Advances in Mechanical Problems of Functionally Graded Materials and Structures

Advances in Mechanical Problems of Functionally Graded Materials and Structures

Special Issue Editors

Tinh Quoc Bui

Le Van Lich

Tiantang Yu

Indra Vir Singh

MDPI • Basel • Beijing • Wuhan • Barcelona • Belgrade

MDPI

Special Issue Editors
Tinh Quoc Bui
Tokyo Institute of Technology
Japan

Le Van Lich
Kyoto University
Japan

Tiantang Yu
Hohai University
China

Indra Vir Singh
Indian Institute of Technology Roorkee
India

Editorial Office
MDPI
St. Alban-Anlage 66
4052 Basel, Switzerland

This is a reprint of articles from the Special Issue published online in the open access journal *Materials* (ISSN 1996-1944) from 2018 to 2019 (available at: https://www.mdpi.com/journal/materials/special_issues/AMPFGMS).

For citation purposes, cite each article independently as indicated on the article page online and as indicated below:

LastName, A.A.; LastName, B.B.; LastName, C.C. Article Title. *Journal Name* **Year**, *Article Number*, Page Range.

ISBN 978-3-03921-658-1 (Pbk)
ISBN 978-3-03921-659-8 (PDF)

Contents

About the Special Issue Editors

Tinh Quoc Bui (Associate Professor), Prof. Bui is currently Associate Professor at the Department of Civil and Environmental Engineering, Tokyo Institute of Technology, Japan. He is also Visiting Associate Professor at the National Taiwan University of Science and Technology. Prof. Bui held successive postdoctoral positions in France, Germany, and Japan. He earned his BSc degree in Mathematics and Computer Science at VNU-HCMC University of Science (2002), MSc in Mechanics of Construction, University of Liege, Belgium (2005), and PhD degree in Mechanical Engineering, Vienna University of Technology, Austria (2009). He was awarded the 2018 JACM Award for Young Investigators in Computational Mechanics from the Japan Association for Computational Mechanics. Prof. Bui is Subject Editor for Applied Mathematical Modelling (Elsevier), and editorial board member for several journals including Thin-Walled Structures (Elsevier). He is has authored and co-authored of over 150 ISI papers. His research interests lie in the areas of computational mechanics, fracture mechanics, nonlinear plate/shell structures, composites, computational intelligence, stochastic high-performance computing, and numerical methods.

Le Van Lich (Ph.D), Dr. Lich received his PhD degree in Engineering Mechanics from Kyoto University in September 2016. He then worked as a Postdoctoral Research Associate at Kyoto University (2016–2017). Since October 2017, he joined Hanoi University of Science and Technology as a Researcher and Lecturer. His research mainly focuses on the phase field modeling and finite element analysis of the multifield coupling properties of ferroic materials and fracture mechanics.

Tiantang Yu is Professor of Engineering Mechanics at Hohai University, China. He obtained his PhD in 2000 from Hohai University. Between September 2000 and February 2001, he was Postdoctoral Research Associate at Lille University of Science and Technology, France. His research interests include advanced numerical methods, structure optimization, composites, and damage and fracture mechanics. Prof. Yu has authored over 70 papers in numerous international journals.

Indra Vir Singh is currently Professor at the Department of Mechanical and Industrial Engineering, Indian Institute of Technology Roorkee, India. Before joining IIT Roorkee, he completed his postdoctoral research work on nanocomposite modeling and simulation from Shinshu University, Japan. He has published more than 125 research articles in international journals of repute. His research areas of interest include FEM, XFEM, meshfree methods, isogeometric analysis, phase field modeling, fracture and damage mechanics, mechanical behavior of materials, modeling and simulations of composite, nonlinear and multiscale modeling.

Preface to "Advances in Mechanical Problems of Functionally Graded Materials and Structures"

All of the papers published in this Special Issue have been peer-reviewed according to the journal standard. Our special thanks and sincere gratitude go to all the authors and reviewers for their hard work, timely responses, and careful revisions. All authors have made tremendous contributions and offered generous support, thus ensuring the success of this Special Issue.

Tinh Quoc Bui, Le Van Lich, Tiantang Yu, Indra Vir Singh
Special Issue Editors

materials

MDPI

Article

Quadratic Solid–Shell Finite Elements for Geometrically Nonlinear Analysis of Functionally Graded Material Plates

Hocine Chalal and Farid Abed-Meraim *

Laboratory LEM3, Université de Lorraine, CNRS, Arts et Métiers ParisTech, F-57000 Metz, France; hocine.chalal@ensam.eu

* Correspondence: farid.abed-meraim@ensam.eu; Tel.: +33-3-87375-479

Received: 30 May 2018; Accepted: 17 June 2018; Published: 20 June 2018

Abstract: In the current contribution, prismatic and hexahedral quadratic solid–shell (SHB) finite elements are proposed for the geometrically nonlinear analysis of thin structures made of functionally graded material (FGM). The proposed SHB finite elements are developed within a purely 3D framework, with displacements as the only degrees of freedom. Also, the in-plane reduced-integration technique is combined with the assumed-strain method to alleviate various locking phenomena. Furthermore, an arbitrary number of integration points are placed along a special direction, which represents the thickness. The developed elements are coupled with functionally graded behavior for the modeling of thin FGM plates. To this end, the Young modulus of the FGM plate is assumed to vary gradually in the thickness direction, according to a volume fraction distribution. The resulting formulations are implemented into the quasi-static ABAQUS/Standard finite element software in the framework of large displacements and rotations. Popular nonlinear benchmark problems are considered to assess the performance and accuracy of the proposed SHB elements. Comparisons with reference solutions from the literature demonstrate the good capabilities of the developed SHB elements for the 3D simulation of thin FGM plates.

Keywords: quadratic solid–shell elements; finite elements; functionally graded materials; thin structures; geometrically nonlinear analysis

---

1. Introduction

Over the last decades, the concept of functionally graded materials (FGMs) has emerged, and FGMs were introduced in the industrial environment due to their excellent performance compared to conventional materials. This new class of materials was first introduced in 1984 by a Japanese research group, who made a new class of composite materials (i.e., FGMs) for aerospace applications dealing with very high temperature gradients [1,2]. These heterogeneous materials are made from several isotropic material constituents, which are usually ceramic and metal. Among the many advantages of FGMs, their mechanical and thermal properties change gradually and continuously from one surface to the other, which allows for overcoming delamination between interfaces as compared to conventional composite materials. In addition, FGMs can resist severe environment conditions (e.g., very high temperatures), while maintaining structural integrity.

Thin structures are widely used in the automotive industry, especially through sheet metal forming into automotive components. In this context, the finite element (FE) method is considered nowadays as a practical numerical tool for the simulation of thin structures. Traditionally, shell and solid elements are used in the simulation of linear and nonlinear problems. However, the simulation results require very fine meshes to obtain accurate solutions due to the various locking phenomena that are inherent to these elements, which lead to high computational costs. To overcome these issues, many researchers have devoted their works to the development of locking-free finite elements. More specifically, the technology of solid–shell elements has become an interesting alternative to traditional solid and shell elements for the efficient modeling of thin structures (see, e.g., [3–8]). Solid-shell elements are based on a fully 3D formulation with only nodal displacements as degrees of freedom. They can be easily combined with various fully 3D constitutive models (e.g., orthotropic elastic behavior, plastic behavior), without any further assumptions, such as plane-stress assumptions. Based on the reduced-integration technique (see, e.g., [9]), they are often combined with advanced strategies to alleviate locking phenomena, such as the assumed-strain method (ASM) (see, e.g., [4]), the enhanced assumed strain (EAS) formulation (see, e.g., [10]), and the assumed natural strain (ANS) approach (see, e.g., [11]). Several FE formulations for the analysis of thin FGM structures have been developed in the literature. They can be classified into three main formulations: The shell-based FGM FE formulation, the solid-based FGM FE formulation, and the solid–shell-based FE formulation. The first formulation is considered as the most widely adopted approach for the modeling of 2D thin FGM structures. However, this approach requires specific kinematic assumptions in the FE formulation, such as the classical Kirchhoff plate theory, first and high-order shear theories, plane-stress assumption (see, e.g., [12–15]). The second approach is based on a 3D formulation of solid elements, in which a fully 3D elastic behavior for FGMs is adopted. In such an approach, some specific kinematic assumptions for thin plates, such as the classical Kirchhoff plate theory and the von Karman theory, are also adopted in the FE formulation (see, e.g., [16–18]). The third approach is based on the concept of solid–shell elements, which are combined with FGM behavior. Few works in the literature have investigated the behavior of thin FGM plates with this approach. Among them, the work of Zhang et al. [19], who investigated the piezo-thermo-elastic behavior of FGM shells with EAS-ANS solid–shell elements. Recently, Hajlaoui et al. [20,21] studied the buckling and nonlinear dynamic analysis of FGM shells using an EAS solid–shell element based on the first-order shear deformation concept.

In this work, quadratic prismatic and hexahedral shell-based (SHB) continuum elements, namely SHB15 and SHB20, respectively, are proposed for the modeling of thin FGM plates. SHB15 is a fifteen-node prismatic solid-shell element with a user-defined number of through-thickness integration points, while SHB20 is a twenty-node hexahedral solid-shell element with a user-defined number of through-thickness integration points. These solid-shell elements have been first developed in the framework of isotropic elastic materials and small strains (see [22]), and recently coupled with anisotropic elastic–plastic behavior models within the framework of large strains for the modeling of sheet metal forming processes [23]. In this paper, the formulations of the quadratic SHB15 and SHB20 elements are combined with functionally graded behavior for the modeling of thin FGM plates. To achieve this, the elastic properties of the proposed elements are assumed to vary gradually in the thickness direction according to a power-law volume fraction. The resulting formulations are implemented into the quasi-static ABAQUS/Standard software. The performance of the proposed elements is assessed through the simulation of various nonlinear benchmark problems taken from the literature.

2. SHB15 and SHB20 Solid-Shell Elements

2.1. Element Reference Geometries

The proposed SHB elements are based on a 3D formulation, with displacements as the only degrees of freedom. Figure 1 shows the reference geometry of the quadratic prismatic SHB15 and quadratic hexahedral SHB20 elements and the position of the associated integration points. Within the reference frame of each element, direction ζ represents the thickness, along which several integration points can be arranged.

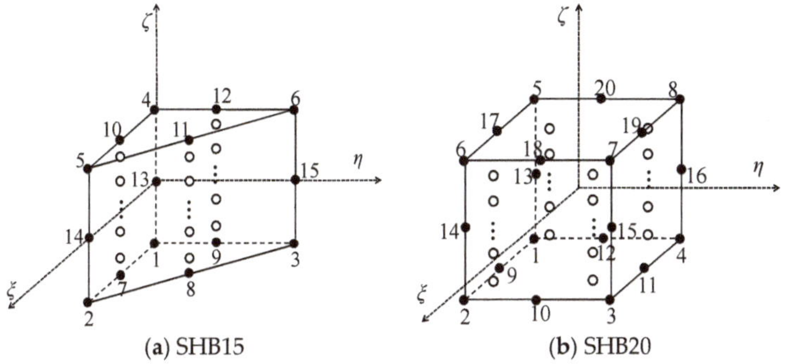

Figure 1. Reference geometry of (**a**) quadratic prismatic SHB15 element and (**b**) quadratic hexahedral SHB20 element, and position of the associated integration points.

2.2. Quadratic Approximation for the SHB Elements

Conventional quadratic interpolation functions for traditional continuum prismatic and hexahedral elements are used in the formulation of the SHB elements. According to this formulation, the spatial coordinates x_i and the displacement field u_i are approximated using the following interpolation functions:

$$x_i = x_{iI}N_I(\xi,\eta,\zeta) = \sum_{I=1}^{K} x_{iI}N_I(\xi,\eta,\zeta), \tag{1}$$

$$u_i = d_{iI}N_I(\xi,\eta,\zeta) = \sum_{I=1}^{K} d_{iI}N_I(\xi,\eta,\zeta), \tag{2}$$

where d_{iI} are the nodal displacements, $i = 1, 2, 3$ correspond to the spatial coordinate directions, and I varies from 1 to K, with K being the number of nodes per element, which is equal to 15 for the SHB15 element and 20 for the SHB20 element.

2.3. Strain Field and Gradient Operator

Using the above approximation for the displacement within the element, the linearized strain tensor $\boldsymbol{\varepsilon}$ can be derived as:

$$\varepsilon_{ij} = \frac{1}{2}(u_{i,j} + u_{j,i}) = \frac{1}{2}(d_{iI}N_{I,j} + d_{jI}N_{I,i}). \tag{3}$$

By combining Equations (1) and (2) with the help of the interpolation functions, the nodal displacement vectors $\mathbf{d}_i$ write:

$$\mathbf{d}_i = a_{0i}\mathbf{s} + a_{1i}\mathbf{x}_1 + a_{2i}\mathbf{x}_2 + a_{3i}\mathbf{x}_3 + \sum_{\alpha} c_{\alpha i}\mathbf{h}_\alpha, i = 1, 2, 3, \tag{4}$$

where $\mathbf{x}_i^T = (x_{i1}, x_{i2}, x_{i3}, \cdots, x_{iK})$ are the nodal coordinate vectors. In Equation (4), index α goes from 1 to 11 for the SHB15 element, and from 1 to 16 for the SHB20 element. In addition, vector $\mathbf{s}^T = (1, 1, \cdots, 1)$ has fifteen constant components in the case of the SHB15 element, and twenty constant components vector for the SHB20 element. Vectors $\mathbf{h}_\alpha$ are composed of h_α functions, which are evaluated at the element nodes, and the full details of their expressions can be found in [23].

With the help of some well-known orthogonality conditions and of the Hallquist [24] vectors $\mathbf{b}_i = \frac{\partial \mathbf{N}}{\partial x_i}_{|\xi=\eta=\zeta=0}$, where vector $\mathbf{N}$ contains the expressions of the interpolation functions N_I, the unknown constants a_{ji} and $c_{\alpha i}$ in Equation (4) can be derived as:

$$a_{ji} = \mathbf{b}_j^T \cdot \mathbf{d}_i, \quad c_{\alpha i} = \gamma_\alpha^T \cdot \mathbf{d}_i, \tag{5}$$

where the complete details on the expressions of vectors γ_α can be found in [22].

By introducing the discrete gradient operator **B**, the strain field in Equation (3) writes:

$$\nabla_s(\boldsymbol{u}) = \begin{bmatrix} u_{x,x} \\ u_{y,y} \\ u_{z,z} \\ u_{x,y} + u_{y,x} \\ u_{y,z} + u_{z,y} \\ u_{x,z} + u_{z,x} \end{bmatrix} = \mathbf{B} \cdot \mathbf{d} = \mathbf{B} \cdot \begin{bmatrix} \mathbf{d}_x \\ \mathbf{d}_y \\ \mathbf{d}_z \end{bmatrix}, \tag{6}$$

where the expression of the discrete gradient operator **B** is:

$$\mathbf{B} = \begin{bmatrix} \mathbf{b}_x^T + h_{\alpha,x}\gamma_\alpha^T & 0 & 0 \\ 0 & \mathbf{b}_y^T + h_{\alpha,y}\gamma_\alpha^T & 0 \\ 0 & 0 & \mathbf{b}_z^T + h_{\alpha,z}\gamma_\alpha^T \\ \mathbf{b}_y^T + h_{\alpha,y}\gamma_\alpha^T & \mathbf{b}_x^T + h_{\alpha,x}\gamma_\alpha^T & 0 \\ 0 & \mathbf{b}_z^T + h_{\alpha,z}\gamma_\alpha^T & \mathbf{b}_y^T + h_{\alpha,y}\gamma_\alpha^T \\ \mathbf{b}_z^T + h_{\alpha,z}\gamma_\alpha^T & 0 & \mathbf{b}_x^T + h_{\alpha,x}\gamma_\alpha^T \end{bmatrix}. \tag{7}$$

2.4. Hu–Washizu Variational Principle

The SHB solid-shell elements are based on the assumed-strain method, which is derived from the simplified form of the Hu–Washizu variational principle [25]. In terms of assumed-strain rate $\dot{\bar{\boldsymbol{\varepsilon}}}$, interpolated stress $\boldsymbol{\sigma}$, nodal velocities $\dot{\mathbf{d}}$, and external nodal forces $\mathbf{f}^{ext}$, this principle writes

$$\pi\left(\dot{\bar{\boldsymbol{\varepsilon}}}\right) = \int_{\Omega_e} \delta\dot{\bar{\boldsymbol{\varepsilon}}}^T \cdot \boldsymbol{\sigma}\, d\Omega \; - \; \delta\dot{\mathbf{d}}^T \cdot \mathbf{f}^{ext} = 0. \tag{8}$$

The assumed-strain rate is expressed in terms of the discrete gradient operator **B** as:

$$\dot{\bar{\boldsymbol{\varepsilon}}}(x,t) = \mathbf{B} \cdot \dot{\mathbf{d}}. \tag{9}$$

Substituting the expression of the assumed-strain rate given by Equation (9) into the variational principle (Equation (8)), the expressions of the stiffness matrix $\mathbf{K}_e$ and the internal forces $\mathbf{f}^{\text{int}}$ for the SHB elements are

$$\mathbf{K}_e = \int_{\Omega_e} \mathbf{B}^T \cdot \mathbf{C}^e(\zeta) \cdot \mathbf{B}\, d\Omega \; + \; \mathbf{K}_{GEOM}, \tag{10}$$

$$\mathbf{f}^{\text{int}} = \int_{\Omega_e} \mathbf{B}^T \cdot \boldsymbol{\sigma}\, d\Omega, \tag{11}$$

where $\mathbf{K}_{GEOM}$ is the geometric stiffness matrix. As to the fourth-order tensor $\mathbf{C}^e(\zeta)$, it describes the functionally graded elastic behavior of the FGM material. Its expression is given hereafter.

2.5. Description of Functionally Graded Elastic Behavior

In the framework of large displacements and rotations, the formulation of the SHB elements requires the definition of a local frame with respect to the global coordinate system, as illustrated in Figure 2. The local frame, which is designated as the "element frame" in Figure 2, is defined for each element using the associated nodal coordinates. In such an element frame, where the ζ-coordinate represents the thickness direction, the fourth-order elasticity tensor $\mathbf{C}^e(\zeta)$ for the FGM is specified.

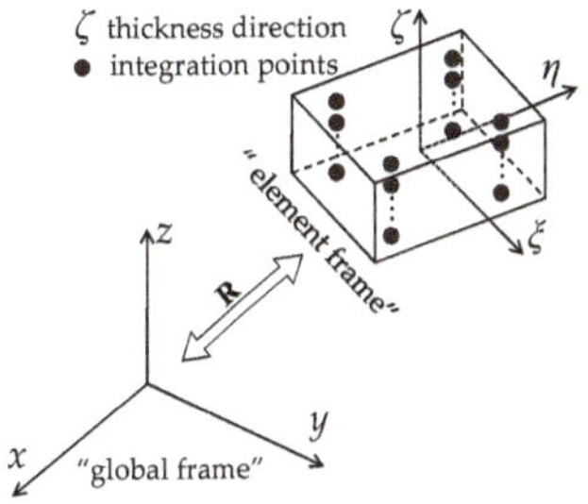

Figure 2. Element frame and global frame for the proposed SHB elements.

In this work, a two-phase FGM is considered, which consists of two constituent mixtures of ceramic and metal. The ceramic phase of the FGM can sustain very high temperature gradients, while the ductility of the metal phase prevents the onset of fracture due to the cyclic thermal loading. In such FGMs, the material at the bottom surface of the plate is fully metal and at the top surface of the plate is fully ceramic, as illustrated in Figure 3. Between these bottom and top surfaces, the elastic properties vary continuously through the thickness from metal to ceramic properties, respectively, according to a power-law volume fraction. The corresponding volume fractions for the ceramic phase f_c and the metal phase f_m are expressed as (see, e.g., [26,27]):

$$f_c = \left(\frac{z}{\mathrm{t}} + \frac{1}{2}\right)^n \text{ and } f_m = 1 - f_c, \tag{12}$$

where n is the power-law exponent, which is greater than or equal to zero, and $z \in [-\mathrm{t}/2,\ \mathrm{t}/2]$, with t the thickness of the plate. For $n = 0$, the material is fully ceramic, while when $n \to \infty$ the material is fully metal (see Figure 4).

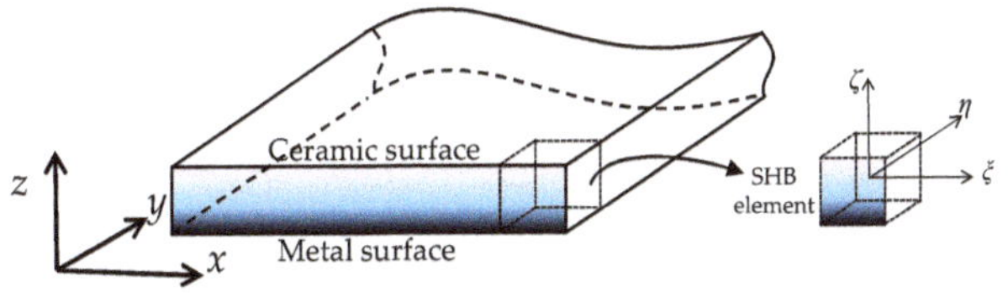

Figure 3. Schematic representation of the functionally graded thin plate.

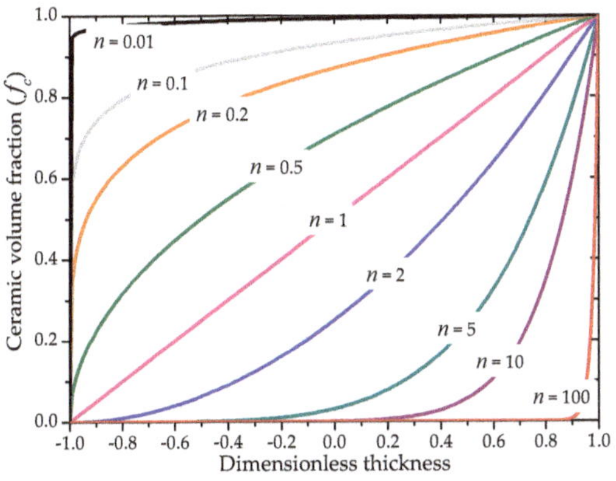

Figure 4. Volume fraction distribution of the ceramic phase as function of the power-law exponent n.

For an isotropic elastic behavior, the constitutive equations are governed by the Hooke elasticity law, which is expressed by the following relationship:

$$\boldsymbol{\sigma} = \lambda tr(\boldsymbol{\varepsilon})\mathbf{1} + 2\mu\boldsymbol{\varepsilon}, \tag{13}$$

where **1** denotes the second-order unit tensor, λ and μ are the Lamé constants given by:

$$\lambda = \frac{\nu E}{(1-2\nu)(1+\nu)} \text{ and } \mu = \frac{E}{2(1+\nu)}, \tag{14}$$

with E and ν the Young modulus and the Poisson ratio, respectively.

For FGMs that are made of ceramic and metal constituents, it is commonly assumed that only the Young modulus E varies in the thickness direction, while the Poisson ratio ν is kept constant. Therefore, the constant Young modulus in Equation (14) is replaced by $E(z)$, whose value evolves according to the following power-law distribution:

$$E(z) = (E_c - E_m)f_c + E_m, \tag{15}$$

where E_c and E_m are the Young modulus of the ceramic and metal, respectively.

3. Nonlinear Benchmark Problems

In this section, the performance of the proposed elements is assessed through the simulation of several popular nonlinear benchmark problems. The static ABAQUS/Standard solver has been used to solve the following static benchmark problems. More specifically, the classical Newton method is considered for most benchmark problems, aside from limit-point buckling problems for which the Riks arc-length method is used.

To accurately describe the variation of the Young modulus through the thickness of the FGM plates, only five integration points within a single element layer is used in the simulations. For each benchmark problem, the simulation results given by the proposed elements are compared to the reference solutions taken from the literature. In the subsequent simulations, it is worth noting that the elastic properties of the metal and ceramic constituents of the FGM plates do not reflect a real metallic or ceramic material. Indeed, the terms metal and ceramic are commonly used in the literature to emphasize the difference between the properties of the FGM constituents (see, e.g., [15,26,27]).

Regarding the meshes used in the simulations, the following mesh strategy is adopted: $(N_1 \times N_2) \times N_3$ for the hexahedral SHB20 element, where N_1 is the number of elements along the length, N_2 is the number of elements along the width, and N_3 is the number of elements along the thickness direction. As to the prismatic SHB15 element, the mesh strategy consists of $(N_1 \times N_2 \times 2) \times N_3$, due to the in-plane subdivision of a rectangular element into two triangles.

3.1. Cantilever Beam Sujected to End Shear Force

Figure 5a shows a simple cantilever FGM beam with a bending load at its free end. This is a classical popular benchmark problem, which has been widely considered in many works for the analysis of cantilever beams with isotropic material (see, e.g., [28,29]). The Poisson ratio of the FGM beam is assumed to be $\nu = 0.3$, while the Young modulus of the metal and ceramic constituents are $E_m = 2.1 \times 10^5$ Mpa and $E_c = 3.8 \times 10^5$ Mpa, respectively. Figure 5b illustrates the final deformed shape of the cantilever beam with respect to its undeformed shape, as discretized with SHB20 elements, in the case of fully metallic material. Figure 6 shows the load–deflection curves obtained with the quadratic SHB elements, along with the reference solutions taken from [15], for various values of the power-law exponent n. One recalls that fully metallic material is obtained when $n \to \infty$, and fully ceramic material for $n = 0$. Overall, the SHB elements show excellent agreement with the reference solutions corresponding to the various values of exponent n. More specifically, it can be observed that the bending behavior of the FGM beam lies between that of the fully ceramic and fully metal beam, which is consistent with the power-law distribution of the Young modulus in the thickness direction. Another advantage of the proposed SHB elements is that, using the same in-plane mesh discretization as in reference [15], only five integration points through the thickness are sufficient for the SHB elements, while ten integration points have been considered in [15] to simulate this benchmark problem.

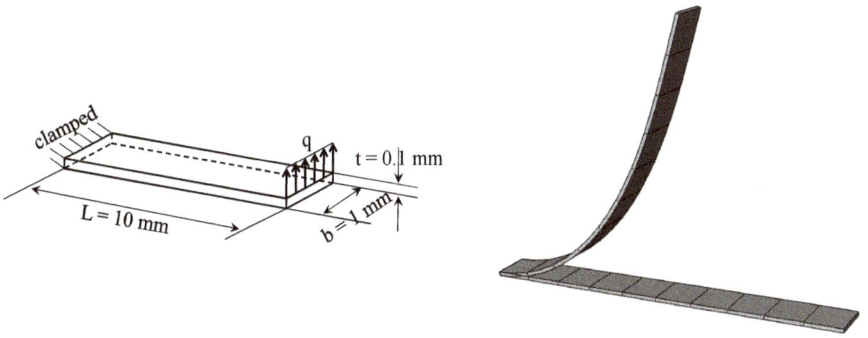

(**a**) (**b**)

Figure 5. Cantilever beam: (**a**) geometry and (**b**) undeformed and deformed configurations.

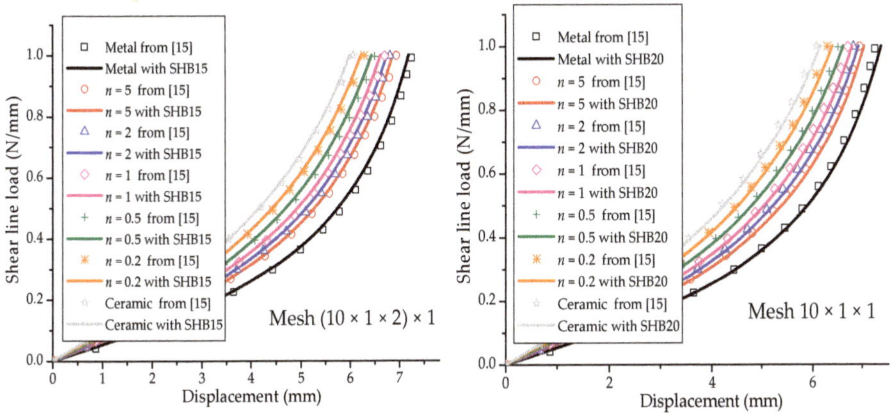

(**a**) (**b**)

Figure 6. Load–deflection curves for the cantilever beam. (**a**) Prismatic SHB15 element; (**b**) hexahedral SHB20 element.

3.2. Slit Annular Plate

In this section, the well-known slit annular plate problem is considered (see, e.g., [29–31]). The annular plate is clamped at one end and loaded by a line shear force P, as illustrated in Figure 7a. The inner and outer radius of the annular plate are $R_i = 6$ m and $R_o = 10$ m, respectively, while the thickness is t = 0.03 m. The Poisson ratio of the annular plate is $\nu = 0.3$, while the Young modulus of the metal and ceramic constituents are $E_m = 21$ Gpa and $E_c = 38$ Gpa, respectively. Figure 7b illustrates the undeformed and deformed shapes of the annular plate, as discretized with SHB20 elements, in the case of fully metallic material. Figure 8 reports the load–out-of-plane vertical deflection curves at the outer point A of the annular plate as obtained with the SHB elements, along with the reference solutions taken from [15]. One can observe that the SHB elements perform very well with respect to the reference solutions for all considered values of exponent n. Similar to the previous benchmark problem, the same in-plane mesh discretization as in [15] with only five integration points through the thickness has been adopted by the proposed SHB elements for this nonlinear test, while ten integration points have been considered in [15].

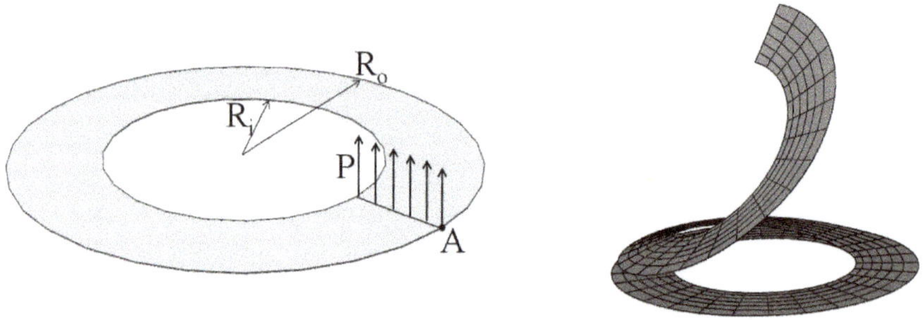

(**a**) (**b**)

Figure 7. Slit annular plate: (**a**) geometry and (**b**) undeformed and deformed configurations.

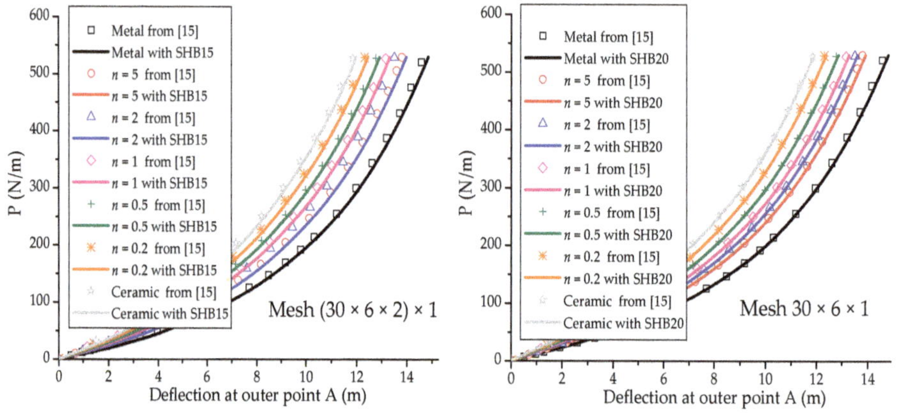

(**a**) (**b**)

Figure 8. Load–deflection curves at the outer point A for the slit annular plate. (**a**) Prismatic SHB15 element; (**b**) hexahedral SHB20 element.

3.3. Clamped Square Plate under Pressure

Figure 9a illustrates a fully clamped square plate, which is loaded by a uniformly distributed pressure. The length and thickness of the square plate are L = 1000 mm and t = 2 mm, respectively. The Poisson ratio is $\nu = 0.3$, while the Young modulus of the metal and ceramic constituents are $E_m = 2 \times 10^5$ Mpa and $E_c = 3.8 \times 10^5$ MPa, respectively. Considering the problem symmetry, a quarter of the plate is discretized. Figure 9b illustrates the undeformed and deformed shapes of the square plate, as discretized with SHB20 elements, in the case of fully metallic material. The pressure–displacement curves for the SHB elements (where the displacement is computed at the center of the plate), along with the reference solutions taken from [15], are all depicted in Figure 10. The results obtained with the SHB elements, by adopting only five integration points in the thickness direction and the same in-plane mesh discretization as in [15], are in excellent agreement with the reference solutions that required ten through-thickness integration points.

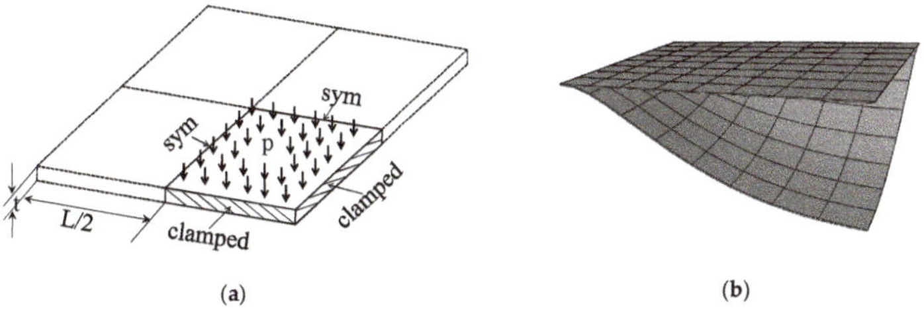

Figure 9. Clamped square plate: (**a**) geometry and (**b**) undeformed and deformed configurations.

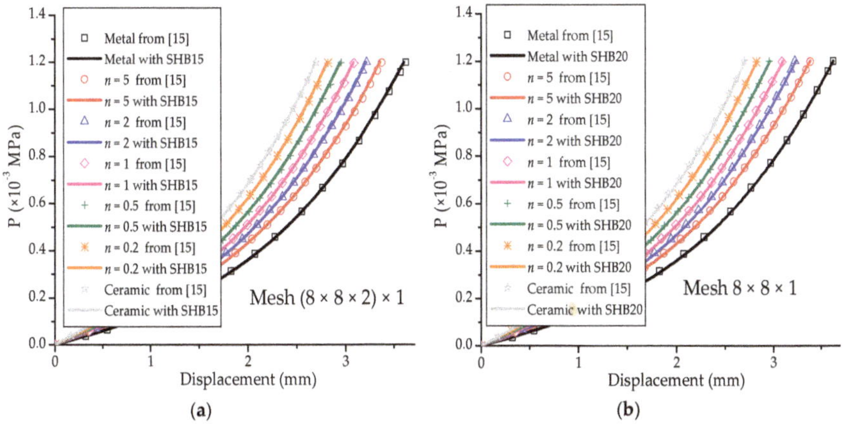

Figure 10. Load–deflection curves at the center point for the square plate. (**a**) Prismatic SHB15 element; (**b**) hexahedral SHB20 element.

3.4. Hinged Cylindrical Roof

Figure 11a shows a hinged cylindrical roof subjected to a concentrated force at its center. Two types of roofs are considered, thick and thin, with thicknesses t = 12.7 mm and t = 6.35 mm, respectively. Because this nonlinear benchmark test involves geometric-type instabilities (limit-point buckling), the Riks path-following method is used to follow the load–displacement curves beyond the limit points. The Poisson ratio of the cylindrical roof is $\nu = 0.3$, while the Young modulus of the metal and ceramic constituents are $E_m = 70 \times 10^3$ Mpa and $E_c = 151 \times 10^3$ Mpa, respectively. Owing to the symmetry, only one quarter of the cylindrical roof is modeled. Figure 11b illustrates the undeformed and deformed shapes of the hinged cylindrical roof, as discretized with SHB20 elements, in the case of fully metallic material. The load–vertical displacement curves at the central point A of the thick and thin hinged cylindrical roofs are shown in Figures 12 and 13, and compared with the reference solutions taken from [30]. From these figures, it can be seen that the results obtained with the proposed quadratic SHB elements are in good agreement with the reference solutions for the different values of exponent n, corresponding to different volume fractions (from fully metal to fully ceramic). More specifically, the snap-through and snap-back phenomena, which are typically exhibited in such limit-point buckling problems, are very well reproduced by the proposed SHB elements. Note that, for the thick roof (i.e., t = 12.7 mm), the converged solutions in Figure 12 are obtained by using a mesh of $(8 \times 8 \times 2) \times 1$ in the case of prismatic SHB15 elements, and a mesh of $8 \times 8 \times 1$ with hexahedral SHB20 elements. As to the thin roof (i.e., t = 6.35 mm), finer meshes of $(16 \times 16 \times 2) \times 1$ for the prismatic SHB15 elements, and $16 \times 16 \times 1$ for the hexahedral SHB20 elements are required to obtain converged results (see Figure 13). These mesh refinements are similar to those used by Sze et al. [29] for the thick and thin roof in the case of an isotropic material as well as for multilayered composite materials.

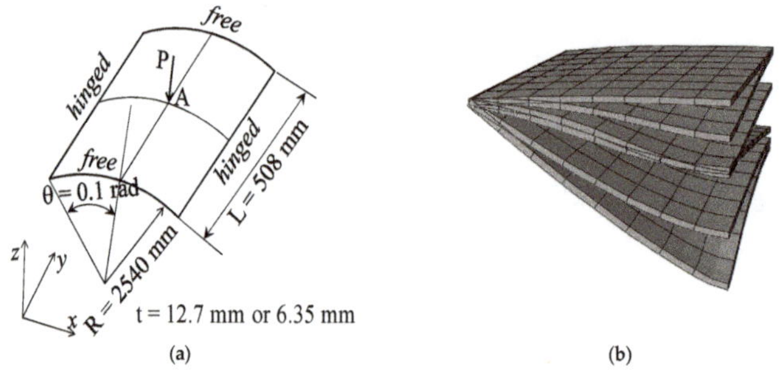

Figure 11. Hinged cylindrical roof: (**a**) geometry and (**b**) undeformed and deformed configurations.

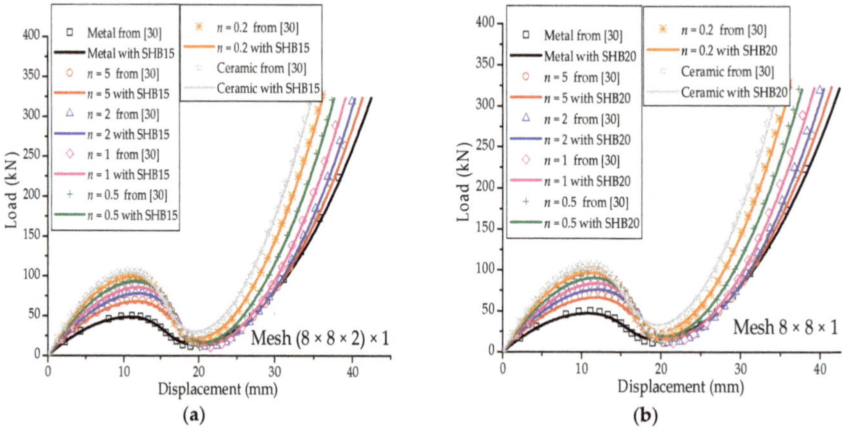

Figure 12. Deflection at the central point A under concentrated force for the thick hinged roof. (**a**) prismatic SHB15 element; (**b**) hexahedral SHB20 element.

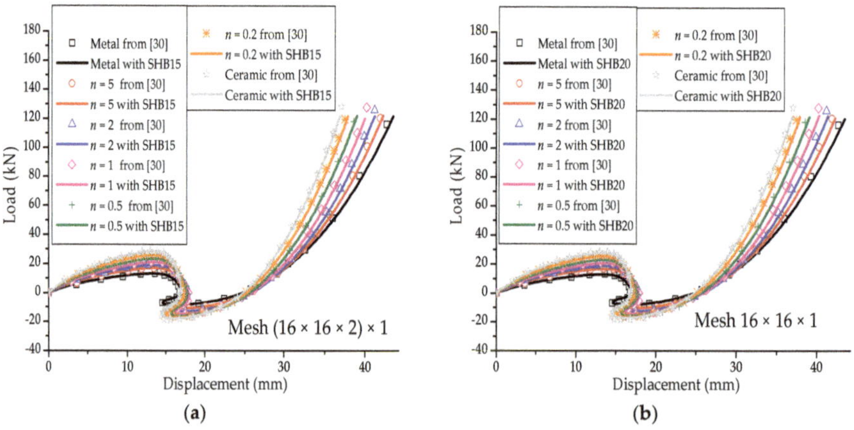

Figure 13. Deflection at the central point A under concentrated force for the thin hinged roof. (**a**) prismatic SHB15 element; (**b**) hexahedral SHB20 element.

3.5. Pull-Out of an Open-Ended Cylinder

The well-known pull-out test for an open-ended cylinder is considered in this section. As illustrated in Figure 14a, the cylinder is pulled by two opposite radial forces, which results in the deformed shape shown in Figure 14b. The isotropic material case as well as the laminated composite material case have been considered by many authors in the literature (see, e.g., [29,30,32]). The Poisson ratio of the cylinder is $\nu = 0.3$, while the Young modulus of the metal and ceramic constituents are $E_m = 0.7 \times 10^9$ Pa and $E_c = 1.51 \times 10^9$ Pa, respectively. Owing to the symmetry of the problem, only one eighth of the cylinder is modeled. The force–radial displacement curves at points A, B and C (as depicted in Figure 14a), obtained with the SHB elements, are shown in Figures 15–17, respectively, along with the reference solutions taken from [13]. It can be observed that the developed SHB elements successfully pass this benchmark test as compared to the reference solutions. More specifically, the transition zone in the load–radial displacement curves, which is marked by the snap-through point, is well reproduced by both prismatic and hexahedral SHB elements for the various values of the power-law exponent n. Note that the converged solutions in Figures 15–17 are obtained with the proposed elements by using only five integration points in the thickness direction, and meshes of $(24 \times 36 \times 2) \times 1$ and $12 \times 18 \times 1$ in the case of the prismatic SHB15 element and hexahedral SHB20 element, respectively. Hence, the required meshes for convergence are coarser than those used by Sze et al. [29] in the case of an isotropic material.

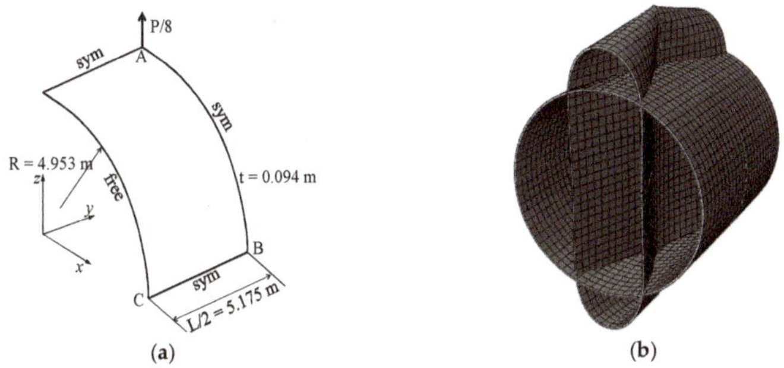

Figure 14. Pull-out of an open-ended cylinder: (**a**) geometry and (**b**) undeformed and deformed configurations.

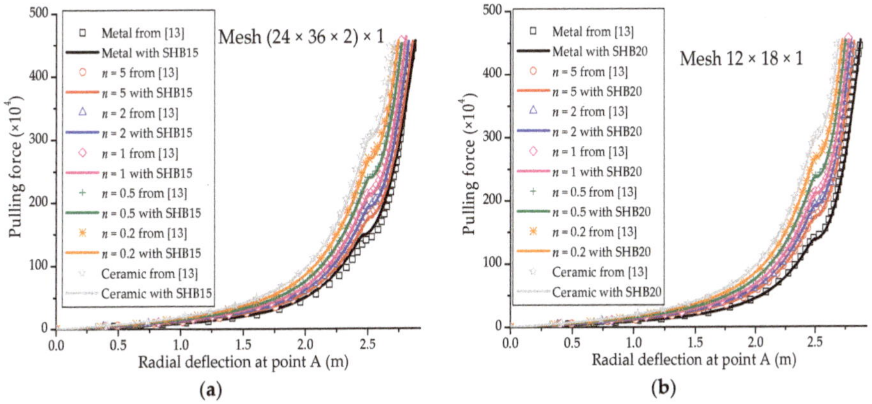

Figure 15. Radial displacement at point A under concentrated force for the open-ended cylinder. (**a**) Prismatic SHB15 element; (**b**) hexahedral SHB20 element.

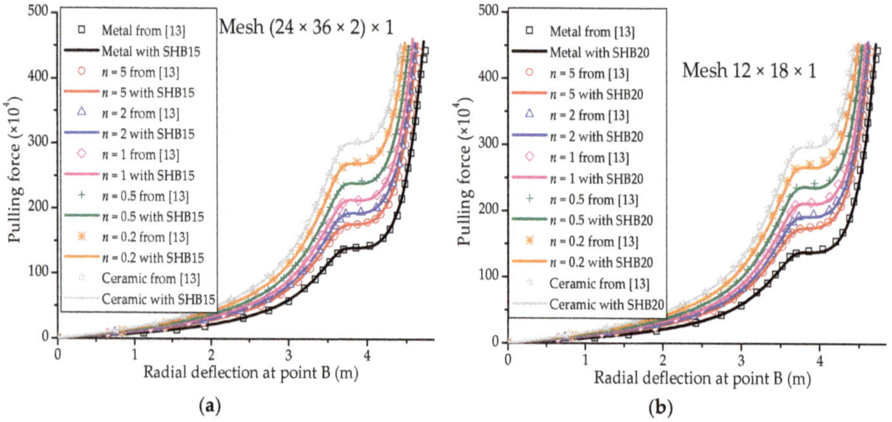

Figure 16. Radial displacement at point B under concentrated force for the open-ended cylinder. (**a**) prismatic SHB15 element; (**b**) hexahedral SHB20 element.

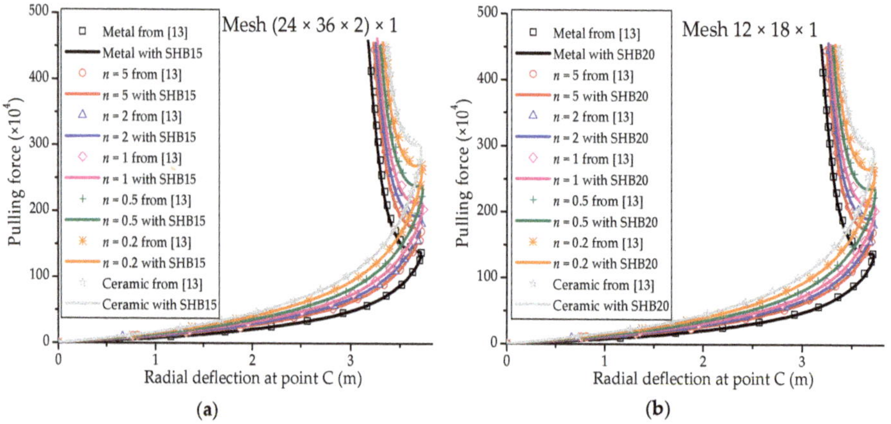

Figure 17. Radial displacement at point C under concentrated force for the open-ended cylinder. (**a**) prismatic SHB15 element; (**b**) hexahedral SHB20 element.

3.6. Pinched Hemispherical Shell

It is worth noting that although the performance of the prismatic SHB15 element is similar to that of the hexahedral SHB20 element, as demonstrated in the above nonlinear benchmark problems, the main motivation in developing the prismatic solid–shell element is to use it for the mesh discretization of complex geometries. Indeed, it is well-known that complex geometries cannot be discretized with only hexahedral elements, and require either an irregular mesh with prismatic elements, or a mixture based on a combination of prismatic and hexahedral elements. In this section, a hemispherical shell is loaded by alternating radial forces as shown in Figure 18a. Note that this benchmark problem has been considered in the literature for an isotropic material as well as a laminated composite material (see, e.g., [30]), while the case of FGMs has not been considered yet. Consequently, only the simulation results obtained with the proposed SHB elements corresponding to the fully metallic shell can be compared to the reference solution taken from [30].

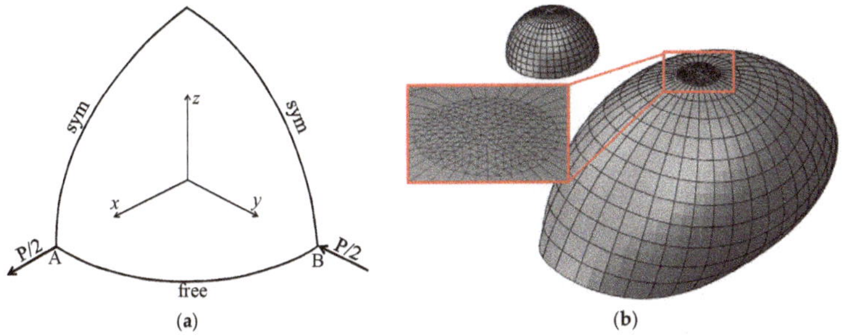

Figure 18. Pinched hemisphere: (**a**) geometry and (**b**) undeformed and deformed configurations.

The radius and thickness of the hemispherical shell are R = 10 m and t = 0.04 m, respectively. The Poisson ratio of the hemispherical shell is $\nu = 0.3$, while the Young modulus of the metal and ceramic constituents are $E_m = 6.825 \times 10^7$ Pa and $E_c = 1.46 \times 10^8$ Pa, respectively. Due to the symmetry, a quarter of the structure is discretized. The hemispherical shell is discretized with a mixture of prismatic and hexahedral elements, which consists of 90 SHB15 elements located at the top of the hemisphere (far from the load points, see Figure 18b) and 110 SHB20 elements for the remaining area.

The simulation results in terms of force–radial deflection at point A, for various values of the power-law exponent n, are plotted in Figure 19. This figure shows that the results corresponding to a fully metallic shell (i.e., $n \to \infty$), obtained by the combination of prismatic and hexahedral SHB elements, are in excellent agreement with those provided in [30] for an isotropic shell. Note that an equivalent in-plane mesh discretization has been used in [30], where a fully integrated shell element with several integration points has been considered. Moreover, Figure 19 reveals that for all values of the exponent n, the simulated load–radial deflection curves lie between that of the fully ceramic shell and that of the fully metal shell, which is consistent with the numerical results found in the previous benchmark problems.

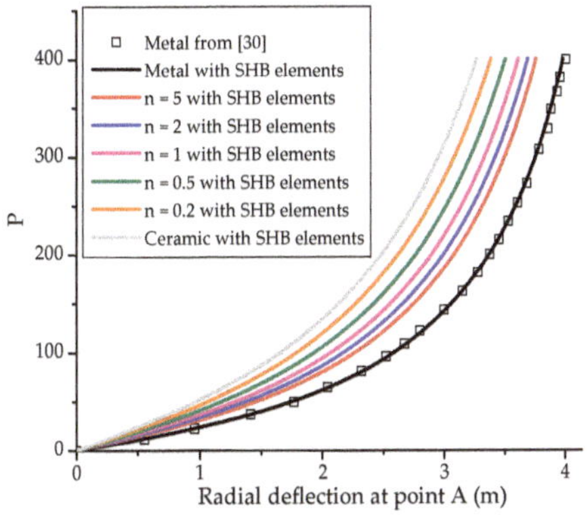

Figure 19. Load–displacement curves at point A for the pinched hemispherical shell, obtained with a mixture of prismatic SHB15 and hexahedral SHB20 elements.

4. Conclusions

In this work, quadratic prismatic and hexahedral solid–shell SHB elements have been proposed for the 3D modeling of thin FGM structures. The formulation of the SHB elements adopts the in-plane reduced-integration technique along with the assumed-strain method to alleviate various locking phenomena. A local (element) frame has been defined for each element, in which the thickness direction is specified. In this local frame, elastic properties of the thin structure are assumed to vary gradually through the thickness according to a power-law volume fraction distribution. The resulting formulations are implemented into the finite element software ABAQUS/Standard in the framework of large displacements and rotations. A series of selective and representative benchmark problems, involving FGM thin structures, has been performed to evaluate the performance of the SHB elements in geometrically nonlinear analysis. The results obtained with the SHB elements have been compared with reference solutions. Note that the state-of-the-art ABAQUS solid and shell elements have not been considered in the simulations, because these elements do not allow modeling of FGM behavior with only a single layer of elements. Overall, the numerical results obtained with the SHB elements showed excellent agreement with the available reference solutions. More specifically, the load–displacement curves for each benchmark test lie between that of the fully ceramic and fully metal behavior, which is consistent with the power-law distribution of the Young modulus in the thickness direction of the FGM plates. This good performance of the proposed SHB elements only requires a few integration points in the thickness direction (i.e., only five integration points), as compared to the number of integration points used in the literature to model thin FGM structures. Furthermore, it has been shown that the prismatic SHB15 element can be naturally combined with the hexahedral SHB20 element, within the same simulation, to help discretize complex geometries. Overall, the proposed SHB elements showed good capabilities in 3D modeling of thin FGM structures with only a single layer of elements and few integration points, which is not the case of traditional solid and shell elements.

Author Contributions: H.C. conceived and performed the simulations. H.C. and F.A.-M. analyzed and discussed the results. Both authors contributed to the writing of the manuscript.

Funding: This research received no external funding.

Conflicts of Interest: The authors declare no conflicts of interest.

References

1. Yamanoushi, M.; Koizumi, M.; Hiraii, T.; Shiota, I. FGM-90. In Proceedings of the First International Symposium on Functionally Gradient Materials, Sendai, Japan, 8–9 October 1990.
2. Koizumi, M. The concept of FGM. *Ceram. Trans. Funct. Gradient Mater.* **1993**, *34*, 3–10.
3. Hauptmann, R.; Schweizerhof, K. A systematic development of solid–shell element formulations for linear and nonlinear analyses employing only displacement degrees of freedom. *Int. J. Numer. Methods Eng.* **1998**, *42*, 49–70. [CrossRef]
4. Abed-Meraim, F.; Combescure, A. SHB8PS—A new adaptive, assumed-strain continuum mechanics shell element for impact analysis. *Comput. Struct.* **2002**, *80*, 791–803. [CrossRef]
5. Reese, S. A large deformation solid-shell concept based on reduced integration with hourglass stabilization. *Int. J. Numer. Methods Eng.* **2007**, *69*, 1671–1716. [CrossRef]
6. Abed-Meraim, F.; Combescure, A. An improved assumed strain solid–shell element formulation with physical stabilization for geometric non-linear applications and elastic–plastic stability analysis. *Int. J. Numer. Methods Eng.* **2009**, *80*, 1640–1686. [CrossRef]
7. Salahouelhadj, A.; Abed-Meraim, F.; Chalal, H.; Balan, T. Application of the continuum shell finite element SHB8PS to sheet forming simulation using an extended large strain anisotropic elastic-plastic formulation. *Arch. Appl. Mech.* **2012**, *82*, 1269–1290. [CrossRef]
8. Pagani, M.; Reese, S.; Perego, U. Computationally efficient explicit nonlinear analyses using reduced integration-based solid–shell finite elements. *Comput. Methods Appl. Mech. Eng.* **2014**, *268*, 141–159. [CrossRef]
9. Zienkiewicz, O.C.; Taylor, R.L.; Too, J.M. Reduced integration technique in general analysis of plates and shells. *Int. J. Numer. Methods Eng.* **1971**, *3*, 275–290. [CrossRef]
10. Alves de Sousa, R.J.; Cardoso, R.P.R.; Fontes Valente, R.A.; Yoon, J.W.; Grácio, J.J.; Natal Jorge, R.M. A new one-point quadrature enhanced assumed strain (EAS) solid-shell element with multiple integration points along thickness: Part I—Geometrically linear applications. *Int. J. Numer. Methods Eng.* **2005**, *62*, 952–977. [CrossRef]

11. Edem, I.B.; Gosling, P.D. Physically stabilised displacement-based ANS solid–shell element. *Finite Elem. Anal. Des.* **2013**, *74*, 30–40. [CrossRef]
12. Reddy, J.N. Analysis of functionally graded plates. *Int. J. Numer. Methods Eng.* **2000**, *47*, 663–684. [CrossRef]
13. Arciniega, R.A.; Reddy, J.N. Large deformation analysis of functionally graded shells. *Int. J. Solids Struct.* **2007**, *44*, 2036–2052. [CrossRef]
14. Cao, Z.-Y.; Wang, H.-N. Free vibration of FGM cylindrical shells with holes under various boundary conditions. *J. Sound Vib.* **2007**, *306*, 227–237. [CrossRef]
15. Beheshti, A.; Ramezani, S. Nonlinear finite element analysis of functionally graded structures by enhanced assumed strain shell elements. *Appl. Math. Model.* **2015**, *39*, 3690–3703. [CrossRef]
16. Asemi, K.; Salami, S.J.; Salehi, M.; Sadighi, M. Dynamic and Static analysis of FGM Skew plates with 3D Elasticity based Graded Finite element Modeling. *Lat. Am. J. Solids Struct.* **2014**, *11*, 504–533. [CrossRef]
17. Nguyen, K.D.; Nguyen-Xuan, H. An isogeometric finite element approach for three-dimensional static and dynamic analysis of functionally graded material plate structures. *Compos. Struct.* **2015**, *132*, 423–439. [CrossRef]
18. Vel, S.S.; Batra, R.C. Three dimensional exact solution for the vibration of FGM rectangular plates. *J. Sound Vib.* **2004**, *272*, 703–730. [CrossRef]
19. Zheng, S.J.; Dai, F.; Song, Z. Active control of piezothermoelastic FGM shells using integrated piezoelectric sensor/actuator layers. *Int. J. Appl. Electromagn.* **2009**, *30*, 107–124. [CrossRef]
20. Hajlaoui, A.; Jarraya, A.; El Bikri, K.; Dammak, F. Buckling analysis of functionally graded materials structures with enhanced solid-shell elements and transverse shear correction. *Compos. Struct.* **2015**, *132*, 87–97. [CrossRef]
21. Hajlaoui, A.; Triki, E.; Frikha, A.; Wali, M.; Dammak, F. Nonlinear dynamics analysis of FGM shell structures with a higher order shear strain enhanced solid-shell element. *Lat. Am. J. Solids Struct.* **2017**, *14*, 72–91. [CrossRef]
22. Abed-Meraim, F.; Trinh, V.D.; Combescure, A. New quadratic solid–shell elements and their evaluation on linear benchmark problems. *Computing* **2013**, *95*, 373–394. [CrossRef]
23. Wang, P.; Chalal, H.; Abed-Meraim, F. Quadratic solid-shell elements for nonlinear structural analysis and sheet metal forming simulation. *Comput. Mech.* **2017**, *59*, 161–186. [CrossRef]
24. Hallquist, J.O. *Theoretical Manual for DYNA3D*; Report UC1D-19041; Lawrence Livermore National Laboratory: Livermore, CA, USA, 1983.
25. Simo, J.C.; Hughes, T.J.R. On the variational foundations of assumed strain methods. *J. Appl. Mech.* **1986**, *53*, 51–54. [CrossRef]
26. Chi, S.H.; Chung, Y.L. Mechanical behavior of functionally graded material plates under transverse load—Part I: Analysis. *Int. J. Solids Struct.* **2006**, *43*, 3657–3674. [CrossRef]
27. Chi, S.H.; Chung, Y.L. Mechanical behavior of functionally graded material plates under transverse load—Part II: Numerical results. *Int. J. Solids Struct.* **2006**, *43*, 3675–3691. [CrossRef]
28. Betsch, P.; Gruttmann, F.; Stein, E. A 4-node finite shell element for the implementation of general hyperelastic 3D-elasticity at finite strains. *Comput. Methods Appl. Mech. Eng.* **1996**, *130*, 57–79. [CrossRef]
29. Sze, K.Y.; Liu, X.H.; Lo, S.H. Popular benchmark problems for geometric nonlinear analysis of shells. *Finite Elem. Anal. Des.* **2004**, *40*, 1551–1569. [CrossRef]
30. Arciniega, R.A.; Reddy, J.N. Tensor-based finite element formulation for geometrically nonlinear analysis of shell structures. *Comput. Methods Appl. Mech. Eng.* **2007**, *196*, 1048–1073. [CrossRef]
31. Andrade, L.G.; Awruch, A.M.; Morsch, I.B. Geometrically nonlinear analysis of laminate composite plates and shells using the eight-node hexahedral element with one-point integration. *Compos. Struct.* **2007**, *79*, 571–580. [CrossRef]
32. Sansour, C.; Kollmann, F.G. Families of 4-node and 9-node finite elements for a finite deformation shell theory. An assessment of hybrid stress, hybrid strain and enhanced strain elements. *Comput. Mech.* **2000**, *24*, 435–447. [CrossRef]

materials

MDPI

Article

On the Finite Element Implementation of Functionally Graded Materials

Emilio Martínez-Pañeda

Department of Engineering, Cambridge University, Cambridge CB2 1PZ, UK; mail@empaneda.com

Received: 24 December 2018; Accepted: 15 January 2019; Published: 17 January 2019

Abstract: We investigate the numerical implementation of functionally graded properties in the context of the finite element method. The macroscopic variation of elastic properties inherent to functionally graded materials (FGMs) is introduced at the element level by means of the two most commonly used schemes: (i) nodal based gradation, often via an auxiliary (non-physical) temperature-dependence, and (ii) Gauss integration point based gradation. These formulations are extensively compared by solving a number of paradigmatic boundary value problems for which analytical solutions can be obtained. The nature of the notable differences revealed by the results is investigated in detail. We provide a user subroutine for the finite element package ABAQUS to overcome the limitations of the most popular approach for implementing FGMs in commercial software. The use of reliable, element-based formulations to define the material property variation could be key in fracture assessment of FGMs and other non-homogeneous materials.

Keywords: functionally graded materials; finite element analysis; graded finite elements

1. Introduction

There is an emerging interest in the analysis of the mechanical response of materials with spatially varying properties. New manufacturing technologies make it possible to engineer materials with functionally graded microstructures, so-called functionally graded materials (FGMs). The resulting spatial variation of material properties eliminates stress discontinuities at material interfaces and optimizes material performance under non-uniform service conditions. For example, the performance of coatings subjected to large thermal gradients can be significantly improved by using metal-ceramic FGMs [1], which combine the thermal and corrosive resistance of ceramics with the mechanical strength and high tenacity of metals. In addition, FGMs are now employed in a host of commercial applications, ranging from cutting tools to biomedical devices [2]. This widespread use of FGMs is largely due to their capacity to reduce residual stresses [3], increase the strength of joints [4], and tailor material microstructure to specific service requirements [5].

The complexity and cost associated with the manufacture and testing of functionally graded specimens has intensified the use of numerical tools to analyse their mechanical response. Although a variety of numerical techniques have been used, including mesh-free methods [6,7] and enriched formulations [8,9], the finite element method is by far the most popular approach [10–15]. Several formulations have been proposed to accurately capture a smooth material gradient by defining the material property variation at the element level [11–13]. While these formulations are been widely and indistinctly used, a performance assessment of the different types of *graded* elements has not been conducted yet. We investigate the performance of different types of functionally *graded* elements by comparing with analytical solutions of paradigmatic boundary value problems. We show that notable differences can be attained and that the most common approach in commercial software entails a number of limitations. An alternative implementation is presented in the context of the commercial finite element package ABAQUS.

2. Numerical Formulation

The assignment of material properties in the numerical model must reflect the material property distribution in the functionally graded specimen under consideration. However, an accurate characterization of the material gradient is not a straightforward task. Typically, the information available is the spatial variation of the volume fractions of constituent materials, which is provided as input to the production technique [16]. The macroscopic material property variation does not tend to mirror the volume fraction profile, but one can estimate the former from the latter by using homogenization laws [17]. However, the micromechanical assumptions upon which these theoretical mixing laws are built may hinder an accurate characterization of the macroscopic variation of material properties. An alternative approach is to determine the material property variation directly by experimentation. For example, by producing and testing individual homogeneous specimens with a range of volume fractions [18], by testing the graded material through indentation or ultrasonic techniques [19], or by cutting and testing small samples from a larger graded specimen [20]. Capturing this material gradation profile in the numerical model is key to designing optimal FGM specimens, as well as reproducing and gaining insight into experimental results.

From the numerical perspective, material properties can vary between elements or between nodes and integration points. Numerical works in the mechanics of functionally graded materials can be classified into two large groups depending on their approach to the implementation of the material gradient, see Figure 1, using either *homogeneous* elements (see, e.g., [10]) or *graded* elements (see [15] and references therein). The former is appropriate for layered functionally graded composites, but it constitutes a poor approximation otherwise. Assuming constant material properties within each element leads to a discontinuous step-type variation and requires uniform meshing along the material gradation direction. A material property variation at the element level is generally more appropriate and different graded elements formulations have been proposed.

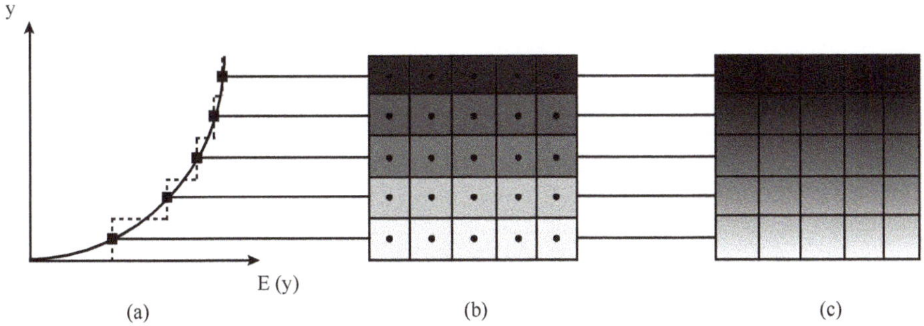

Figure 1. Sketch outlining the (**a**) gradual variation of Young's modulus E along the y-coordinate, as captured by (**b**) *homogeneous* elements and (**c**) *graded* elements.

2.1. Gauss Integration Point-Based Variation

Consider an isoparametric finite element with n number of nodes, the displacement field $\boldsymbol{u}(\boldsymbol{x})$ is interpolated from the nodal values $\hat{\boldsymbol{u}}_i$ as

$$\boldsymbol{u} = \sum_{i=1}^{n} N_i(\xi, \eta, \zeta)\, \hat{\boldsymbol{u}}_i, \tag{1}$$

where i is a given node and N_i are the shape functions. For example, in an eight-node quadrilateral element, the shape functions read, for the corner nodes

$$N_i = \frac{1}{4}\left(1+\xi\xi_i\right)\left(1+\eta\eta_i\right)\left(\xi\xi_i+\eta\eta_i-1\right) \tag{2}$$

and for the mid-side nodes,

$$N_i = \frac{1}{2}\left(1-\xi^2\right)\left(1+\eta\eta_i\right), \qquad \text{for } \xi_i = 0, \tag{3}$$

$$N_i = \frac{1}{2}\left(1+\xi\xi_i\right)\left(1-\eta^2\right), \qquad \text{for } \eta_i = 0, \tag{4}$$

with (ξ, η) denoting the intrinsic coordinates in the interval $[-1, 1]$, and (ξ_i, η_i) denoting the local coordinates of node i. Accordingly, the strain field $\boldsymbol{\varepsilon}(\boldsymbol{x})$ is computed from the nodal displacements by means of the strain-displacement matrix

$$\boldsymbol{\varepsilon} = \sum_{i=1}^{n} \boldsymbol{B}_i\left(\xi, \eta, \zeta\right) \hat{\boldsymbol{u}}_i \tag{5}$$

with the matrix $\boldsymbol{B}_i$ containing the appropriate derivatives of the shape functions N_i. For example, in a plane strain element, the strain-displacement matrix for a node i reads

$$\boldsymbol{B}_i\left(\xi, \eta\right) = \begin{bmatrix} \partial N_i\left(\xi, \eta\right)/\partial x & 0 \\ 0 & \partial N_i\left(\xi, \eta\right)/\partial y \\ 0 & 0 \\ \partial N_i\left(\xi, \eta\right)/\partial y & \partial N_i\left(\xi, \eta\right)/\partial x \end{bmatrix} \tag{6}$$

so as to compute the strain components $\varepsilon_{xx}, \varepsilon_{yy}, \varepsilon_{zz}, \gamma_{xy}$ from the nodal displacements.

Let us assume linear elastic behaviour, which is arguably appropriate for ceramic-based FGMs. The Cauchy stress field $\boldsymbol{\sigma}$ is related to the strain tensor $\boldsymbol{\varepsilon}$ through a *spatially varying* constitutive matrix $\boldsymbol{C}\left(\boldsymbol{x}\right)$ as

$$\boldsymbol{\sigma} = \boldsymbol{C}\left(\boldsymbol{x}\right)\boldsymbol{\varepsilon}. \tag{7}$$

The principle of virtual work yields the relation between the deformation work given by the element and the elemental nodal force vector $\boldsymbol{F}^e$ as

$$\boldsymbol{K}^e \hat{\boldsymbol{u}}_i = \boldsymbol{F}^e, \tag{8}$$

where $\boldsymbol{K}^e$ is the element strain-displacement matrix—see, for example, Ref. [21]. Accordingly, the element stiffness matrix over the volume of the element V^e reads

$$\boldsymbol{K}^e = \int_{V_e} \boldsymbol{B}^{e\,T} \boldsymbol{C}\left(\boldsymbol{x}\right) \boldsymbol{B}^e \, \mathrm{d}V, \tag{9}$$

where $\boldsymbol{B}^e$ is the element stiffness matrix, given by the assembly of $\boldsymbol{B}_i$ over n nodes. Therefore, the linear elastic stiffness matrix is defined to match the material gradation profile at the Gauss integration points. This basic finite element formulation for functionally graded solids was presented by Santare and Lambros [12].

2.2. Nodal-Based Variation via Temperature Dependence

An alternative approach to develop a formulation for graded finite elements was proposed by Kim and Paulino [13]. They propose a generalized isoparametric finite element formulation where the same shape functions are employed to interpolate the nodal displacements, the geometry, and the material properties. Thus, consider a standard isoparametric formulation where the spatial coordinates (x, y, z) are interpolated as

$$x = \sum_{i=1}^{n} N_i x_i, \quad y = \sum_{i=1}^{n} N_i y_i, \quad z = \sum_{i=1}^{n} N_i z_i. \tag{10}$$

The isoparametric concept can be generalized to interpolate the spatially varying Young's modulus $E(\boldsymbol{x})$ and Poisson's ratio $\nu(\boldsymbol{x})$ as

$$E = \sum_{i=1}^{n} N_i E_i, \qquad \nu = \sum_{i=1}^{n} N_i \nu_i, \tag{11}$$

where E_i and ν_i are the elastic properties defined at each node i. Hence, the material gradient is defined precisely at the nodes and subsequently interpolated to the Gauss integration points to compute the stresses through Equation (7).

A generalized isoparametric graded finite element can be easily implemented into a commercial finite element package by taking advantage of the possibility of defining temperature-dependent material properties [11,15]. For example, one can define E as a function of the temperature and provide the specimen with an initial temperature distribution that matches the Young's modulus variation desired. Here, the temperature has no physical meaning and unwanted thermal strains are suppressed by assigning a zero thermal expansion coefficient. Since the temperature field is defined at the nodes and subsequently interpolated to the Gauss integration points, this technique constitutes a straightforward implementation of a generalized isoparametric graded element, enjoying great popularity. Evident drawbacks are the inability to (i) model thermomechanical problems, and (ii) define different profiles for Poisson's ratio and Young's modulus. Furthermore, to obtain a consistent variation of mechanical and thermal strains, many commercial codes interpolate nodal temperature values using shape functions one order lower than those used for the nodal displacements. Consequently, there is an inherent error in the presence of a nonlinear material gradation profile, as sketched in Figure 2. The implications of adopting this technique, relative to the Gauss-based approach defined in Section 2.1, are explored here.

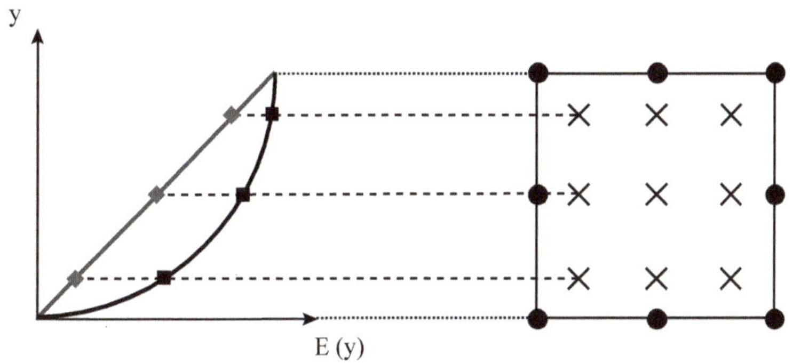

Figure 2. Sketch outlining the gradual variation of Young's modulus E and its associated interpolation by means of temperature-based generalized isoparametric graded element for an equivalent interpolation of thermal and mechanical strains.

3. Results

The variation in elastic properties inherent to FGMs is implemented at the element level by making use of user subroutines within the commercial finite element package ABAQUS. The graded elements described in Section 2 can be readily implemented by using a USDFLD subroutine, for a Gauss points-based implementation, or a UFIELD subroutine, for a nodal-based graded element. In addition, as discussed above, the temperature can be used as an auxiliary field to effectively implement a generalized isoparametric graded element. The user must provide the material properties as a function of a user defined field (or temperature). Then, a suitable field (or temperature distribution) is defined to match the material property variation desired. A direct comparison between the two approaches in terms of computational time is hindered by their different implementations; nevertheless, the sampling of material properties at integration points or nodes is achieved at a negligible computational cost. The performance of different types of graded elements will be benchmarked by considering a Gauss point-based implementation (Section 2.1) and, via temperature dependent properties, a generalized isoparametric approach (Section 2.2). For the sake of simplicity, we will consider bi-dimensional problems and four quadrilateral element types (see Figure 3): linear elements with reduced integration (Q4R), linear elements with full integration (Q4), quadratic elements with reduced integration (Q8R), and quadratic elements with full integration (Q8).

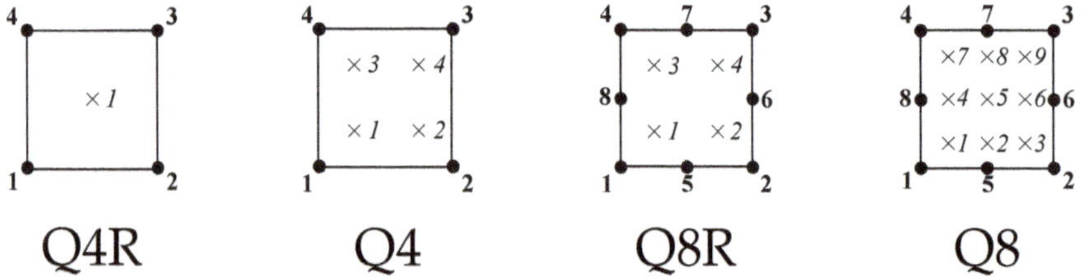

Figure 3. Graded finite elements employed and notation used.

Three plane problems for which analytical solutions can be obtained will be addressed, as shown in Figure 4. Young's modulus will be varied along the x-direction and Poisson's ratio will be assumed to be constant. We assume plane stress conditions. As sketched in Figure 4, the three boundary value problems considered involve a functionally graded plate being subjected to (i) uniform displacement perpendicular to the material gradient direction, (ii) uniform traction perpendicular to the material gradient direction, and (iii) uniform traction in the direction parallel to material gradation.

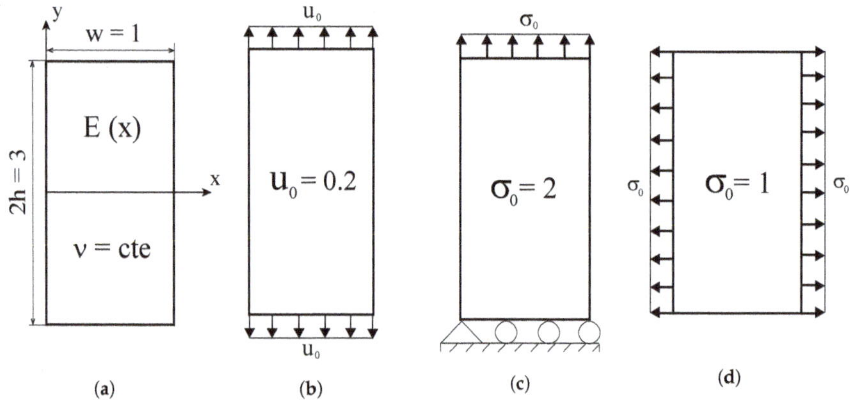

Figure 4. Boundary value problems under consideration: (**a**) functionally graded plate with spatially varying Young's modulus subjected to (**b**) uniform displacement perpendicular to the material gradient direction, (**c**) uniform traction perpendicular to the material gradient direction, and (**d**) uniform traction in the direction parallel to material gradation—consistent units.

In all cases, we assume that the Young's modulus varies exponentially as

$$E(x) = E_0 \exp(\beta x) \tag{12}$$

with E_0 and β being material constants. This choice is motivated by the existence of analytical solutions for functionally graded solids exhibiting an exponential variation of the elastic properties; see, for example, Refs. [13,22]. Many other functions have been employed in the literature (see Ref. [23] for a review), but our choice is appropriate for our aim: comparing on equal footing different graded finite element implementations. Consistent units are throughout the manuscript and, therefore, units will be omitted subsequently. We consider a width $w = 1$, a total height of $2h = 3$, and choose $E_0 = 1$ and $\beta = \ln 8$ so as to vary E gradually in the x-direction from $E(0) = 1$ to $E(w) = 8$.

3.1. *Uniform Displacement Perpendicular to the Material Gradient Direction*

Consider first the case of an FGM plate subjected to a remote strain $\varepsilon_0 = u_0/h$, where u_0 denotes the displacement in the remote boundary and h denotes half the height of the plate. A constant Poisson's ratio of $\nu = 0.3$ throughout the plate is assumed. The relevant stress component is given by

$$\sigma_{yy}(x, y) = E(x)u_0/h. \tag{13}$$

This analytical solution is compared in Figure 5 with the finite element results obtained for the element types and graded element formulations described above. A remote displacement of $u_0 = 0.2$ is prescribed in all cases. As shown in the insets of Figure 5, a uniform mesh of 4 by 12 elements is employed.

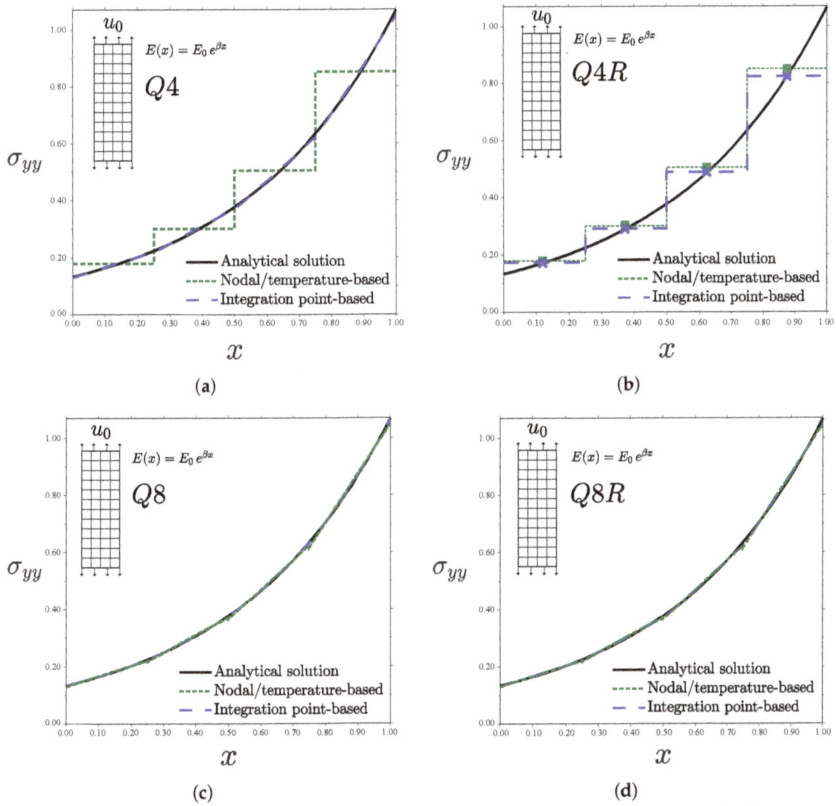

Figure 5. Uniform displacement perpendicular to the material gradient for different kinds of elements: (**a**) Q4; (**b**) Q4R; (**c**) Q8; and (**d**) Q8R—consistent units.

Results reveal differences between the different types of graded elements. Consider first the case of the linear element with full integration Q4—see Figure 5a. The Gauss integration point-based approach accurately captures the material gradient depicted by the analytical solution. In fact, the numerical result is exact at the integration points since the displacement field is linear; a single Q4 element will suffice to capture the FGM response. However, the generalized isoparametric formulation (via temperature-dependent properties) exhibits a step-type variation with constant stress in each element. This behaviour is inherently related to how ABAQUS interpolates nodal temperature values. As many other finite element packages, ABAQUS interpolates nodal temperatures with shape functions that are one order lower than those used for the displacements, so as to obtain an equivalent distribution of mechanical and thermal strains. An average value of the temperature in the nodes is passed to the integration points when using linear elements and a linear variation is assumed in quadratic elements. In agreement with expectations, the use of linear elements with reduced integration (Q4R, see Figure 5b) exhibits the response inherent to homogeneous elements for both cases. However, the constant value of σ_{yy} attained in each element depends on the implementation approach. The integration point-based scheme computes the exact σ_{yy} at the element centroid, where E is sampled. On the other hand, the nodal-based approach averages nodal temperatures, introducing a source of error when the elastic properties vary in a nonlinear manner. The use of quadratic elements leads to a good agreement with the analytical solution for both schemes, although differences are observed (Figure 5c,d). We plot the error obtained with both schemes for the element Q8 in Figure 6. It is evident that the integration point-based approach reproduces the analytical result more accurately.

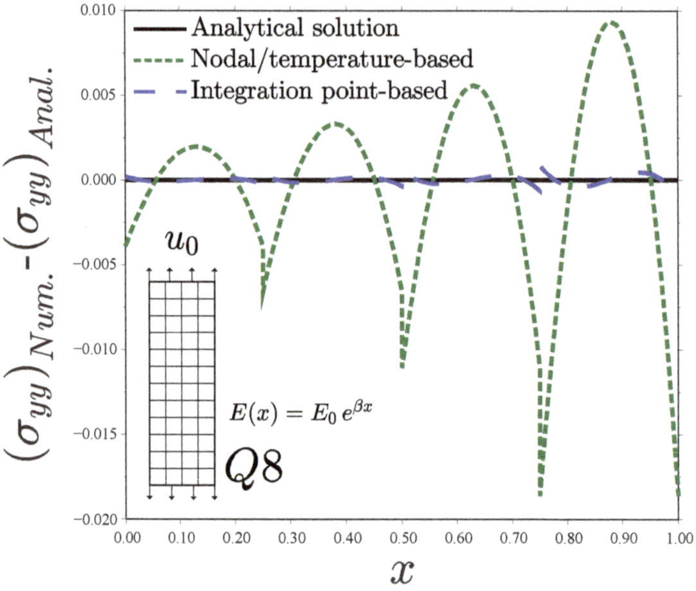

Figure 6. Uniform displacement perpendicular to the material gradient. Error analysis for the Q8 element. Consistent units.

Further insight is gained by reproducing the analysis with a single element in the x-direction—see Figure 7. Inspection of Figure 7a reveals substantial differences between Gauss points-based and nodal-based implementations. The former reproduces precisely the analytical stress distribution, with the exact result being obtained at the integration points. Contrarily, the approximation is much poorer when the material gradient is implemented via the temperature. Differences are particularly significant at the left side of the specimen, where the error is on the order of 50%.

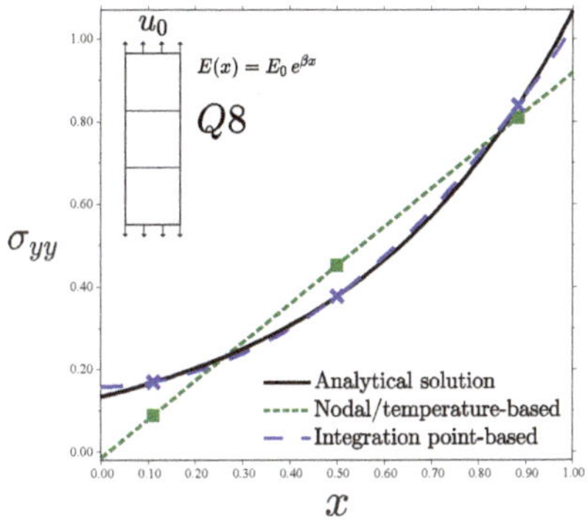

(**a**)

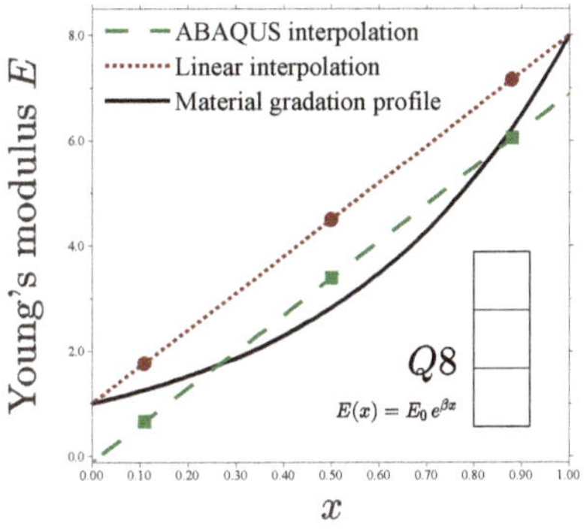

(**b**)

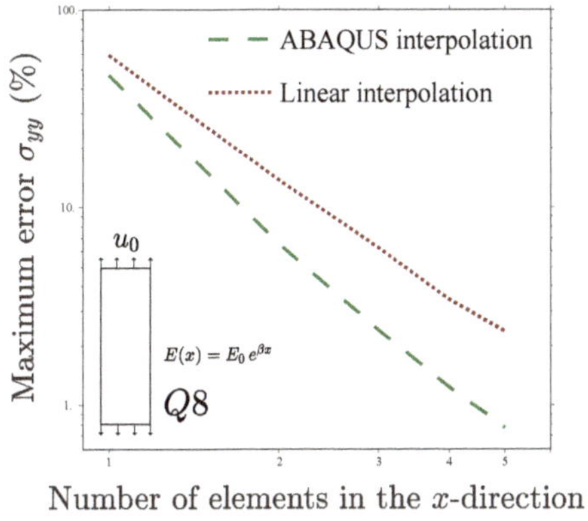

(**c**)

Figure 7. Uniform displacement perpendicular to the material gradient, (**a**) tensile stress for one Q8 element in the x-direction; (**b**) Young's modulus interpolation through different schemes; and (**c**) mesh-sensitivity error analysis—consistent units.

Remarkably, *negative* stresses are predicted for $x = 0$. These non-physical compressive stresses arise as a consequence of the particularities of ABAQUS' criterion for interpolating nodal temperatures, which does not correspond to the linear interpolation outlined in Figure 2. In ABAQUS, the nodal temperature values are multiplied by certain weights, such that the temperature T at an integration point i is given by

$$T_i = \sum_{j=1}^{m} T_j W_{ij} \quad \text{with}\, i = 1, \cdots, n. \tag{14}$$

Here, T_j is the temperature in node j, W_{ij} the weight associated with the nodal temperature j and integration point i, and n and m respectively denote the total number of nodes and integration points. The specific values of W_{ij} depend on a number of numerical considerations and can be easily obtained by means of a one-element model. This criterion is motivated by numerical convergence in thermomechanical problems, as it smoothens localized temperature peaks. The resulting variation in the elastic properties within the element is shown in Figure 7b. Differences with a direct linear interpolation are evident. The weighting procedure implemented in ABAQUS brings non-physical values of E, but it shows a better agreement with the material gradation profile at the Gauss integration points. In turn, this better approximation of $E(x)$ reduces the error in the computation of the stresses, as quantified in Figure 7c as a function of the number of elements. The log–log plot of Figure 7c shows that the weighted interpolation of ABAQUS exhibits a smaller maximum error in the computation of σ_{yy} at the Gauss points, as well as a faster convergence rate. Consequently, the error intrinsic to a temperature-based graded element is magnified if a standard linear interpolation is used and, therefore, the conclusions of the present study are even more relevant to finite element codes that employ non-weighted interpolations of nodal temperatures. Recall that the integration point-based scheme presented in Section 2.1 captures the analytical solution at the Gauss points with a single element.

3.2. Uniform Traction Perpendicular to the Material Gradient Direction

Consider now the case where the remote load is prescribed as a traction perpendicular to the elastic gradient—see Figure 4c. The Dirichlet boundary conditions of the problem read

$$u_x(0,0) = 0, \tag{15}$$

$$u_y(x,0) = 0. \tag{16}$$

In the case of a plate with infinite height, the only non-zero component of the Cauchy stress tensor is σ_{yy}. Following Refs. [13,22], a membrane resultant N along the $x = w/2$ line can be defined as a function of the remote stress σ_0 and the width,

$$N = \sigma_0 w. \tag{17}$$

The compatibility condition $\partial^2\varepsilon_{yy}/\partial x^2 = 0$ requires the strain component to be of the form

$$\varepsilon_{yy}(x) = Ax + B, \tag{18}$$

and, consequently, one can readily obtain the stress field by considering the exponential elastic modulus variation assumed (12) and making use of Hooke's law as

$$\sigma_{yy}(x) = E_0 e^{\beta x}(Ax + B). \tag{19}$$

The coefficients A and B are obtained by solving

$$\int_0^w \sigma_{yy}(x)\mathrm{d}x = N, \tag{20}$$

$$\int_0^w \sigma_{yy}(x)x\mathrm{d}x = N\frac{w}{2}, \tag{21}$$

such that

$$A = \frac{\beta N}{2E_0}\left(\frac{w\beta^2 e^{\beta w} - 2\beta e^{\beta w} + w\beta^2 + 2\beta}{e^{\beta w}\beta^2 w^2 - e^{2\beta w} + 2e^{\beta w} - 1}\right), \tag{22}$$

$$B = \frac{\beta N}{2E_0} \cdot \frac{e^{\beta w}[e^{\beta w}(-w^2\beta^2 + 3\beta w - 4) + w^2\beta^2 - 2\beta w + 8] - \beta w - 4}{(e^{\beta w} - 1)(e^{\beta w}\beta^2 w^2 - e^{2\beta w} + 2e^{\beta w} - 1)}. \tag{23}$$

The displacement solution can be readily obtained by making use of the strain-displacement relations and applying the boundary conditions (15) and (16),

$$u_x(x,y) = \nu\left(\frac{A}{2}x^2 + Bx\right) - \frac{A}{2}y^2, \tag{24}$$

$$u_y(x,y) = (Ax + B)\,y. \tag{25}$$

The analytical solution in the middle line $y = h/2$ is compared with the numerical predictions for $\sigma_0 = 2$. Results are shown in Figure 8 for a uniform mesh of 48 plane stress elements.

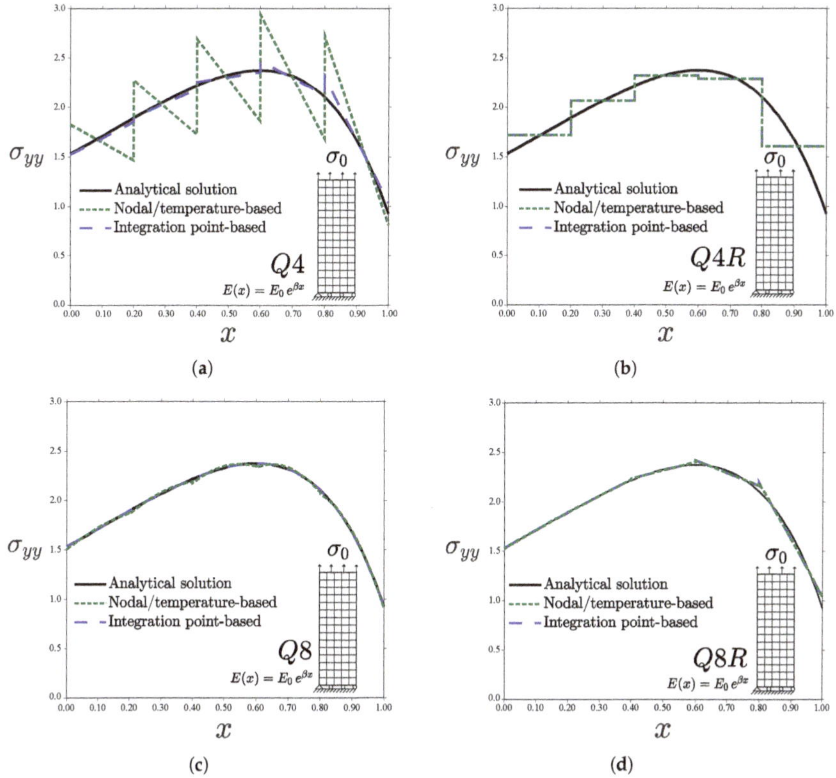

Figure 8. Uniform traction perpendicular to the material gradient for different kinds of elements: (**a**) Q4; (**b**) Q4R; (**c**) Q8; and (**d**) Q8R—consistent units.

Differences between the integration point-based scheme and the nodal/temperature-based implementation are particularly significant for the case of linear elements with full integration (Q4, Figure 8a). Sampling the material gradient directly at the Gauss points leads to a good agreement with the analytical solution. However, using temperature-based properties renders the homogeneous element solution. Both the analytical and Gauss point-based solutions show an increasing σ_{yy} along x within those elements close to the left edge. Contrarily, the inverse response is observed when using temperature-dependent properties, as the strain field decreases with x and E is constant element-wise. The Q4 element predicts in all cases a linear variation of σ_{yy} within each element for the quadratic displacement solution under consideration. On the other hand, identical predictions between graded element schemes are obtained when using linear elements with reduced integration (Q4R, Figure 8b). This is unlike the case of a prescribed displacement (see Figure 5b), as the error in the approximation of the strain field ε_{yy} (exact at the integration points only for the Gauss points-based scheme) is compensated. Quadratic elements (Q8 and Q8R) introduce an element-wise variation of E in both approaches and, therefore, differences appear to be smaller than in linear elements—see Figure 8c,d. The error in the approximation is shown in Figure 9 for the Q8 element case. When using full integration, the Gauss point-based approach outperforms the temperature-based, generalized isoparameteric graded element.

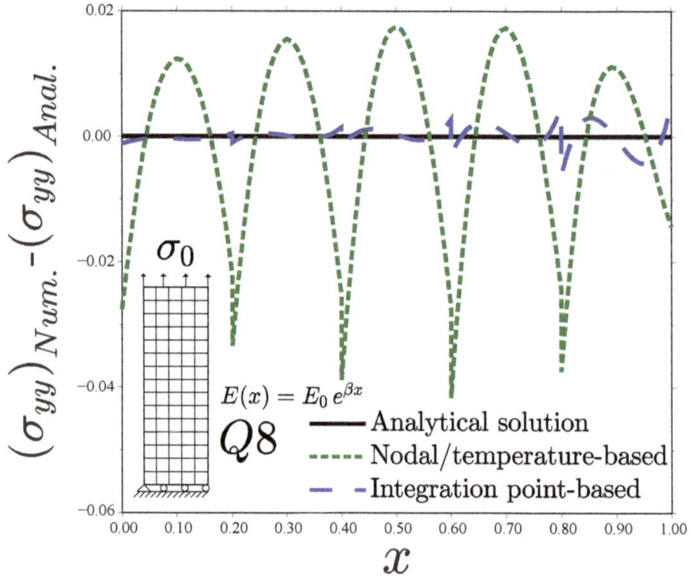

Figure 9. Uniform traction perpendicular to the material gradient. Error analysis for the Q8 element—consistent units.

Further insight is gained by analysing a very coarse mesh with a single element in the x-direction; results are shown in Figure 10. Regarding the stress (Figure 10a), the prediction obtained from the integration point-based implementation of graded elements agrees well with the analytical solution, being exact at the Gauss points (symbols). On the other hand, the approximation via a nodal/temperature-based approach introduces a significant source of error (larger than 20% in all the integration points). Moreover, when the load is prescribed as a traction, the approximation of the material gradient also influences the strain field, see Equations (8) and (9). As shown in Figure 10b, a better approximation is attained if the material properties are sampled directly at the Gauss points. Both schemes differ at the edges with the analytical solution for a plate of infinite height.

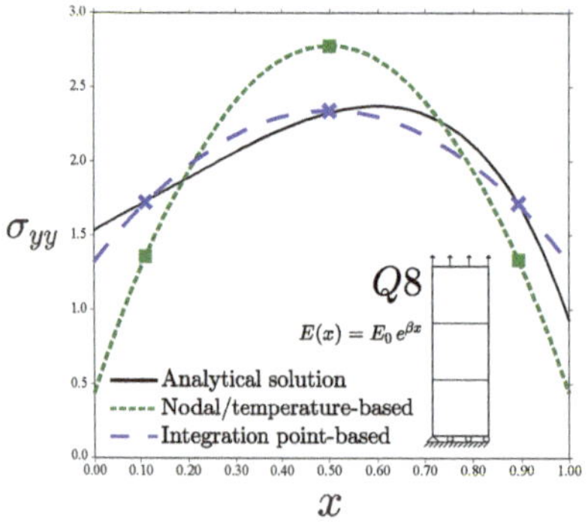

(a)

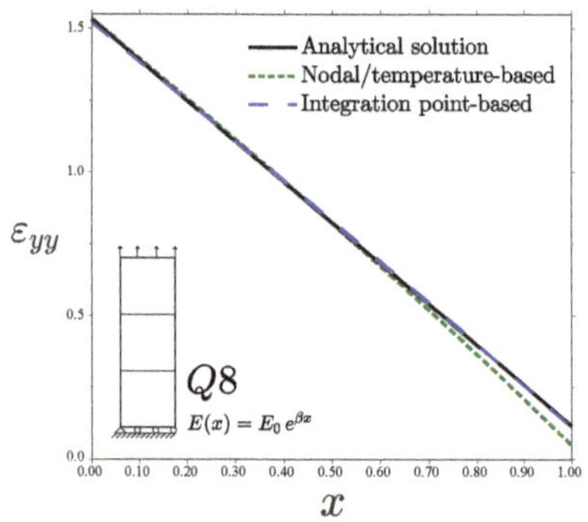

(b)

Figure 10. Uniform traction perpendicular to the material gradient, tensile (**a**) stress and (**b**) strain for one Q8 element in the x-direction—consistent units.

3.3. *Uniform Traction Parallel to the Material Gradient Direction*

The last case study involves a functionally graded plate subjected to traction in the x-direction, parallel to the elastic gradient, see Figure 4d. Under those conditions, the normal stress component equals the applied stress

$$\sigma_{xx}(x, y) = \sigma_0 \tag{26}$$

if Poisson's ratio is made equal to zero, $\nu = 0$. The results obtained for a uniform mesh of 75 plane stress elements are shown in Figure 11.

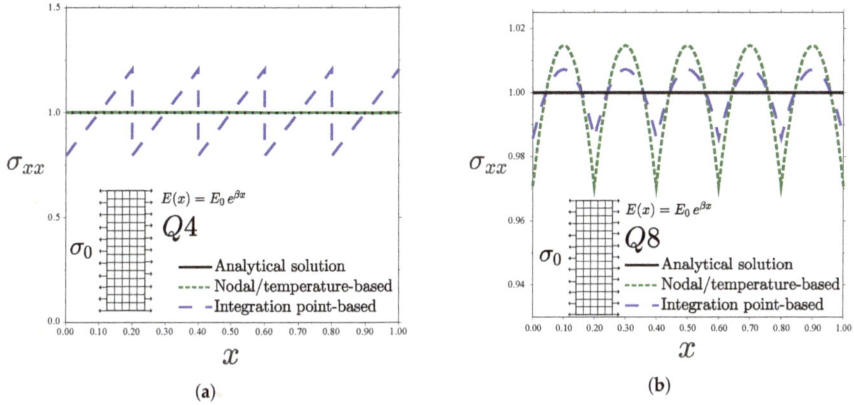

Figure 11. Uniform traction parallel to the material gradient for different kinds of elements: (**a**) Q4 and (**b**) Q8—consistent units.

Contrarily to what has been observed so far, the Q4 results (Figure 11a) show that the nodal/temperature-based approach outperforms the integration point-based counterpart. Hooke's law requires the strains to vary according to an inverse exponential distribution to obtain a constant stress for an exponentially varying E. However, linear elements predict a constant strain field and, consequently, an effectively homogeneous element will predict a constant stress. This trend is inverted for the case of a quadratic element with full integration (Q8). As shown in Figure 11b, again the use of a Gauss point-based graded element formulation approximates the analytical solution better. The results pertaining to reduced integration elements (Q4R and Q8R), not shown for brevity, reveal a perfect agreement with the analytical solution in all cases. Thus, reduced integration improves precision in this specific case study as the resulting stress is constant—either because there is a single integration point (Q4R), leading to a constant ε_{xx} and E, or because both ε_{xx} and E are element-wise linear (Q8R).

4. Conclusions

We have explored the influence of element order, integration scheme and graded element formulation in the finite element analysis of functionally graded materials (FGMs). Two graded element formulations are presented to account for the variation in space of material properties: nodal and integration point based gradations. The nodal based variation is implemented by defining temperature-dependent properties with a zero thermal expansion coefficient, a simple approach that enables the use of this scheme in commercial finite element packages. Important insight is gained by solving, analytically and numerically, three boundary value problems involving remote tractions and displacements, applied parallel and perpendicular to the material gradation direction.

Results reveal that integration point-based graded elements generally outperform a nodal-based implementation through temperature-dependent properties. The former approximates better the analytical solution in all the boundary value problems considered if quadratic shape functions are used. A much finer mesh is needed to attain a similar degree of precision with a nodal-based approach.

However, the temperature-based generalized isoparametric graded element is more accurate when linear elements are employed and the traction is applied parallel to the direction of material gradation. These observations are inherent to the interpolation of nodal temperatures with shape functions that are one order lower than those employed for the displacement field, as done in most finite element codes to ensure an equivalent variation of thermal and mechanical strains. In addition, in the case of the commercial package ABAQUS, the nodal temperature averaging criterion employed can lead to non-physical results.

The results presented have implications in the analysis of functionally graded structures, both in terms of computation time and local precision, particularly relevant for fracture studies. A user subroutine for ABAQUS is presented to overcome the number of drawbacks identified with the most popular graded finite element implementation. The user subroutine can be downloaded from www.empaneda.com/codes.

Funding: The author gratefully acknowledges financial support from the People Programme (Marie Curie Actions) of the European Union's Seventh Framework Programme (FP7/2007-2013) under REA grant agreement No. 609405 (COFUNDPostdocDTU).

Conflicts of Interest: The author declares no conflict of interest.

References

1. Kawasaki, A.; Watanabe, R. Finite element analysis of thermal stress of the metal/ceramic multi-layer composites with controlled compositional gradients. *J. Jpn. Inst. Metals* **1987**, *51*, 525–529. [CrossRef]
2. Uemura, S. The Activities of FGM on New Application. *Mater. Sci. Forum* **2003**, *423–425*, 1–10. [CrossRef]
3. Lee, Y.D.; Erdogan, F. Residual/thermal stresses in FGM and laminated thermal barrier coatings. *Int. J. Fract.* **1994**, *69*, 145–165. [CrossRef]
4. Ramaswamy, P.; Seetharamu, S.; Varma, K.; Rao, K. Al_2O_3-ZrO_2 Composite coatings for thermal-barrier applications. *Compos. Sci. Technol.* **1997**, *57*, 81–89. [CrossRef]
5. Marur, P.R.; Tippur, H.H.V. Evaluation of mechanical properties of functionally graded materials. *J. Test. Eval.* **1998**, *26*, 539–545.
6. Rao, B.N.; Rahman, S. Mesh-free analysis of cracks in isotropic functionally graded materials. *Eng. Fract. Mech.* **2003**, *70*, 1–27. [CrossRef]
7. Liu, P.; Yu, T.; Bui, T.Q.; Zhang, C. Transient dynamic crack analysis in non-homogeneous functionally graded piezoelectric materials by the X-FEM. *Comput. Mater. Sci.* **2013**, *69*, 542–558. [CrossRef]
8. Comi, C.; Mariani, S. Extended finite element simulation of quasi-brittle fracture in functionally graded materials. *Comput. Methods Appl. Mech. Eng.* **2007**, *196*, 4013–4026. [CrossRef]
9. Natarajan, S.; Baiz, P.M.; Bordas, S.; Rabczuk, T.; Kerfriden, P. Natural frequencies of cracked functionally graded material plates by the extended finite element method. *Compos. Struct.* **2011**, *93*, 3082–3092. [CrossRef]
10. Bao, G.; Wang, L. Multiple cracking in functionally graded ceramic/metal coatings. *Int. J. Solids Struct.* **1995**, *32*, 2853–2871. [CrossRef]
11. Rousseau, C.E.; Tippur, H.V. Compositionally graded materials with cracks normal to the elastic gradient. *Acta Mater.* **2000**, *48*, 4021–4033. [CrossRef]
12. Santare, M.H.; Lambros, J. Use of Graded Finite Elements to Model the Behavior of Nonhomogeneous Materials. *J. Appl. Mech.* **2000**, *67*, 819–822. [CrossRef]
13. Kim, J.H.; Paulino, G.H. Isoparametric Graded Finite Elements for Nonhomogeneous Isotropic and Orthotropic Materials. *J. Appl. Mech.* **2002**, *69*, 502. [CrossRef]
14. Valizadeh, N.; Natarajan, S.; Gonzalez-Estrada, O.A.; Rabczuk, T.; Bui, T.Q.; Bordas, S.P.A. NURBS-based finite element analysis of functionally graded plates: Static bending, vibration, buckling and flutter. *Compos. Struct.* **2013**, *99*, 309–326. [CrossRef]
15. Martínez-Pañeda, E.; Gallego, R. Numerical analysis of quasi-static fracture in functionally graded materials. *Int. J. Mech. Mater. Design* **2015**, *11*, 405–424. [CrossRef]
16. Butcher, R.J.; Rousseau, C.E.; Tippur, H.V. A functionally graded particulate composite: Preparation, measurements and failure analysis. *Acta Mater.* **1998**, *47*, 259–268. [CrossRef]

17. Mathew, T.V.; Natarajan, S.; Martínez-Pañeda, E. Size effects in elastic-plastic functionally graded materials. *Compos. Struct.* **2018**, *204*, 43–51. [CrossRef]
18. Carrillo-Heian, E.M.; Carpenter, R.D.; Paulino, G.H.; Gibeling, J.C.; Munir, Z.A. Dense Layered Molybdenum Disilicide-Silicon Carbide Functionally Graded Composites Formed by Field-Activated Synthesis. *J. Am. Ceram. Soc.* **2001**, *84*, 962–968. [CrossRef]
19. Krumova, M.; Klingshirn, C.; Haupert, F.; Friedrich, K. Microhardness studies on functionally graded polymer composites. *Compos. Sci. Technol.* **2001**, *61*, 557–563. [CrossRef]
20. Abanto-Bueno, J.; Lambros, J. Parameters controlling fracture resistance in functionally graded materials under mode I loading. *Int. J. Solids Struct.* **2006**, *43*, 3920–3939. [CrossRef]
21. Bower, A.F. *Applied Mechanics of Solids*; CRC Press, Taylor & Francis: Boca Raton, FL, USA, 2009.
22. Erdogan, F.; Wu, B.H. The surface crack problem for a plate with functionally graded properties. *J. Appl. Mech.* **1997**, *64*, 449. [CrossRef]
23. Sayyad, A.S.; Ghugal, Y.M. Modeling and analysis of functionally graded sandwich beams: A review. *Mech. Adv. Mater. Struct.* **2018**, doi:10.1080/15376494.2018.1447178. [CrossRef]

materials

MDPI

Article

Description of Residual Stress and Strain Fields in FGM Hollow Disc Subject to External Pressure

Stanislav Strashnov [1,*], Sergei Alexandrov [2,3] and Lihui Lang [2]

1 Department of Civil Engineering, Peoples' Friendship University of Russia (RUDN University), Miklukho-Maklaya st. 6, 117198 Moscow, Russia
2 School of Mechanical Engineering and Automation, Beihang University, No. 37 Xueyuan Road, Beijing 100191, China; sergei_alexandrov@spartak.ru (S.A.); lang@buaa.edu.cn (L.L.)
3 Ishlinsky Institute for Problems in Mechanics, 101-1 Prospect Vernadskogo, 119526 Moscow, Russia
* Correspondence: shtrafnoy@gmail.com

Received: 15 December 2018; Accepted: 25 January 2019; Published: 31 January 2019

Abstract: Elastic/plastic stress and strain fields are obtained in a functionally graded annular disc of constant thickness subject to external pressure, followed by unloading. The elastic modulus and tensile yield stress of the disc are assumed to vary along the radius whereas the Poisson's ratio is kept constant. The flow theory of plasticity is employed. However, it is shown that the equations of the associated flow rule, which are originally written in terms of plastic strain rate, can be integrated with respect to the time giving the corresponding equations in terms of plastic strain. This feature of the solution significantly facilitates the solution. The general solution is given for arbitrary variations of the elastic modulus and tensile yield stress along the radial coordinate. However, it is assumed that plastic yielding is initiated at the inner radius of the disc and that no other plastic region appears in the course of deformation. The solution in the plastic region at loading reduces to two ordinary differential equations. These equations are solved one by one. Unloading is assumed to be purely elastic. This assumption should be verified a posteriori. An illustrative example demonstrates the effect of the variation of the elastic modulus and tensile yield stress along the radius on the distribution of stresses and strains at the end of loading and after unloading. In this case, it is assumed that the material properties vary according to power-law functions.

Keywords: hollow disc; external pressure; residual stress; residual strain; flow theory of plasticity

1. Introduction

Stress and strain analyses of solid and hollow circular discs have long been an important topic in the mechanics of solids. The motivation of doing such analyses is that circular discs subject to mechanical, thermal, and inertial loading are used in many sectors of industry. The performance of such discs under service conditions can be improved by using functionally graded materials. The material may be continuously graded or be piecewise homogeneous. It is assumed in the present paper that the distribution of all material properties is axisymmetric. Discs made of homogeneous materials are not discussed. Discs made of functionally graded materials have been the subject of intense research. A linearly elastic solution under plane stress and plane strain conditions has been given in [1], assuming that the disc is loaded by external or internal pressure. It has been concluded that the stress response of the functionally graded disc is significantly different from that of the homogeneous disc. Another plane stress solution of this boundary value problem has been obtained in [2] and another plane strain solution in [3]. A thermoelastic stress solution for a disc of variable thickness has been found in [2]. A thermoelastic analysis of a disc subject to a steady-state temperature distribution together with external and internal pressures has been provided in [4]. It has been assumed in this work that the material properties are arbitrary smooth functions of the radial coordinate. A similar boundary value problem for a multilayered hollow cylinder has been solved in [5]. A design driven by the minimization of induced stresses in elastic multilayer cylinders under plane stress conditions has been proposed in [6]. All of the solutions above are purely elastic or thermoelastic. The process of autofrettage of a functionally graded cylinder has been studied in [7]. The analysis of this process requires the use of an elastic/plastic model. In [7], the deformation theory of plasticity together with the von Mises yield criterion has been employed. Another elastic/plastic plane strain solution for a functionally graded cylinder has been given in [8]. The solution is based on Tresca's yield criterion, which significantly simplifies the analysis even in the case of the flow theory of plasticity.

There is a vast amount of literature on functionally graded rotating discs. The elastic response of an arbitrary functionally graded polar orthotropic disc has been investigated in [9]. Another purely elastic solution has been given in [10], using the finite difference method. Thermoelastic analyses have been presented in [11–14]. The effect of a non-uniform heat source on thermoelastic behavior of a functionally graded rotating disc has been investigated in [15]. The effect of viscosity on the response of a functionally graded rotating disc of variable thickness has been studied in [16]. The limit of elastic angular velocity has been determined in [17]. The effect of variable angular velocity on the elastic response of a functionally graded rotating disc has been analyzed in [18]. A design driven by weight optimization of a disc subject to thermomechanical loading has been proposed in [19]. Most of the available elastic/plastic solutions fall into three categories. A series of solutions is devoted to discs obeying Tresca's yield criteria [20,21]. As it has been mentioned before, the use of Tresca's yield criterion significantly simplifies the solution. Another category includes the solutions for the deformation theory of plasticity [22–25]. In some cases, using deformation theories of plasticity is justified since the stress path is nearly proportional. However, it has been shown in [26] that it may not be so in thin discs. The third category includes stress solutions [27,28]. In this case, no flow rule is necessary to find the solution.

An advantage of the present elastic/plastic solution is that the flow theory of plasticity in conjunction with the von Mises yield criterion is employed. It is assumed that a hollow disc is subject to external pressure, followed by unloading. First, the general solution is derived under plane stress conditions assuming that the elastic modulus and tensile yield stress are arbitrary smooth functions of the radial coordinate. It is, however, assumed that plastic yielding initiates at the inner radius of the disc and that there is one plastic region throughout the process of deformation. The Poisson's ratio is supposed to be constant. This is a typical assumption for functionally graded discs [1,3,10,19]. Second, a numerical example is given assuming that the material properties vary according to power-law functions. This is also a typical assumption for functionally graded discs [1,9–13]. The solution found can be considered as an extension of the solution provided in [1] to the plastic range.

2. Statement of the Problem

Consider a thin hollow disc of functionally graded material subject to uniform pressure p_0 over its outer radius b_0, followed by unloading. The inner radius of the disc is denoted as a_0. The thickness of the disc is constant. The mechanical properties of the disc are classified in terms of the yield stress tension σ_Y, Poisson's ratio ν, and Young's modulus E. It is assumed that the Poisson's ratio is constant, whereas the value of both σ_Y and E vary with radius. It is convenient to use a cylindrical coordinate system $(r,\ \theta,\ z)$ whose z-axis coincides with the axis of symmetry of the disc. The normal stresses in this coordinate system are the principal stresses. The state of stress is plane (i.e., the axial stress in the cylindrical coordinate system vanishes). Therefore, Hooke's law can be written as

$$\varepsilon_r^e = \frac{\sigma_r - \nu\sigma_\theta}{E},\ \varepsilon_\theta^e = \frac{\sigma_\theta - \nu\sigma_r}{E},\ \varepsilon_z^e = -\frac{\nu(\sigma_r + \sigma_\theta)}{E} \tag{1}$$

here σ_r is the radial stress, σ_θ is the circumferential stress, ε_r^e, ε_θ^e, and ε_z^e are the elastic strains referred to the cylindrical coordinate system. Plastic yielding is controlled by the von Mises yield criterion. Under a plane stress condition, this criterion reads

$$\sigma_r^2 + \sigma_\theta^2 - \sigma_r\sigma_\theta = \sigma_Y^2 \tag{2}$$

The flow theory of plasticity is adopted. The flow rule associated with the yield criterion (2) is

$$\xi_r^p = \lambda(2\sigma_r - \sigma_\theta),\ \xi_\theta^p = \lambda(2\sigma_\theta - \sigma_r),\ \xi_z^p = -\lambda(\sigma_r + \sigma_\theta) \tag{3}$$

here ξ_r^p, ξ_θ^p, and ξ_z^p are the plastic strain rates referred to the cylindrical coordinate system and λ is a non-negative multiplier. The total strain components in the cylindrical coordinate system are given by

$$\varepsilon_r = \varepsilon_r^e + \varepsilon_r^p,\ \varepsilon_\theta = \varepsilon_\theta^e + \varepsilon_\theta^p,\ \varepsilon_z = \varepsilon_z^e + \varepsilon_z^p \tag{4}$$

here ε_r^p, ε_θ^p, and ε_z^p are the plastic strains referred to the cylindrical coordinate system. The constitutive equations should be complemented with the equilibrium equation

$$\frac{\partial \sigma_r}{\partial r} + \frac{\sigma_r - \sigma_\theta}{r} = 0 \tag{5}$$

and the equation of strain compatibility of the form

$$r\frac{\partial \varepsilon_\theta}{\partial r} = \varepsilon_r - \varepsilon_\theta \tag{6}$$

The boundary conditions at the stage of loading are

$$\sigma_r = -p_0 \tag{7}$$

for $r = b_0$ and

$$\sigma_r = 0 \tag{8}$$

for $r = a_0$. The boundary conditions at the stage of unloading will be formulated in Section 6.

It is convenient to introduce the following dimensionless quantities:

$$\rho = \frac{r}{b_0},\ a = \frac{a_0}{b_0},\ k = \frac{\sigma_0}{E_0},\ p = \frac{p_0}{\sigma_0} \tag{9}$$

here σ_0 is the value of σ_Y at $r = b_0$ and E_0 is the value of E at $r = b_0$. Then, the variation of σ_Y and E with ρ can be represented as

$$\sigma_Y = \sigma_0\Phi(\rho) \text{ and } E = E_0\eta(\rho) \tag{10}$$

where $\Phi(\rho)$ and $\eta(\rho)$ are arbitrary functions of ρ satisfying the conditions $\Phi(\rho) = 1$ and $\eta(\rho) = 1$ at $\rho = 1$. In what follows, it is assumed that these functions are such that plastic yielding initiates at the inner radius of the disc and no other plastic region appears in continued deformation.

3. Purely Elastic Solution

When p is small enough, the entire disc is elastic. In this case, the total strains are equal to the elastic strains. The system of equations comprises Hooke' law, the equilibrium equation, and the equation of strain compatibility. Using (9) it is possible to rewrite Equations (5) and (6) as

$$\frac{\partial \sigma_r}{\partial \rho} + \frac{\sigma_r - \sigma_\theta}{\rho} = 0 \tag{11}$$

and

$$\rho \frac{\partial \varepsilon_\theta^e}{\partial \rho} = \varepsilon_r^e - \varepsilon_\theta^e \tag{12}$$

Eliminating the strains in this equation by means of (1) and using (9) results in

$$\frac{\partial \sigma_\theta}{\partial \rho} + \frac{\nu}{\rho}(\sigma_r - \sigma_\theta) + \frac{\eta}{\rho}(\sigma_\theta - \nu\sigma_r)\frac{\partial(\rho/\eta)}{\partial \rho} - \frac{(\sigma_r - \nu\sigma_\theta)}{\rho} = 0 \tag{13}$$

Equations (11) and (13) comprise the system for determining the distribution of stresses in the purely elastic disc. Then, the distribution of strains can be readily found from (1) and (9). However, the purely elastic solution is not of interest in the case under consideration. Therefore, the solution to Equations (11) and (13) is only necessary to determine the value of p at which plastic yielding is initiated. This value of p is denoted as p_e. By assumption, plastic yielding is initiated at $\rho = a$. It follows from the boundary condition (8) and the yield criterion (2) that $\sigma_\theta = -\sigma_Y$ at $\rho = a$ at the initiation of plastic yielding. Using (10), this condition can be rewritten as $\sigma_\theta = -\sigma_0\Phi(a)$ at $\rho = a$. This is one of the boundary conditions of the boundary value problem to be solved. The other boundary condition is given by (8). Equations (11) and (13) should be solved together with these boundary conditions. The value of p_e is readily found from this solution as $p_e = -\sigma_r(1)/\sigma_0$. In what follows, it is assumed that $p > p_e$.

4. Elastic/Plastic Stress Solution

If $p > p_e$, then the disc consists of two regions, elastic and plastic. The elastic region occupies the domain $\rho_c \le \rho \le 1$ and the plastic region the domain $a \le \rho \le \rho_c$. Here, ρ_c is the elastic/plastic boundary. Consider the plastic region. The yield criterion (2) is satisfied by the following substitution:

$$\frac{\sigma_r}{\sigma_0} = \frac{2\Phi(\rho)\sin\psi}{\sqrt{3}} \text{ and } \frac{\sigma_\theta}{\sigma_0} = \Phi(\rho)\left(\frac{\sin\psi}{\sqrt{3}} + \cos\psi\right) \tag{14}$$

here ψ is a new function of ρ. This function should be found from the solution. Substituting (14) into (11) yields

$$\frac{d\psi}{d\rho} + \frac{\tan\psi}{\Phi}\frac{d\Phi}{d\rho} + \frac{\left(\tan\psi - \sqrt{3}\right)}{2\rho} = 0 \tag{15}$$

The boundary condition to this equation follows from (8) and (14). In particular, $\sigma_r = 0$ if $\psi = 0$ or $\psi = \pi$. It is evident that $\sigma_\theta < 0$ at $\rho = a$. Then, the boundary condition to Equation (15) is

$$\psi = \pi \tag{16}$$

for $\rho = a$. Solving Equation (15) together with this boundary condition supplies the variation of ψ with ρ. This solution and (14) determine the distribution of the stresses in the plastic region. Let p_p be the value of p at which the entire disc becomes plastic. Putting in the solution for the radial stress $\rho = 1$ gives the value of p_p as $p_p = -\sigma_r(1)/\sigma_0$. In what follows, it is assumed that $p < p_p$. In this case, $a < \rho_c < 1$. The value of σ_r on the plastic side of the elastic/plastic boundary is denoted as σ_r^c and the value of σ_θ on the plastic side of the elastic/plastic boundary as σ_θ^c. These values are readily found from the solution of (15) and (14). Equations (11) and (13) are valid in the elastic region. The radial and circumferential stresses must be continuous across the elastic/plastic boundary. Therefore, the boundary conditions to Equations (11) and (13) are

$$\sigma_r = \sigma_r^c \text{ and } \sigma_\theta = \sigma_\theta^c \tag{17}$$

for $\rho = \rho_c$. Solving Equations (11) and (13) together with these boundary conditions supplies the distribution of the radial and circumferential stresses in the elastic region. In particular, the value of p

involved in (7) is determined from the equation $p = -\sigma_r(1)/\sigma_0$. Therefore, the solution found connects p and ρ_c. One of these parameters should be prescribed. Then, the other parameter is determined from the solution.

5. Elastic/Plastic Strain Solution

Consider the plastic region, $a \le \rho \le \rho_c$. Eliminating λ between the equations in (3) gives

$$\frac{\xi_r^p}{\xi_\theta^p} = \frac{(2\sigma_r - \sigma_\theta)}{(2\sigma_\theta - \sigma_r)}, \ \frac{\xi_z^p}{\xi_\theta^p} = -\frac{(\sigma_r + \sigma_\theta)}{(2\sigma_\theta - \sigma_r)} \tag{18}$$

Using (14), the stresses in these equations can be expressed in terms of ψ. Then, taking into account that $\xi_r^p = \partial\varepsilon_r^p/\partial t$, $\xi_\theta^p = \partial\varepsilon_\theta^p/\partial t$, and $\xi_z^p = \partial\varepsilon_z^p/\partial t$ Equation (18) is transformed to

$$\frac{\partial\varepsilon_r^p}{\partial t} = \frac{\left(\sqrt{3}\tan\psi - 1\right)}{2}\frac{\partial\varepsilon_\theta^p}{\partial t}, \ \frac{\partial\varepsilon_z^p}{\partial t} = -\frac{\left(\sqrt{3}\tan\psi + 1\right)}{2}\frac{\partial\varepsilon_\theta^p}{\partial t} \tag{19}$$

here t is the time. It is seen from the structure of Equation (15) and the boundary condition (16) that ψ is independent of t. Therefore, the coefficients of $\partial\varepsilon_\theta^p/\partial t$ in (19) are also independent of t, and the equations in (19) can be immediately integrated with respect to the time to give

$$\varepsilon_r^p = \frac{\left(\sqrt{3}\tan\psi - 1\right)}{2}\varepsilon_\theta^p, \ \varepsilon_z^p = -\frac{\left(\sqrt{3}\tan\psi + 1\right)}{2}\varepsilon_\theta^p \tag{20}$$

It has been taken into account here that $\varepsilon_r^p = \varepsilon_\theta^p = \varepsilon_z^p = 0$ at the elastic/plastic boundary. The elastic strains in the plastic region, ε_r^{ep}, ε_θ^{ep} and ε_z^{ep}, are determined from (1) and (14) with the use of (9) and (10). As a result,

$$\begin{aligned} \varepsilon_r^{ep} &= k\Lambda\left[\tfrac{(2-\nu)}{\sqrt{3}}\sin\psi - \nu\cos\psi\right], \ \varepsilon_\theta^{ep} = k\Lambda\left[\tfrac{(1-2\nu)}{\sqrt{3}}\sin\psi + \cos\psi\right], \\ \varepsilon_z^{ep} &= -\nu k\Lambda\left(\sqrt{3}\sin\psi + \cos\psi\right) \end{aligned} \tag{21}$$

here $\Lambda = \Phi/\eta$. The total strains are found from (4), (20), and (21) as

$$\begin{aligned} \varepsilon_r &= k\Lambda\left[\tfrac{(2-\nu)}{\sqrt{3}}\sin\psi - \nu\cos\psi\right] + \tfrac{\left(\sqrt{3}\tan\psi - 1\right)}{2}\varepsilon_\theta^p, \\ \varepsilon_\theta &= k\Lambda\left[\tfrac{(1-2\nu)}{\sqrt{3}}\sin\psi + \cos\psi\right] + \varepsilon_\theta^p, \\ \varepsilon_z &= -\nu k\Lambda\left(\sqrt{3}\sin\psi + \cos\psi\right) - \tfrac{\left(\sqrt{3}\tan\psi + 1\right)}{2}\varepsilon_\theta^p \end{aligned} \tag{22}$$

It follows from these equations that

$$\begin{aligned} \varepsilon_r - \varepsilon_\theta &= \left(\tan\psi - \sqrt{3}\right)\left[\tfrac{k(1+\nu)\cos\psi}{\sqrt{3}}\Lambda + \tfrac{\sqrt{3}}{2}\varepsilon_\theta^p\right], \\ \tfrac{\partial\varepsilon_\theta}{\partial\rho} &= k\tfrac{d\Lambda}{d\rho}\left[\tfrac{(1-2\nu)}{\sqrt{3}}\sin\psi + \cos\psi\right] + k\Lambda\left[\tfrac{(1-2\nu)}{\sqrt{3}}\cos\psi - \sin\psi\right]\tfrac{d\psi}{d\rho} + \tfrac{\partial\varepsilon_\theta^p}{\partial\rho}. \end{aligned}$$

.

These equations and Equation (6), in which r should be replaced with ρ by means of (9), combine to give

$$\begin{aligned} &\tfrac{\partial\varepsilon_\theta^p}{\partial\rho} - \tfrac{\sqrt{3}}{2}\tfrac{\left(\tan\psi - \sqrt{3}\right)}{\rho}\varepsilon_\theta^p + k\tfrac{d\Lambda}{d\rho}\left[\tfrac{(1-2\nu)}{\sqrt{3}}\sin\psi + \cos\psi\right] + \\ &k\Lambda\left[\tfrac{(1-2\nu)}{\sqrt{3}}\cos\psi - \sin\psi\right]\tfrac{d\psi}{d\rho} - \tfrac{k(1+\nu)\Lambda\left(\sin\psi - \sqrt{3}\cos\psi\right)}{\sqrt{3}\rho} = 0. \end{aligned}$$

The derivative $d\psi/d\rho$ in this equation can be eliminated by means of (15). Then,

$$\begin{aligned}&\frac{\partial\varepsilon_\theta^p}{\partial\rho}-\frac{\sqrt{3}}{2}\frac{(\tan\psi-\sqrt{3})}{\rho}\varepsilon_\theta^p+k\frac{d\Lambda}{d\rho}\left[\frac{(1-2\nu)}{\sqrt{3}}\sin\psi+\cos\psi\right]-\\&k\Lambda\left[\frac{(1-2\nu)}{\sqrt{3}}\cos\psi-\sin\psi\right]\left[\frac{\tan\psi}{\Phi}\frac{d\Phi}{d\rho}+\frac{(\tan\psi-\sqrt{3})}{2\rho}\right]-\\&\frac{k(1+\nu)\Lambda(\sin\psi-\sqrt{3}\cos\psi)}{\sqrt{3}\rho}=0\end{aligned} \tag{23}$$

Since ψ has already been determined as a function of ρ in Section 4, (23) is a linear differential equation for ε_θ^p. The boundary condition to this equation is

$$\varepsilon_\theta^p=0 \tag{24}$$

for $\rho=\rho_c$. Once Equation (23) has been solved, the distribution of ε_r^p and ε_z^p in the plastic region is found from (20) and the distribution of the total strains from (22).

The distribution of strains in the elastic region is determined from the solution for stress found in Section 4 and Hooke's law.

6. Unloading

Let p_f be the value of p at the end of loading. Then, the boundary conditions for the stage of unloading are

$$\Delta\sigma_r=0 \tag{25}$$

for $\rho=a$ and

$$\Delta\sigma_r=\sigma_0 p_f \tag{26}$$

for $\rho=1$. Here, $\Delta\sigma_r$ is the increment of the radial stress after unloading ($\Delta\sigma_\theta$ will stand for the increment of the circumferential stress). It is assumed that unloading is purely elastic. This assumption should be verified a posteriori. Equations (11) and (13), in which σ_r should be replaced with $\Delta\sigma_r$ and σ_θ with $\Delta\sigma_\theta$, are valid. An iterative procedure should be used for solving this system of equations together with the boundary conditions (25) and (26). Once this boundary value problem has been solved, the increment of strains is determined from Hooke's law as

$$\Delta\varepsilon_r=\frac{\Delta\sigma_r-\nu\Delta\sigma_\theta}{E},\ \Delta\varepsilon_\theta=\frac{\Delta\sigma_\theta-\nu\Delta\sigma_r}{E},\ \Delta\varepsilon_z=-\frac{\nu(\Delta\sigma_r+\Delta\sigma_\theta)}{E} \tag{27}$$

The distribution of residual stresses, σ_r^{res} and σ_θ^{res}, is given by

$$\sigma_r^{res}=\sigma_r^{(f)}+\Delta\sigma_r,\ \sigma_\theta^{res}=\sigma_\theta^{(f)}+\Delta\sigma_\theta \tag{28}$$

Here, $\sigma_r^{(f)}$ is the distribution of the radial stress and $\sigma_\theta^{(f)}$ is the distribution of the circumferential stress at the end of loading. These distributions have been found in Section 4. Substituting (28) into (2) provides the condition to verify that the process of unloading is purely elastic in the form

$$(\sigma_r^{res})^2+(\sigma_\theta^{res})^2-\sigma_r^{res}\sigma_\theta^{res}-\sigma_0^2\Phi^2\le 0 \tag{29}$$

Here, Equation (10) has been taken into account.

The distribution of residual strains, ε_r^{res}, ε_θ^{res} and ε_z^{res}, is given by

$$\varepsilon_r^{res}=\varepsilon_r^{(f)}+\Delta\varepsilon_r,\ \varepsilon_\theta^{res}=\varepsilon_\theta^{(f)}+\Delta\varepsilon_\theta,\ \varepsilon_z^{res}=\varepsilon_z^{(f)}+\Delta\varepsilon_z \tag{30}$$

Here, $\varepsilon_r^{(f)}$ is the distribution of the radial strain, $\varepsilon_\theta^{(f)}$ is the distribution of the circumferential strain and $\varepsilon_z^{(f)}$ is the distribution of the axial strain at the end of loading. These distributions have been found in Section 5.

7. Illustrative Example

It is often assumed that material properties vary according to a power law along the radius of the disc [1,9–13]. In the case under consideration one possible variant of this law reads

$$\Phi(\rho) = \rho^m \text{ and } \eta(\rho) = \rho^n \tag{31}$$

In all calculations, $\nu = 0.3$, $a = 0.3$, and $n = 0.3$. The value of m varies in the range $0 \leq m \leq 0.3$. It is worthy of note that there is no need to prescribe the values of σ_0 and E_0 for numerical analysis since the stress components are proportional to σ_0, and the strain components are proportional to k. It is assumed that $\rho_c = 0.8$. Then, the value of p_f has been found from the stress solution given in Section 4. The dependence of p_f on m is presented in Table 1. The variation of the radial and circumferential stresses with ρ at $p = p_f$ is depicted in Figures 1 and 2, respectively. It has been verified that the yield criterion (2) is not violated in the elastic region. The variation of the radial, circumferential, and axial strains with ρ at $p = p_f$ is depicted in Figures 3–5, respectively.

Using the values of p_f found (Table 1) the system of Equations (11) and (13) together with the boundary conditions (25) and (26) has been solved for $\Delta\sigma_r/\sigma_0$ and $\Delta\sigma_\theta/\sigma_0$. Having this solution and the radial distribution of the radial and circumferential stresses at the end of loading (Figures 1 and 2) the radial distribution of the residual stresses is determined from (28). The variation of the residual radial and circumferential stresses with ρ is shown in Figures 6 and 7, respectively. Then, it has been verified that the inequality (29) is satisfied in the range $a \leq \rho \leq 1$. Having the solution for $\Delta\sigma_r/\sigma_0$ and $\Delta\sigma_\theta/\sigma_0$, the increment of the strains is determined from (27). It is evident from (10) and (27) that these increments are proportional to k introduced in (9). The radial distribution of the residual strains is determined from (30). The variation of the residual radial, circumferential, and axial strains with ρ is shown in Figures 8–10, respectively.

Table 1. Dependence of the value of pressure at the end of loading on the value of m introduced in (31).

m	p_f
0	0.79
0.1	0.75
0.2	0.71
0.3	0.68

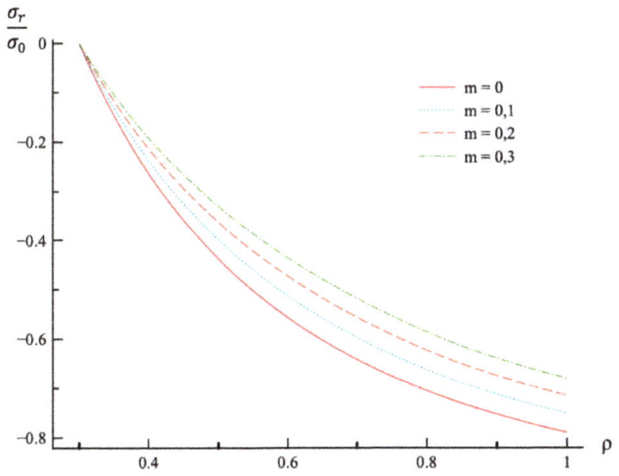

Figure 1. Variation of the radial stress, σ_r, with the dimensionless radius, ρ, at $p = p_f$.

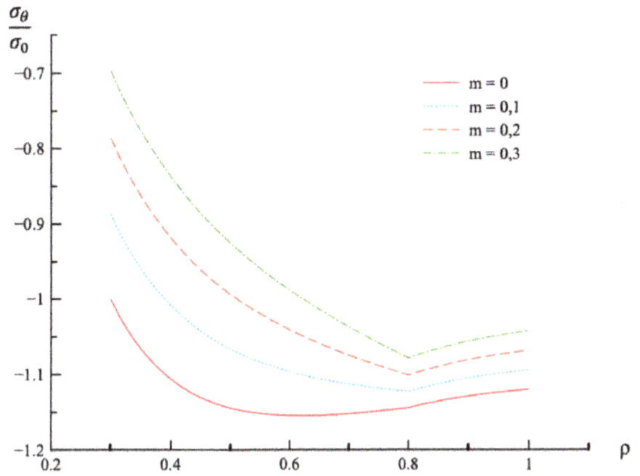

Figure 2. Variation of the circumferential stress, σ_θ, with the dimensionless radius, ρ, at $p = p_f$.

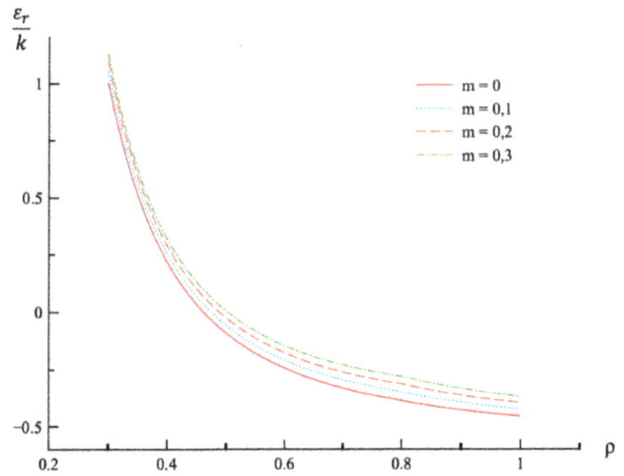

Figure 3. Variation of the radial strain, ε_r, with the dimensionless radius, ρ, at $p = p_f$.

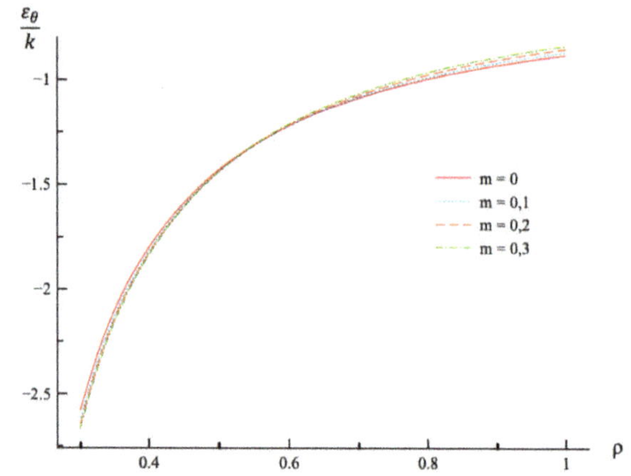

Figure 4. Variation of the circumferential strain, ε_θ, with the dimensionless radius, ρ, at $p = p_f$.

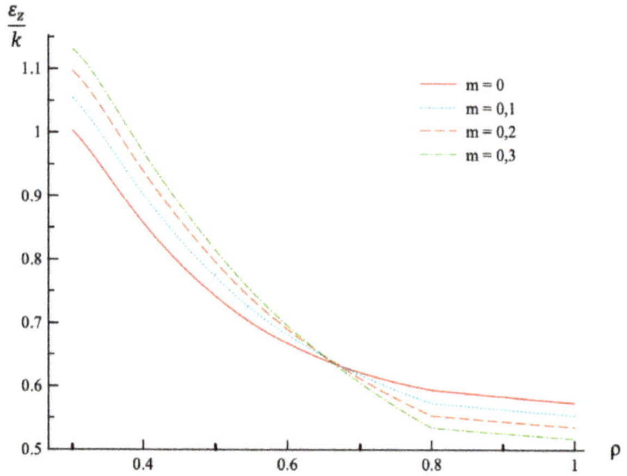

Figure 5. Variation of the axial strain, ε_z, with the dimensionless radius, ρ, at $p = p_f$.

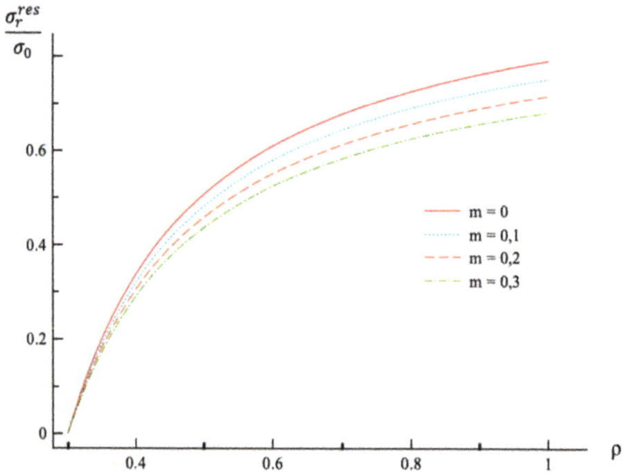

Figure 6. Variation of the residual radial stress, σ_r^{res}, with the dimensionless radius, ρ.

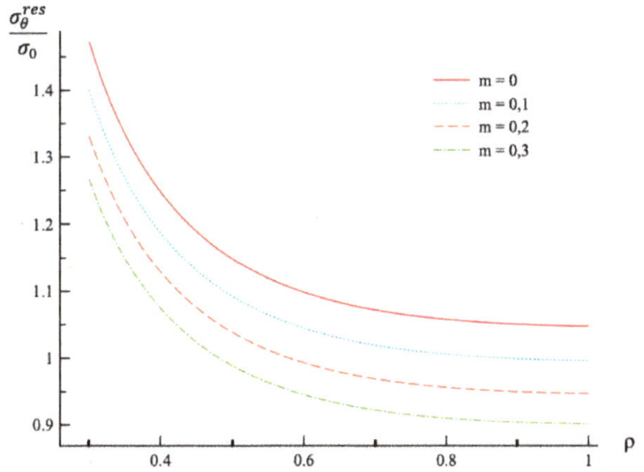

Figure 7. Variation of the residual circumferential stress, σ_θ^{res}, with the dimensionless radius, ρ.

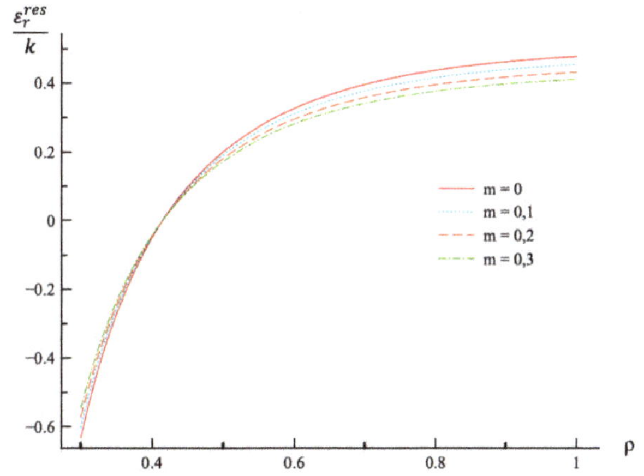

Figure 8. Variation of the residual radial strain, ε_r^{res}, with the dimensionless radius, ρ.

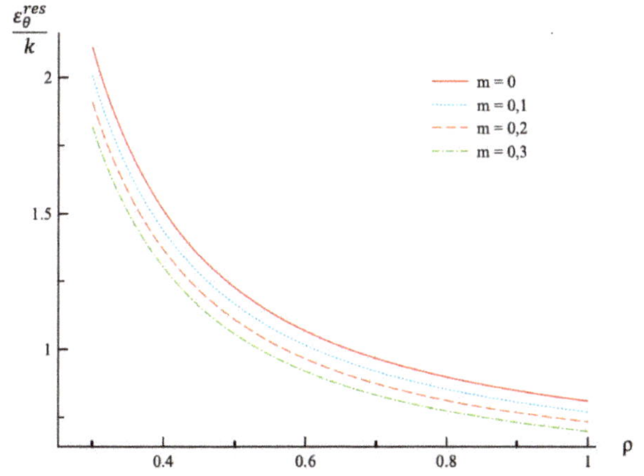

Figure 9. Variation of the residual circumferential strain, ε_θ^{res}, with the dimensionless radius, ρ.

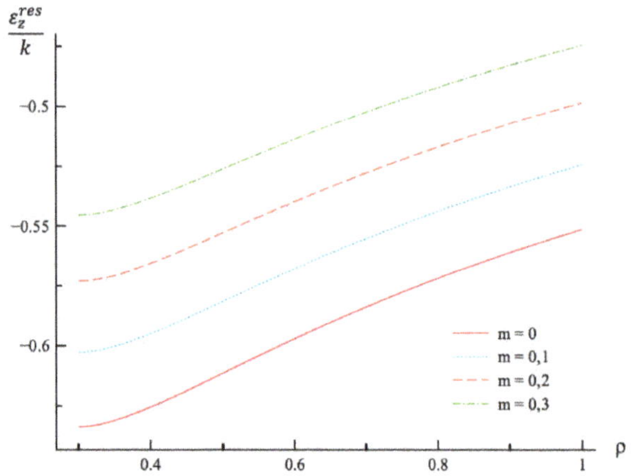

Figure 10. Variation of the residual axial strain, ε_z^{res}, with the dimensionless radius, ρ.

8. Discussion

The paper presents a general solution for the distribution of stress and strain in an functionally graded hollow disc subject to external pressure, followed by unloading. The solution is valid for any variation of the yield stress and Young's modulus with radius if the plastic region initiates at, and then propagates from, the inner radius of the disc. The purely elastic solution is valid if $p \leq p_e$. This solution is known, and therefore is not considered in the present paper. However, the general elastic solution is used in the elastic region, $\rho_c \leq \rho \leq 1$. The stress solution follows from the solution of Equation (13). Then, the distribution of strains is determined from Hooke's law shown in (1).

The constitutive equations of the classical flow theory of plasticity are adopted. In particular, Hooke's law shown in (1) is used to connect the stress components and the elastic strain components. This law is valid in the elastic region. The von Mises yield criterion (2) is adopted in the plastic region. In this region, the stress components are connected to the strain rate components rather than to the strain components. The corresponding constitutive equation is the associated flow rule (3). The total strain components in the plastic region are given by (4). A detailed description of this material model can be found in any textbook on plasticity theory (for example, [29,30]).

For any given functions $\Phi(\rho)$ and $\eta(\rho)$ involved in (10), the distribution of stress and strain in the plastic region, $a \leq \rho \leq \rho_c$, can be calculated as follows. Two ordinary differential equations, Equations (15) and (23), should be solved numerically. These equations can be solved one by one. In particular, the dependence of ψ on ρ is found from (15). Then, this numerical function is substituted into (23). As a result, a linear differential equation for the circumferential plastic strain, ε_θ^p, is obtained. The solution of Equation (15) supplies the distribution of stresses in the plastic region according to (14). A remarkable feature of the strain solution is that the equations in (18), which are derived from the associated flow rule written in terms of plastic strain rates, can be immediately integrated with respect to the time to result in the equations in terms of plastic strains (Equation (20)). This feature of the solution significantly facilitates the solution. Another remarkable feature of the strain solution is that all strain components are proportional to k introduced in (9). It is seen from Equations (22) and (23). Therefore, simple scaling of any strain solution provides the strain solution for similar discs of material with the same other properties and geometry, but any value of k. To calculate the distribution of strains, it is first necessary to solve Equation (23) for the circumferential plastic strain. Then, the radial and axial plastic strains are readily found from (20) and the solution of (15). The elastic portions of strain in the plastic region are determined from (14) and Hooke's law shown in (1). Finally, the total strains follow from (4).

The general solution found is used to find a numerical solution, assuming that the power laws shown in (31) are valid. The effect on m-value on the radial and circumferential stresses in an $a = 0.3$ disc at $\nu = 0.3$, $n = 0.3$, and $\rho_c = 0.8$ is illustrated in Figures 1 and 2, respectively. The corresponding values of the external pressure are depicted in Table 1. It is seen from these figures that this effect is quite significant, especially on the circumferential stress. The associated distribution of strains is illustrated in Figures 3–5 (the radial strain is depicted in Figure 3, the circumferential strain in Figure 4 and the axial strain in Figure 5). The effect of m-value on these distributions is not so pronounced as compared to the stress distributions. This is associated with the imposed condition that $\rho_c = 0.8$ in all cases. If the value of p_f were fixed, then the effect of m-value on the strain distributions would be more significant as compared to its effect on the stress distribution. In particular, it is seen from Table 1 that the value of p_f is quite sensitive to the value of m.

Residual stress and strain fields after purely elastic unloading are also obtained. These distributions are given by Equations (28) and (30). The validity of the pure elastic solution at unloading should be verified by means of Equation (29). The distribution of residual radial and circumferential stresses is illustrated in Figures 6 and 7, respectively. As in the case of the stress distributions at the end of loading, the effect of m-value is most significant on the circumferential stress. The effect of this value on the distribution of the residual radial strain is small (Figure 8). It is moderate in the case of the residual circumferential strain (Figure 9) and large in the case of the residual axial strain (Figure 10).

9. Conclusions

Stress and strain fields in an elastic/plastic functionally graded annular disc of constant thickness subject to external pressure are obtained under plane stress conditions. Residual stress and strain fields after purely elastic unloading are also obtained. From this work, the following conclusions can be drawn.

1. A remarkable feature of the strain solution is that the equations in (18), which are derived from the associated flow rule written in terms of plastic strain rates, can be immediately integrated with respect to the time to result in the equations in terms of plastic strains (Equation (20)). This significantly facilitates the solution.
2. Another remarkable feature of the strain solution is that all strain components are proportional to k introduced in (9). Therefore, simple scaling of any strain solution provides the strain solution for similar discs of material with the same other properties and geometry, but any value of k.
3. In the case of the stress solution, the effect of m-value involved in (31) is most significant on the distribution of the circumferential stress and the residual circumferential stress.
4. In the case of the strain solution, the effect of m-value is most significant on the distribution of the axial strain and the residual axial strain.

The method used in the present paper is a generalization of the method developed in [31] for homogeneous discs. It is evident from the solutions provided in [31] that the method can be more successfully adopted for disc subject to other loading conditions than those used in the present paper. In particular, the basic equations derived are independent of boundary conditions. Therefore, the solution of these equations used in conjunction with any other boundary conditions (of course, the boundary value problem should be axisymmetric) supplies the distribution of stress and strain. This will be the subject of a subsequent investigation.

Author Contributions: The general method has been developed by S.A., analytical treatment has been carried out by S.S., numerical results have been obtained by L.L.

Funding: S.A. acknowledges support from the Russian Ministry of Science and Higher Education (Project AAAA-A17-117021310373-3).

Acknowledgments: This work was performed while S.V. Strashnov was a visiting research fellow at Beihang University, Beijing, China.

Conflicts of Interest: The authors declare no conflict of interest.

References

1. Horgan, C.O.; Chan, A.M. The Pressurized Hollow Cylinder or Disk Problem for Functionally Graded Isotropic Linearly Elastic Materials. *J. Elast.* **1999**, *55*, 43–59. [CrossRef]
2. You, L.H.; Wang, J.X.; Tang, B.P. Deformations and stresses in annular disks made of functionally graded materials subjected to internal and/or external pressure. *Meccanica* **2009**, *44*, 283–292. [CrossRef]
3. Tutuncu, N. Stresses in thick-walled FGM cylinders with exponentially-varying properties. *Eng. Struct.* **2007**, *29*, 2032–2035. [CrossRef]
4. Gönczi, D.; Ecsedi, I. Thermoelastic Analysis of Functionally Graded Hollow Circular Disc. *Arch. Mech. Eng.* **2015**, *62*, 5–18. [CrossRef]
5. Yeo, W.H.; Purbolaksono, J.; Aliabadi, M.H.; Ramesh, S.; Liew, H.L. Exact solution for stresses/displacements in a multilayered hollow cylinder under thermo-mechanical loading. *Int. J. Press. Vessels Pip.* **2017**, *151*, 45–53. [CrossRef]
6. Shabana, Y.M.; Elsawaf, A.; Khalaf, H.; Khalil, Y. Stresses minimization in functionally graded cylinders using particle swarm optimization technique. *Int. J. Press. Vessels Pip.* **2017**, *154*, 1–10. [CrossRef]
7. Haghpanah Jahromi, B.; Farrahi, G.H.; Maleki, M.; Nayeb-Hashemi, H.; Vaziri, A. Residual stresses in autofrettaged vessel made of functionally graded material. *Eng. Struct.* **2009**, *31*, 2930–2935. [CrossRef]
8. Eraslan, A.N.; Akis, T. Plane strain analytical solutions for a functionally graded elastic–plastic pressurized tube. *Int. J. Press. Vessels Pip.* **2006**, *83*, 635–644. [CrossRef]
9. Yildirim, V. Numerical/analytical solutions to the elastic response of arbitrarily functionally graded polar orthotropic rotating discs. *J. Braz. Soc. Mech. Sci. Eng.* **2018**, *40*, 320. [CrossRef]
10. Jalali, M.H.; Shahriari, B. Elastic Stress Analysis of Rotating Functionally Graded Annular Disk of Variable Thickness Using Finite Difference Method. *Math. Probl. Eng.* **2018**, *2018*, 1–11. [CrossRef]
11. Hosseini Kordkheili, S.A.; Naghdabadi, R. Thermoelastic analysis of a functionally graded rotating disk. *Compos. Struct.* **2007**, *79*, 508–516. [CrossRef]
12. Hassani, A.; Hojjati, M.H.; Farrahi, G.; Alashti, R.A. Semi-exact elastic solutions for thermo-mechanical analysis of functionally graded rotating disks. *Compos. Struct.* **2011**, *93*, 3239–3251. [CrossRef]
13. Dai, T.; Dai, H.-L. Thermo-elastic analysis of a functionally graded rotating hollow circular disk with variable thickness and angular speed. *Appl. Math. Model.* **2016**, *40*, 7689–7707. [CrossRef]
14. Allam, M.N.M.; Tantawy, R.; Zenkour, A.M. Thermoelastic stresses in functionally graded rotating annular disks with variable thickness. *J. Theor. Appl. Mech.* **2018**, *56*, 1029–1041. [CrossRef]
15. Leu, S.-Y.; Chien, L.-C. Thermoelastic analysis of functionally graded rotating disks with variable thickness involving non-uniform heat source. *J. Therm. Stress.* **2015**, *38*, 415–426. [CrossRef]
16. Allam, M.N.M.; Tantawy, R.; Yousof, A.; Zenkour, A.M. Elastic and Viscoelastic Stresses of Nonlinear Rotating Functionally Graded Solid and Annular Disks with Gradually Varying Thickness. *Arch. Mech. Eng.* **2017**, *64*, 423–440. [CrossRef]
17. Sondhi, L.; Sanyal, S.; Saha, K.; Bhowmick, S. Limit Elastic Speeds of Functionally Graded Annular Disks. *FME Trans.* **2018**, *46*, 603–611.
18. Zheng, Y.; Bahaloo, H.; Mousanezhad, D.; Vaziri, A.; Nayeb-Hashemi, H. Displacement and Stress Fields in a Functionally Graded Fiber-Reinforced Rotating Disk with Nonuniform Thickness and Variable Angular Velocity. *J. Eng. Mater. Technol.* **2017**, *139*, 1–10. [CrossRef]
19. Khorsand, M.; Tang, Y. Design functionally graded rotating disks under thermoelastic loads: Weight optimization. *Int. J. Press. Vessels Pip.* **2018**, *161*, 33–40. [CrossRef]
20. Nejad, M.Z.; Rastgoo, A.; Hadi, A. Exact elasto-plastic analysis of rotating disks made of functionally graded materials. *Int. J. Eng. Sci.* **2014**, *85*, 47–57. [CrossRef]
21. Nejad, M.Z.; Fatehi, P. Exact elasto-plastic analysis of rotating thick-walled cylindrical pressure vessels made of functionally graded materials. *Int. J. Eng. Sci.* **2015**, *86*, 26–43. [CrossRef]
22. Hassani, A.; Hojjati, M.H.; Farrahi, G.H.; Alashti, R.A. Semi-exact solution for thermo-mechanical analysis of functionally graded elastic-strain hardening rotating disks. *Commun. Nonlinear Sci. Numer. Simulat.* **2012**, *17*, 3747–3762. [CrossRef]

23. Haghpanah Jahromi, B.; Nayeb-Hashemi, H.; Vaziri, A. Elasto-Plastic Stresses in a Functionally Graded Rotating Disk. *J. Eng. Mater. Technol.* **2012**, *134*, 1–11.
24. Mahdavi, E.; Ghasemi, A.; Alashti, R.A. Elastic–plastic analysis of functionally graded rotating disks with variable thickness and temperature-dependent material properties under mechanical loading and unloading. *Aerosp. Sci. Technol.* **2016**, *59*, 57–68. [CrossRef]
25. Kalali, A.T.; Hadidi-Moud, S.; Hassani, B. Elasto-Plastic Stress Analysis in Rotating Disks and Pressure Vessels Made of Functionally Graded Materials. *Latin Am. J. Solids Struct.* **2016**, *13*, 819–834. [CrossRef]
26. Pirumov, A.; Alexandrov, S.; Jeng, Y.-R. Enlargement of a Circular Hole in a Disc of Plastically Compressible Material. *Acta Mech.* **2013**, *224*, 2965–2976. [CrossRef]
27. Çallıoglu, H.; Sayer, M.; Demir, E. Elastic–plastic stress analysis of rotating functionally graded discs. *Thin-Walled Struct.* **2015**, *94*, 38–44. [CrossRef]
28. Zheng, Y.; Bahaloo, H.; Mousanezhad, D.; Mahdi, E.; Vaziri, A.; Nayeb-Hashemi, H. Stress analysis in functionally graded rotating disks with non-uniform thickness and variable angular velocity. *Int. J. Mech. Sci.* **2016**, *116*, 283–293. [CrossRef]
29. Chakrabarty, J. *Theory of Plasticity*; McGraw-Hill Book Company: New York, NY, USA, 1987.
30. Rees, D.W.A. *Basic Engineering Plasticity*; Elsevier: Amsterdam, The Netherlands, 2006.
31. Alexandrov, S. *Elastic/Plastic Discs under Plane Stress Conditions*; Springer: Berlin, Germany, 2015.

materials

MDPI

Article

One-Dimensional and Two-Dimensional Analytical Solutions for Functionally Graded Beams with Different Moduli in Tension and Compression

Xue Li [1], Jun-yi Sun [1,2,*], Jiao Dong [1] and Xiao-ting He [1,2]

1 School of Civil Engineering, Chongqing University, Chongqing 400045, China; lixuecqu@126.com (X.L.); dongjiaocqu@126.com (J.D.); hexiaoting@cqu.edu.cn (X.H.)

2 Key Laboratory of New Technology for Construction of Cities in Mountain Area (Chongqing University), Ministry of Education, Chongqing 400045, China

* Correspondence: sunjunyi@cqu.edu.cn; Tel.: +86-(0)23-6512-0720

Received: 17 April 2018; Accepted: 15 May 2018; Published: 17 May 2018

Abstract: The material considered in this study not only has a functionally graded characteristic but also exhibits different tensile and compressive moduli of elasticity. One-dimensional and two-dimensional mechanical models for a functionally graded beam with a bimodular effect were established first. By taking the grade function as an exponential expression, the analytical solutions of a bimodular functionally graded beam under pure bending and lateral-force bending were obtained. The regression from a two-dimensional solution to a one-dimensional solution is verified. The physical quantities in a bimodular functionally graded beam are compared with their counterparts in a classical problem and a functionally graded beam without a bimodular effect. The validity of the plane section assumption under pure bending and lateral-force bending is analyzed. Three typical cases that the tensile modulus is greater than, equal to, or less than the compressive modulus are discussed. The result indicates that due to the introduction of the bimodular functionally graded effect of the materials, the maximum tensile and compressive bending stresses may not take place at the bottom and top of the beam. The real location at which the maximum bending stress takes place is determined via the extreme condition for the analytical solution.

Keywords: functionally graded beams; different moduli in tension and compression; bimodulus; analytical solution; neutral layer

1. Introduction

Most materials may exhibit different elastic responses in a state of tension and compression, but these characteristics are often neglected due to the complexity of their analysis. Materials that have apparently different moduli in tension and compression are known as bimodular materials [1], for example, ceramics, graphite, concrete, and some biological materials (nacre, for example [2]). During recent decades, many studies have described useful material models for studying bimodular materials. One is Bert's model [3] based on the criterion of positive-negative signs of the strains in longitudinal fibers. This model is widely used in laminated composites [4–8]. Another is Ambartsumyan's bimodular model [9] for isotropic materials, which has attracted the most attention in the engineering community. This model assesses different moduli in terms of tension and compression based on the positive-negative signs of principal stresses, which is especially important for the analysis and design of structures. It is well-known that the cracking direction of a concrete beam is always normal to the direction of principal tensile stresses in the beam. The difficulty in applying Ambartsumyan's bimodular model is that the stress state of a point must be known in advance. However, with the exception of some fundamental problems, we must resort to finite element analysis to acquire the states of the stresses in a structure [10–14].

In addition to the bimodular effect in materials, it is also interesting to consider the functionally graded characteristic of materials. Functionally graded materials (FGMs) possess properties that vary gradually with the location within the material. The use of FGMs has many advantages in aerospace, automotive, and biomedical applications. There are many approximations that may be used to model the variation of properties in FGMs. One is the exponential variation, where the elastic constants vary according to the form of the exponential function. Many researchers have found this functional form to be convenient in solving elasticity problems. Sankar [15] obtained an elasticity solution for a functionally graded beam subjected to transverse loads in which the Young's modulus is assumed to vary exponentially through the thickness and the Poisson ratio is held constant. Sankar and co-workers studied the relative issues of functionally graded beams, including thermal stresses [16], a sandwich beam with a functionally graded core [17], and a combined Fourier series–Galerkin method [18]. Without specifying the gradient variations of a material property, Zhong and co-workers presented a general solution of a functionally graded beam by the Airy stress function method [19] and a displacement function approach [20]. Daouadji et al. [21] employed the stress function approach to study the problem of a functionally graded cantilever beam subjected to a linearly distributed load, in which the Young's modulus along the thickness direction varies with power-law functions or with exponential functions. Considering that there are many research works in this field, we do not review them in detail.

Recently, analytical studies of bimodular beams and plates have been performed. Among these works, the determination of the unknown neutral layer is a key issue because it opens up the possibility for the establishment of a mechanical model based on a subarea in tension and compression. Under the assumption that shearing stresses have no contribution to the neutral axis, Yao and Ye [22] obtained a one-dimensional analytical solution of a bimodular shallow beam. He et al. adopted the stress function method to find the elasticity solution of a bimodular straight beam [23] and curved beams [24]. Later, the classical Kirchhoff hypothesis was used to assess the existence of the elastic neutral layers of a thin plate during bending with a small deflection [25]. Consequently, a series of analytical solutions of plates is derived in rectangular and polar coordinate systems. More recently, He et al. [26] presented an elasticity solution of a bimodular FGM beam under uniformly distributed loads and discussed several concrete numerical examples. However, some basic problems are still unclear, which include the consistency between a one-dimensional solution and a two-dimensional solution, the validity of the plane section assumption, the corresponding relation among a classical beam, a standard FGM beam, and a bimodular FGM beam, as well as the bimodular effect on stresses and deformations in a general sense.

In this study, we will adopt a bimodular FGM beam theory to derive the one-dimensional and two-dimensional solutions. Theoretically speaking, any FGM beams may be suitable for this theory provided that the bimodular effect in tension and compression needs to be emphasized for a refined analysis; or, in other words, a certain constituent that forms functionally graded materials presents a relatively obvious bimodular effect which can not be ignored otherwise it will introduce much error into the analysis. The article is organized as follows. The corresponding analytical solutions under pure bending and lateral-force bending will be obtained in Sections 2 and 3, respectively. Specifically, a perturbation method is adopted to solve the transcendental equation for the determination of the unknown neutral layer. The validity of the plane section assumption is discussed and some important physical quantities among a classical beam, a standard FGM beam, and a bimodular FGM beam are compared in Section 4. Besides this, without specifying the real magnitude of the external load and the geometrical dimension of the beam, the bimodular effect on the stress and deformation in a general sense will be investigated in Section 4. Some important conclusions and subsequent studies are given in the concluding remarks.

2. Functionally Graded Beams under Pure Bending

2.1. One-Dimensional Solution

2.1.1. Bending Stress

A bimodular functionally graded beam with a rectangular section dimension of $h \times b$ is subjected to a bending moment M alone as shown in Figure 1. This causes a bending of the beam in the plane coordinate system xoz. Note that due to the introduction of the bimodular effect in tension and compression as well as the functionally graded characteristic of the material, the neutral layer of the beam generally does not locate on the half height of the section. The x axis is established on the unknown neutral layer as shown in Figure 1. It is obvious that the zone below the neutral layer is in tension while the zone up the layer is in compression. Let the tensile and compressive section heights of the beam be h_1 and h_2, respectively. Also, let the modulus of elasticity of the material in the tensile and compressive zones be $E^+(z)$ and $E^-(z)$, respectively, while the Poisson's ratios remain the same.

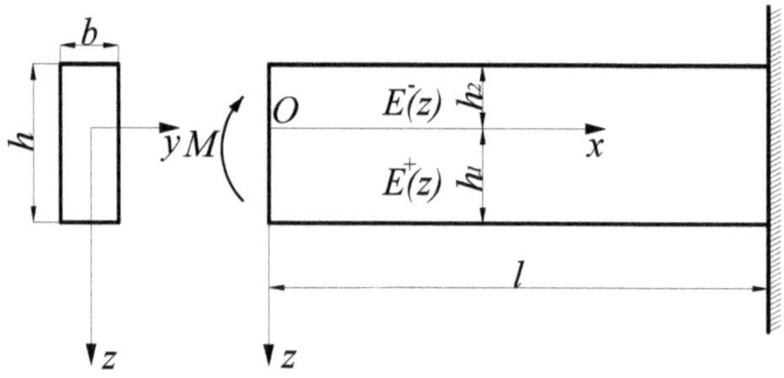

Figure 1. Scheme of a bimodular functionally graded material (FGM) beam under pure bending.

If an exponential function is used to express the function grade of the material, $E^+(z)$ and $E^-(z)$ may be expressed as

$$E^+(z) = E_0 e^{\alpha_1 z/h}, E^-(z) = E_0 e^{\alpha_2 z/h}, \tag{1}$$

where α_1 and α_2 are two grade indexes. $E^+(z) = E^-(z) = E_0$ when $z = 0$, that is, at the neutral layer the tensile modulus is equal to the compressive one. Let the curvature radius of the neutral layer be ρ; then, the bending strain along the x axis in the whole beam will be the uniform expression $\varepsilon_x = z/\rho$ if the plane section assumption holds. Thus, according to Ambartsumyan's bimodular model the bending stress in the tensile and compressive zones, σ_x^+ and σ_x^-, are also the tensile and compressive principal stress and they are, respectively,

$$\sigma_x^+ = \frac{E^+(z)}{\rho} z, \text{ for } 0 \le z \le h_1, \tag{2}$$

and

$$\sigma_x^- = \frac{E^-(z)}{\rho} z, \text{ for } -h_2 \le z \le 0. \tag{3}$$

Let the normal resultant at any section be N, thus $N = 0$ yields

$$\int_0^{h_1} \sigma_x^+ bdz + \int_{-h_2}^0 \sigma_x^- bdz = 0. \tag{4}$$

Substituting Equations (2) and (3) into Equation (4), we have

$$\int_0^{h_1} \frac{E_0 e^{\alpha_1 z/h} z b}{\rho} dz + \int_{-h_2}^{0} \frac{E_0 e^{\alpha_2 z/h} z b}{\rho} dz = 0. \tag{5}$$

If we let

$$\begin{aligned} \int_0^{h_1} e^{\alpha_1 z/h} z dz &= \left(\frac{hh_1}{\alpha_1} - \frac{h^2}{\alpha_1^2}\right) e^{\alpha_1 h_1/h} + \frac{h^2}{\alpha_1^2} = A_1^+ \\ \int_{-h_2}^{0} e^{\alpha_2 z/h} z dz &= \left(\frac{hh_2}{\alpha_2} + \frac{h^2}{\alpha_2^2}\right) e^{-\alpha_2 h_2/h} - \frac{h^2}{\alpha_2^2} = A_1^- \end{aligned}, \tag{6}$$

Equation (4) will lead to the following relation

$$A_1^+ + A_1^- = 0, \tag{7}$$

which is used for the determination of the unknown neutral layer later.

Similarly, the bending moment at any section will give

$$\int_0^{h_1} \sigma_x^+ bzdz + \int_{-h_2}^{0} \sigma_x^- bzdz = M. \tag{8}$$

Substituting Equations (2) and (3) into Equation (8), we have

$$\int_0^{h_1} \frac{E_0 e^{\alpha_1 z/h} z^2 b}{\rho} dz + \int_{-h_2}^{0} \frac{E_0 e^{\alpha_2 z/h} z^2 b}{\rho} dz = M. \tag{9}$$

If we let

$$\begin{aligned} \int_0^{h_1} e^{\alpha_1 z/h} z^2 dz &= \left(\frac{hh_1^2}{\alpha_1} - 2\frac{h^2 h_1}{\alpha_1^2} + 2\frac{h^3}{\alpha_1^3}\right) e^{\alpha_1 h_1/h} - 2\frac{h^3}{\alpha_1^3} = A_2^+ \\ \int_{-h_2}^{0} e^{\alpha_2 z/h} z^2 dz &= -\left(\frac{hh_2^2}{\alpha_2} + 2\frac{h^2 h_2}{\alpha_2^2} + 2\frac{h^3}{\alpha_2^3}\right) e^{-\alpha_2 h_2/h} + 2\frac{h^3}{\alpha_2^3} = A_2^- \end{aligned}, \tag{10}$$

Equation (8) will yield

$$\frac{1}{\rho} = \frac{M}{E_0 b (A_2^+ + A_2^-)}. \tag{11}$$

If D^* is introduced to denote the flexural stiffness of the bimodular functionally graded beam, that is,

$$D^* = E_0 b (A_2^+ + A_2^-), \tag{12}$$

the deformation of the beam will follow the familiar form

$$\frac{1}{\rho} = \frac{M}{D^*}. \tag{13}$$

Substituting the relation (11) into Equations (2) and (3), we obtain the one-dimensional solution of the bending stress in the tensile and compressive zones, respectively,

$$\sigma_x^+ = \frac{M e^{\alpha_1 z/h} z}{b(A_2^+ + A_2^-)}, \text{ for } 0 \le z \le h_1, \tag{14}$$

and

$$\sigma_x^- = \frac{M e^{\alpha_2 z/h} z}{b(A_2^+ + A_2^-)}, \text{ for } -h_2 \le z \le 0. \tag{15}$$

It should be noted here that due to this being the pure bending case, only the bending stress may be obtained and the shearing stress can be derived in the lateral-force bending case, which will be discussed in Section 3.

2.1.2. Deflection Curve

Let the vertical displacement of any point on the neutral layer be w; then, Equation (11) may be expressed in terms of the second-order derivative of w to x as follows

$$\frac{1}{\rho} = -\frac{d^2w}{dx^2} = \frac{M}{E_0 b(A_2^+ + A_2^-)}. \tag{16}$$

Integrating twice with respect to x, we have

$$w = -\frac{Mx^2}{2E_0 b(A_2^+ + A_2^-)} + cx + d, \tag{17}$$

where c and d are two undetermined constants. If a simply-supported beam is considered, the boundary conditions give

$$w = 0, \text{ while } x = 0 \text{ or } l, \tag{18}$$

where l is the span length of the beam. Thus, the deflection curve of the neutral axis is

$$w = \frac{M(l-x)x}{2E_0 b(A_2^+ + A_2^-)}. \tag{19}$$

If a cantilever beam with the right end fixed is considered, as shown in Figure 1, the displacement restriction is

$$w = \frac{dw}{dx} = 0, \text{ while } x = l, \tag{20}$$

and the deflection curve of the neutral axis will be

$$w = -\frac{M(x-l)^2}{2E_0 b(A_2^+ + A_2^-)}. \tag{21}$$

2.1.3. Determination of the Neutral Layer

It should be noted here that the two important parameters h_1 and h_2 have still not been determined. From Equations (6) and (7), we may have

$$\left(\frac{h_1}{\alpha_1} - \frac{h}{\alpha_1^2}\right)e^{\alpha_1 h_1/h} + \left(\frac{h_2}{\alpha_2} + \frac{h}{\alpha_2^2}\right)e^{-\alpha_2 h_2/h} = \frac{h}{\alpha_2^2} - \frac{h}{\alpha_1^2}, \tag{22}$$

where α_1 and α_2 are two indexes concerning the grade function as indicated above. If we introduce the following dimensionless variables

$$H_1 = \frac{h_1}{h}, \ H_2 = \frac{h_2}{h}, \tag{23}$$

and also multiply the two ends of the equation by $\alpha_1^2\alpha_2^2$, Equation (22) may be transformed into a dimensionless form, such that

$$(\alpha_1 H_1 - 1)\alpha_2^2 e^{\alpha_1 H_1} + (\alpha_2 H_2 + 1)\alpha_1^2 e^{-\alpha_2 H_2} = \alpha_1^2 - \alpha_2^2, \tag{24}$$

in which H_1 and H_2 are the basic variables and satisfy $H_1 + H_2 = 1$. It is a transcendental equation and is hard to solve analytically to some extent due to the existence of an exponential function. Next, we will adopt the perturbation idea to solve the transcendental equation.

The exponential items $e^{\alpha_1 H_1}$ and $e^{-\alpha_2 H_2}$ may be spread with respect to H_1 and H_2, respectively,

$$\begin{aligned} e^{\alpha_1 H_1} &= 1 + \alpha_1 H_1 + \tfrac{1}{2}\alpha_1^2 H_1^2 + \cdots + \tfrac{1}{n!}(\alpha_1 H_1)^n + \cdots, \\ e^{-\alpha_2 H_2} &= 1 - \alpha_2 H_2 + \tfrac{1}{2}\alpha_2^2 H_2^2 + \cdots + \tfrac{1}{n!}(-\alpha_2 H_2)^n + \cdots. \end{aligned} \tag{25}$$

If the linear approximation is adopted, such that

$$e^{\alpha_1 H_1} = 1 + \alpha_1 H_1,\ e^{-\alpha_2 H_2} = 1 - \alpha_2 H_2, \tag{26}$$

substituting it into Equation (24) will yield

$$H_1 = H_2 = \frac{1}{2}, \tag{27}$$

which is exactly the solution of a classical problem without considering the functionally graded property and bimodular effect of the material. We call it the first-order approximation solution of the problem. If the second-order approximation is adopted, such that

$$e^{\alpha_1 H_1} = 1 + \alpha_1 H_1 + \frac{1}{2}\alpha_1^2 H_1^2,\ e^{-\alpha_2 H_2} = 1 - \alpha_2 H_2 + \frac{1}{2}\alpha_2^2 H_2^2, \tag{28}$$

substituting it into Equation (24) and considering $H_2 = 1 - H_1$ yields

$$(\alpha_1 - \alpha_2)H_1^3 + 3\alpha_2 H_1^2 + (2 - 3\alpha_2)H_1 + \alpha_2 - 1 = 0, \tag{29}$$

which is an algebra equation of H_1 and is easily solved either by an analytical method or by a numerical technique once the numerical values of α_1 and α_2 are known. The solution of Equation (29) may be called the second-order approximation solution. Similarly, if more items in Equation (25) are taken, we will obtain a high-order approximation solution according to the procedure indicated above. Thus, based on the perturbation idea, the transcendental equation may be gradually transformed into a nonlinear algebra equation of H_1 and the position of the unknown neutral layer is determined analytically.

2.2. Two-Dimensional Solution

2.2.1. Stress

Let the stress components in the two-dimensional beam problem shown in Figure 1 be σ_x, σ_z, and τ_{xz}, let the strain components be ε_x, ε_z, and γ_{xz}, and also let the displacement components in the same problem be u and w. Then, in the differential equation of equilibrium in which the body forces are neglected, the geometrical relation as well as the consistency equation are the same as those in the classical problem, and they are, respectively,

$$\frac{\partial \sigma_x}{\partial x} + \frac{\partial \tau_{xz}}{\partial z} = 0,\ \frac{\partial \tau_{zx}}{\partial x} + \frac{\partial \sigma_x}{\partial z} = 0, \tag{30}$$

and

$$\left\{ \begin{array}{l} \varepsilon_x = \frac{\partial u}{\partial x},\ \varepsilon_z = \frac{\partial w}{\partial z},\ \gamma_{xz} = \frac{\partial w}{\partial x} + \frac{\partial u}{\partial z} \\ \frac{\partial^2 \varepsilon_x}{\partial z^2} + \frac{\partial^2 \varepsilon_z}{\partial x^2} = \frac{\partial^2 \gamma_{xz}}{\partial x \partial z} \end{array} \right. . \tag{31}$$

The physical equation gives

$$\left\{ \begin{array}{l} \varepsilon_x = s_{11}\sigma_x + s_{13}\sigma_z \\ \varepsilon_z = s_{13}\sigma_x + s_{33}\sigma_z \\ \gamma_{zx} = s_{44}\tau_{zx} \end{array} \right. . \tag{32}$$

After considering the different moduli in tension and compression as well as the functional grade of the material, the physical equation may take the following form

$$\begin{cases} \varepsilon_x^{+/-} = \frac{1}{E_0 e^{\alpha_i z/h}}(\sigma_x^{+/-} - \mu\sigma_z^{+/-}) \\ \varepsilon_z^{+/-} = \frac{1}{E_0 e^{\alpha_i z/h}}(\sigma_z^{+/-} - \mu\sigma_x^{+/-}) \\ \gamma_{zx}^{+/-} = \frac{2(1+\mu)}{E_0 e^{\alpha_i z/h}}\tau_{zx}^{+/-} \end{cases}, \tag{33}$$

where a superscript "$+/-$" denotes a tensile (compressive) quantity and $\alpha_i (i = 1, 2)$ correspond to the cases of tension and compression, respectively. Equation (33) is in essence two sets of equations concerning tension and compression.

Next, the stress function method will be adopted to obtain the solution of this two-dimensional problem. Due to pure bending, here we still consider that the stress function $\varphi^{+/-}(x,z)$ depends only on z, that is

$$\varphi^{+/-}(x,z) = f^{+/-}(z), \tag{34}$$

where $f^{+/-}(z)$ is an unknown function and "$+/-$" still denotes a tensile (compressive) quantity. According to the relation between the stress function and the stress components,

$$\sigma_x^{+/-} = \frac{\partial^2\varphi^{+/-}}{\partial z^2},\ \sigma_z^{+/-} = \frac{\partial^2\varphi^{+/-}}{\partial x^2},\ \tau_{xz}^{+/-} = -\frac{\partial^2\varphi^{+/-}}{\partial x\partial z}. \tag{35}$$

Equation (33) may be changed as

$$\begin{cases} \varepsilon_x^{+/-} = \frac{1}{E_0 e^{\alpha_i z/h}}\frac{d^2 f^{+/-}(z)}{dz^2} \\ \varepsilon_z^{+/-} = \frac{-\mu}{E_0 e^{\alpha_i z/h}}\frac{d^2 f^{+/-}(z)}{dz^2} \\ \gamma_{zx}^{+/-} = 0 \end{cases}. \tag{36}$$

Letting Equation (36) satisfy the consistency relation, we obtain

$$\frac{d^2}{dz^2}\left[\frac{1}{E_0 e^{\alpha_i z/h}}\frac{d^2 f^{+/-}(z)}{dz^2}\right] = 0. \tag{37}$$

Integrating twice with respect to z, we have

$$\frac{d^2 f^{+/-}(z)}{dz^2} = (C_1^{+/-}z + C_2^{+/-})E_0 e^{\alpha_i z/h}, \tag{38}$$

where $C_1^{+/-}$ and $C_2^{+/-}$ are four undetermined constants. Continuously integrating with respect to z, we obtain

$$f^{+/-}(z) = \left(z - \frac{2h}{\alpha_i}\right)\frac{E_0 C_1^{+/-} h^2 e^{\alpha_i z/h}}{\alpha_i^2} + \frac{E_0 C_2^{+/-} h^2 e^{\alpha_i z/h}}{\alpha_i^2} + C_3^{+/-}z + C_4^{+/-}, \tag{39}$$

where $C_3^{+/-}$ and $C_4^{+/-}$ are four undetermined constants and may be neglected. The stress function is simplified as

$$\varphi^{+/-}(x,z) = \left(z - \frac{2h}{\alpha_i}\right)\frac{E_0 C_1^{+/-} h^2 e^{\alpha_i z/h}}{\alpha_i^2} + \frac{E_0 C_2^{+/-} h^2 e^{\alpha_i z/h}}{\alpha_i^2}. \tag{40}$$

Correspondingly, the stress expressions are

$$\sigma_x^{+/-} = (C_1^{+/-}z + C_2^{+/-})E_0 e^{\alpha_i z/h}, \sigma_z^{+/-} = 0,\ \tau_{zx}^{+/-} = 0. \tag{41}$$

Next, we will use the boundary conditions as well as the continuity condition of stress to determine the four unknown constants $C_1^{+/-}$ and $C_2^{+/-}$.

First, the continuity conditions of the stresses on the neutral layer give

$$\sigma_x^+ = \sigma_x^- = 0,\ \sigma_z^+ = \sigma_z^-,\ \tau_{xz}^+ = \tau_{xz}^- \text{ at } z = 0. \tag{42}$$

According to Equation (41), it is easily found that the last two conditions are surely satisfied and the first condition yields

$$C_2^+ = C_2^- = 0. \tag{43}$$

The stress boundary conditions on the two main sides of the beam are, respectively,

$$\begin{cases} \sigma_z^+ = 0,\ \tau_{xz}^+ = 0 \text{ at } z = h_1 \\ \sigma_z^- = 0,\ \tau_{xz}^- = 0 \text{ at } z = -h_2 \end{cases}, \tag{44}$$

which are surely satisfied due to pure bending. At the left end of the beam, the application of Saint-Venant's Principle gives

$$\begin{cases} \int_0^{h_1} \sigma_x^+ bdz + \int_{-h_2}^0 \sigma_x^- bdz = 0, \\ \int_0^{h_1} \sigma_x^+ zbdz + \int_{-h_2}^0 \sigma_x^- zbdz = M \\ \int_0^{h_1} \tau_{xz}^+ bdz + \int_{-h_2}^0 \tau_{xz}^- bdz = 0, \end{cases}, \text{ at } x = 0. \tag{45}$$

It is easily found that the last condition is satisfied and the first two conditions will yield, respectively,

$$C_1^+ \int_0^{h_1} z e^{\alpha_1 z/h} dz + C_1^- \int_{-h_2}^0 z e^{\alpha_2 z/h} dz = 0, \tag{46}$$

and

$$C_1^+ \int_0^{h_1} z^2 e^{\alpha_1 z/h} dz + C_1^- \int_{-h_2}^0 z^2 e^{\alpha_2 z/h} dz = \frac{M}{E_0 b}. \tag{47}$$

Considering the Equations (6), (7), and (10), we solve

$$C_1^+ = C_1^- = \frac{M}{E_0 b(A_2^+ + A_2^-)}. \tag{48}$$

Thus, the final stress components are

$$\sigma_x^{+/-} = \frac{M}{b(A_2^+ + A_2^-)} z e^{\alpha_i z/h}, \sigma_z^{+/-} = 0,\ \tau_{zx}^{+/-} = 0, \tag{49}$$

which is the same as the one-dimensional solution obtained in Section 2.1.1.

2.2.2. Displacement

After the determination of the stress components, the combination of the physical equations and the geometrical equations will give

$$\begin{cases} \varepsilon_x^{+/-} = \frac{M}{E_0 b(A_2^+ + A_2^-)} z = \frac{\partial u}{\partial x} \\ \varepsilon_z^{+/-} = \frac{-\mu M}{E_0 b(A_2^+ + A_2^-)} z = \frac{\partial w}{\partial z} \\ \gamma_{zx}^{+/-} = 0 = \frac{\partial u}{\partial z} + \frac{\partial w}{\partial x} \end{cases}. \tag{50}$$

Integrating the first two expressions with respect to x and z, we have, respectively,

$$u = \frac{M}{E_0 b(A_2^+ + A_2^-)} zx + g_1(z), \tag{51}$$

and

$$w = \frac{-\mu M}{2E_0 b(A_2^+ + A_2^-)} z^2 + g_2(x), \tag{52}$$

where $g_1(z)$ and $g_2(x)$ are two undermined functions. Substituting u and w into the third expression in Equation (50), we have

$$\frac{M}{E_0 b(A_2^+ + A_2^-)} x + \frac{dg_2(x)}{dx} = -\frac{dg_1(z)}{dz} = a, \tag{53}$$

where a is a rigid displacement item. Integrating the above expression with respect to z and x, we have, respectively,

$$g_1(z) = -az + c, \tag{54}$$

and

$$g_2(x) = -\frac{M}{2E_0 b(A_2^+ + A_2^-)} x^2 + ax + d, \tag{55}$$

where and d are still rigid displacement items. Now, the displacement may be expressed as

$$u = \frac{M}{E_0 b(A_2^+ + A_2^-)} zx - az + c, \tag{56}$$

and

$$w = -\frac{M}{2E_0 b(A_2^+ + A_2^-)} (x^2 + \mu z^2) + ax + d. \tag{57}$$

If we consider here a simply-supported beam, the corresponding boundary conditions give

$$\begin{cases} u = w = 0, \text{ while } x = 0,\ z = 0 \\ w = 0, \text{ while } x = l,\ z = 0 \end{cases}, \tag{58}$$

where l is the span length of the beam. Thus, the last displacement components are

$$\begin{cases} u(x,z) = \frac{M}{2E_0 b(A_2^+ + A_2^-)} (2x - l)z \\ w(x,z) = -\frac{M}{2E_0 b(A_2^+ + A_2^-)} (x^2 + \mu z^2 - lx) \end{cases}. \tag{59}$$

The deflection curve of the neutral layer may be obtained by $w(x,z)|_{z=0}$, which is the same as the one-dimensional solution, i.e., Equation (19). If a cantilever beam with the right end fixed is considered, the restriction conditions yield

$$u = w = \frac{\partial w}{\partial x} = 0, \text{ while } x = l,\ z = 0, \tag{60}$$

the last displacement components will be

$$\begin{cases} u(x,z) = \frac{M}{E_0 b(A_2^+ + A_2^-)} (x - l)z \\ w(x,z) = -\frac{M}{2E_0 b(A_2^+ + A_2^-)} (x^2 + \mu z^2 - 2lx + l^2) \end{cases}. \tag{61}$$

Similarly, the deflection curve of the neutral layer is consistent with the result presented in Equation (21).

3. Bimodular Functionally Graded Beams under Latera-Force Bending

Let us consider the lateral-force bending problem of a bimodular functionally graded beam, as shown in Figure 2, in which the left end of the beam is subjected to the action of a concentrated force P and the right end is fixed. Due to the combined action of the bending moment and the shearing force, any point in the beam is in diagonal tension or diagonal compression; so, it is very difficult to determine the position and shape of the unknown neutral layer if the constitutive law defined in the principal stress direction is strictly followed. For this purpose, an important assumption that shearing stresses have no contribution to the neutral axis [22] is used to establish the simplified mechanical model. In the light of the assumption, the beam will deflect and develop a so-called tensile zone and compressive zone under the external load. The tension and compression of any point in the beam depend only on the direction of the bending stress and are independent of the shearing stress. Thus, similar to the case of pure bending shown in Figure 1, the mechanical model based on a subarea in tension and compression is still established in the case of lateral-force bending as shown in Figure 2. The basic equations of the problem are the same as those in Section 2.2.1, that is, Equations (30)–(33). According to the loading conditions, the stress function may be assumed to be

$$\varphi^{+/-} = x f^{+/-}(z), \tag{62}$$

where $f^{+/-}(z)$ is an unknown function, and it may be determined by satisfying the consistency relation. The strain components expressed in term of $f^{+/-}(z)$ are

$$\begin{cases} \varepsilon_x^{+/-} = \frac{x}{E_0 e^{\alpha_i z/h}} \frac{d^2 f^{+/-}(z)}{dz^2} \\ \varepsilon_z^{+/-} = \frac{-\mu x}{E_0 e^{\alpha_i z/h}} \frac{d^2 f^{+/-}(z)}{dz^2} \\ \gamma_{zx}^{+/-} = \frac{-2(1+\mu)}{E_0 e^{\alpha_i z/h}} \frac{d f^{+/-}(z)}{dz} \end{cases}. \tag{63}$$

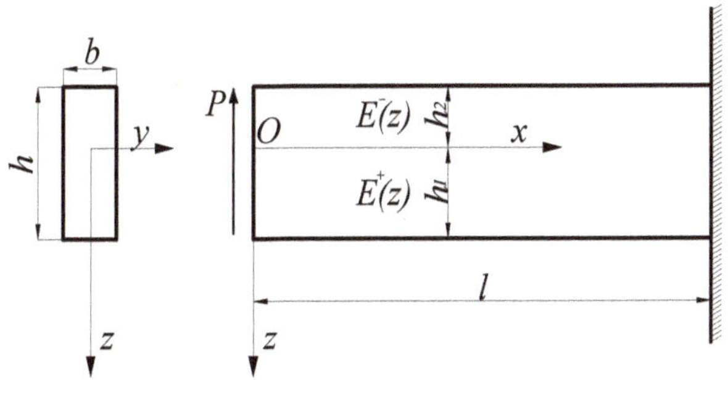

Figure 2. Scheme of a bimodular FGM beam under lateral-force bending.

Satisfying the consistency relation for any x gives

$$\frac{d^2}{dz^2}\left[\frac{1}{E_0 e^{\alpha_i z/h}} \frac{d^2 f^{+/-}(z)}{dz^2}\right] = 0. \tag{64}$$

Continuously integrating with respect to z, we have

$$f^{+/-}(z) = \left(z - \frac{2h}{\alpha_i}\right) \frac{E_0 C_1^{+/-} h^2 e^{\alpha_i z/h}}{\alpha_i^2} + \frac{E_0 C_2^{+/-} h^2 e^{\alpha_i z/h}}{\alpha_i^2} + C_3^{+/-} z, \tag{65}$$

where $C_1^{+/-}$, $C_2^{+/-}$ and $C_3^{+/-}$ are six undetermined constants; the constant item has been neglected. Thus, the stress function now has the following form

$$\varphi^{+/-} = x\left[\left(z - \frac{2h}{\alpha_i}\right)\frac{E_0 C_1^{+/-} h^2 e^{\alpha_i z/h}}{\alpha_i^2} + \frac{E_0 C_2^{+/-} h^2 e^{\alpha_i z/h}}{\alpha_i^2} + C_3^{+/-} z\right]. \tag{66}$$

The stress components expressed in terms of the undetermined constants are

$$\begin{cases} \sigma_x^{+/-} = x(C_1^{+/-} z + C_2^{+/-}) E_0 e^{\alpha_i z/h}, \sigma_z^{+/-} = 0 \\ \tau_{zx}^{+/-} = -(\alpha_i z - h)\frac{E_0 C_1^{+/-} h e^{\alpha_i z/h}}{\alpha_i^2} - \frac{E_0 C_2^{+/-} h e^{\alpha_i z/h}}{\alpha_i} - C_3^{+/-} \end{cases}. \tag{67}$$

The continuity conditions of the stresses on the neutral layer under lateral-force bending are the same as those under pure bending; thus, applying Equation (42) yields

$$C_2^+ = C_2^- = 0, \tag{68}$$

and

$$\frac{E_0 C_1^+ h^2}{\alpha_1^2} - C_3^+ = \frac{E_0 C_1^- h^2}{\alpha_2^2} - C_3^-. \tag{69}$$

Similarly, the stress boundary conditions on the two main sides of the beam are the same as those in Equation (44). Satisfying the conditions in the tensile and compressive zones yields, respectively,

$$\tau_{zx}^+ = -(\alpha_1 h_1 - h)\frac{E_0 C_1^+ h e^{\alpha_1 h_1/h}}{\alpha_1^2} - C_3^+ = 0, \tag{70}$$

and

$$\tau_{zx}^- = -(-\alpha_2 h_2 - h)\frac{E_0 C_1^- h e^{-\alpha_2 h_2/h}}{\alpha_2^2} - C_3^- = 0. \tag{71}$$

At the left end of the beam, the application of Saint-Venant's Principle gives

$$\begin{cases} \int_0^{h_1} \sigma_x^+ b dz + \int_{-h_2}^0 \sigma_x^- b dz = 0, \\ \int_0^{h_1} \sigma_x^+ z b dz + \int_{-h_2}^0 \sigma_x^- z b dz = 0 \\ \int_0^{h_1} \tau_{xz}^+ b dz + \int_{-h_2}^0 \tau_{xz}^- b dz = P, \end{cases}, \text{ at } x = 0. \tag{72}$$

It is easily found that the first two conditions are satisfied and the last condition gives

$$\int_0^{h_1}\left[(\alpha_1 z - h)\frac{E_0 C_1^+ h e^{\alpha_1 z/h}}{\alpha_1^2} + C_3^+\right]dz + \int_{-h_2}^0\left[(\alpha_2 z - h)\frac{E_0 C_1^- h e^{\alpha_2 z/h}}{\alpha_2^2} + C_3^-\right]dz = -\frac{P}{b}. \tag{73}$$

Equations (69), (70), (71), and (73) may be used for the solution of $C_1^{+/-}$and $C_3^{+/-}$. First, substituting Equations (70) and (71) into Equation (69) and also considering A_1^+ and A_1^+ introduced beforehand, we have a simple expression

$$A_1^+ C_1^+ + A_1^- C_1^- = 0, \tag{74}$$

which gives $C_1^+ = C_1^-$ due to $A_1^+ + A_1^- = 0$. Second, integrating Equation (73), substituting Equations (70) and (71) into it, and also considering A_2^+ and A_2^- introduced beforehand, Equation (73) may be simplified as

$$A_2^+ C_1^+ + A_2^- C_1^- = \frac{P}{E_0 b}. \tag{75}$$

Combining Equations (74) and (75) will solve C_1^+ and C_1^-, and substituting them into Equations (70) and (71), we finally obtain

$$\begin{cases} C_1^+ = C_1^- = \frac{P}{E_0 b(A_2^+ + A_2^-)} \\ C_3^+ = \frac{Pe^{\alpha_1 h_1/h}}{b(A_2^+ + A_2^-)}\left(\frac{h^2}{\alpha_1^2} - \frac{hh_1}{\alpha_1}\right) \\ C_3^- = \frac{Pe^{-\alpha_2 h_2/h}}{b(A_2^+ + A_2^-)}\left(\frac{h^2}{\alpha_2^2} + \frac{hh_2}{\alpha_2}\right) \end{cases}, \quad (76)$$

Substituting the determined $C_1^{+/-}$, $C_2^{+/-}$ and $C_3^{+/-}$ into Equation (67), the stress components are obtained as follows

$$\begin{cases} \sigma_x^{+/-} = \frac{Px}{b(A_2^+ + A_2^-)} z e^{\alpha_i z/h}, \sigma_z^{+/-} = 0 \\ \tau_{zx}^+ = \frac{P}{b(A_2^+ + A_2^-)}\left[\left(\frac{h^2}{\alpha_1^2} - \frac{hz}{\alpha_1}\right)e^{\alpha_1 z/h} - \left(\frac{h^2}{\alpha_1^2} - \frac{hh_1}{\alpha_1}\right)e^{\alpha_1 h_1/h}\right] \\ \tau_{zx}^- = \frac{P}{b(A_2^+ + A_2^-)}\left[\left(\frac{h^2}{\alpha_2^2} - \frac{hz}{\alpha_2}\right)e^{\alpha_2 z/h} - \left(\frac{h^2}{\alpha_2^2} + \frac{hh_2}{\alpha_2}\right)e^{-\alpha_2 h_2/h}\right] \end{cases}. \quad (77)$$

It is easily found that the item Px in $\sigma_x^{+/-}$ is exactly the magnitude of the bending moment, which is consistent with Equations (14) and (15).

By use of the physical equation and the geometrical equation, the displacement components may be determined as

$$\begin{cases} u^+ = \frac{P[(1+\mu)h(12h\alpha_1 z - 6\alpha_1^2 z^2) + \mu\alpha_1^3 z^3 + 3\alpha_1^3 x^2 z + 12(1+\mu)h^2 e^{\alpha_1(h_1 - z)/h}(h - h_1\alpha_1)]}{6E_0 b\alpha_1^3(A_2^+ + A_2^-)} - a^+ z - c^+ \\ u^- = \frac{P[(1+\mu)h(12h\alpha_2 z - 6\alpha_2^2 z^2) + \mu\alpha_2^3 z^3 + 3\alpha_2^3 x^2 z + 12(1+\mu)h^2 e^{-\alpha_2(h_2 + z)/h}(h + h_2\alpha_2)]}{6E_0 b\alpha_2^3(A_2^+ + A_2^-)} - a^- z - c^- \\ w^{+/-} = \frac{P}{6E_0 b(A_2^+ + A_2^-)}(-3\mu xz^2 - x^3) + a^{+/-}x + d^{+/-} \end{cases}, \quad (78)$$

where a, d, and c are the items concerning rigid displacement. Using the boundary condition $u = w = \partial w/\partial x = 0$ at $x = l$, $z = 0$, we have

$$\begin{cases} a^{+/-} = \frac{Pl^2}{2E_0 b(A_2^+ + A_2^-)}, d^{+/-} = -\frac{Pl^3}{3E_0 b(A_2^+ + A_2^-)}, \\ c^+ = \frac{2(1+\mu)h^2(h - h_1\alpha_1)Pe^{\alpha_1 h_1/h}}{E_0 b\alpha_1^3(A_2^+ + A_2^-)}, c^- = \frac{2(1+\mu)h^2(h + h_2\alpha_2)Pe^{-\alpha_2 h_2/h}}{E_0 b\alpha_2^3(A_2^+ + A_2^-)} \end{cases}. \quad (79)$$

Thus, the final displacements are determined.

4. Results and Discussions

4.1. Comparision among Three Types of Beam

As indicated before, the material considered in this study not only has a functionally graded characteristic but also presents different mechanical properties in tension and compression. It is valuable to compare physical quantities in a bimodular FGM beam and a standard FGM beam (without bimodular effect) with their counterparts in a classical problem. We should note that in a classical problem, there is no variation of material properties along the thickness direction; thus, the relevant integrals are usually done over the whole section height. The comparisons among the three types of beams are listed in Table 1. It is easily found that when the grade indexes $\alpha_1 = \alpha_2$, the quantities in a bimodular FGM beam regress to the corresponding quantities in a standard FGM beam; when $\alpha_1 = \alpha_2 = 0$, the regression continues up to the classical problem.

Table 1. Comparisons among a classical beam, an FGM beam, and a bimodular FGM beam.

Quantities	A Classical Beam	A FGM Beam	A Bimodular FGM Beam
		Modulus of elasticity	
E	$E = \text{Const.}$	$E(z) = E_0 e^{\alpha z/h}$	$E^+(z) = E_0 e^{\alpha_1 z/h}, E^-(z) = E_0 e^{\alpha_2 z/h}$
		Moment of inertia	
I_y	$\frac{bh^3}{12}$	$\int_A e^{\alpha z/h} z^2 dA$	$\int_0^{h_1} e^{\alpha_1 z/h} z^2 b dz + \int_{-h_2}^{0} e^{\alpha_2 z/h} z^2 b dz$ $= b(A_2^+ + A_2^-)$
		Bending stiffness	
D	EI_y	$E_0 \int_A e^{\alpha z/h} z^2 dA$	$E_0 b(A_2^+ + A_2^-)$
		Curvature	
$\frac{1}{\rho}$	$\frac{M}{EI_y}$	$\frac{M}{E_0 \int_A e^{\alpha z/h} z^2 dA}$	$\frac{M}{E_0 b(A_2^+ + A_2^-)}$
		Bending stress	
σ_x	$\frac{M}{I_y} z$	$\frac{Mz}{\int_A e^{\alpha z/h} dA}$	$\sigma_x^+ = \frac{M}{b(A_2^+ + A_2^-)} z e^{\alpha_1 z/h}$ $\sigma_x^- = \frac{M}{b(A_2^+ + A_2^-)} z e^{\alpha_2 z/h}$
		Static moment when computing shearing stress	
S_y	$\frac{b}{2}\left(\frac{h^2}{4} - z^2\right)$	$\int_A e^{\alpha z/h} z dA$	$S^+ = \int_z^{h_1} e^{a_1 z/h} z b dz = b\left[\left(\frac{h^2}{a_1^2} - \frac{hz}{a_1}\right) e^{a_1 z/h} - \left(\frac{h^2}{a_1^2} - \frac{hh_1}{a_1}\right) e^{a_1 h_1/h}\right]$ for $0 \le z \le h_1$ $S^- = \int_z^{-h_1} e^{a_2 z/h} z b dz = b\left[\left(\frac{h^2}{a_2^2} - \frac{hz}{a_2}\right) e^{a_2 z/h} - \left(\frac{h^2}{a_2^2} + \frac{hh_2}{a_2}\right) e^{a_2 h_2/h}\right]$ for $-h_2 \le z \le 0$
		Shearing stress	
τ_{xz}	$\frac{PS_y}{I_y b}$	$\frac{P\int_A e^{\alpha z/h} z dA}{b\int_A e^{\alpha z/h} z^2 dA}$	$\tau_{zx}^+ = \frac{P}{b(A_2^+ + A_2^-)}\left[\left(\frac{h^2}{a_1^2} - \frac{hz}{a_1}\right) e^{a_1 z/h} - \left(\frac{h^2}{a_1^2} - \frac{hh_1}{a_1}\right) e^{a_1 h_1/h}\right],$ for $0 \le z \le h_1$ $\tau_{zx}^+ = \frac{P}{b(A_2^+ + A_2^-)}\left[\left(\frac{h^2}{a_2^2} - \frac{hz}{a_2}\right) e^{a_2 z/h} - \left(\frac{h^2}{a_2^2} - \frac{hh_2}{a_2}\right) e^{a_2 h_2/h}\right],$ for $-h_2 \le z \le 0$

4.2. Plane Section Assumption

For the pure bending problem, the rotation of a vertical element of the cross section, β, may be obtained from Equation (61),

$$\beta = \frac{\partial u}{\partial z} = \frac{M}{E_0 b(A_2^+ + A_2^-)}(x - l). \tag{80}$$

It is obvious that the rotation is not dependent on z, which shows that for the pure bending problem, the plane section assumption is surely satisfied. However, for the lateral-force bending problem, the rotation may be obtained from Equation (78), respectively, for the tensile area

$$\beta = \frac{\partial u^+}{\partial z} = \frac{P[(1+\mu)h(12h\alpha_1 - 12\alpha_1^2 z) + 3\mu\alpha_1^3 z^2 + 3\alpha_1^3 x^2 - 12(1+\mu)\alpha_1 h e^{\alpha_1 (h_1 - z)/h}(h - h_1\alpha_1)]}{6E_0 b\alpha_1^3 (A_2^+ + A_2^-)} - a^+, \tag{81}$$

and for the compressive area

$$\beta = \frac{\partial u^-}{\partial z} = \frac{P[(1+\mu)h(12h\alpha_2 - 12\alpha_2^2 z) + 3\mu\alpha_2^3 z^2 + 3\alpha_2^3 x^2 - 12(1+\mu)\alpha_2 h e^{-\alpha_2 (h_2 + z)/h}(h + h_2\alpha_2)]}{6E_0 b\alpha_2^3 (A_2^+ + A_2^-)} - a^-. \tag{82}$$

It is readily seen that the rotation is now the function of z. This means that on any cross section, a vertical element under bending will deviate from the original vertical direction and the deviated value varies with the distance from the neutral layer, i.e., z. Consequently, for the lateral-force bending problem the plane section assumption no longer holds. Moreover, unlike the pure bending problem, the rotation will not continuously develop at the neutral layer due to the difference in tension and compression.

4.3. Bimodular Effect on Stress and Displacement

The bimodular effect on stress and displacement may be analyzed by the use of the analytical results obtained. To avoid the inconvenience introduced by the dimension of physical quantities, besides Equation (23), we adopt the following dimensionless manner:

$$\begin{aligned} m = \frac{M}{E_0 h^3},\ p = \frac{P}{E_0 h^2},\ a = \frac{A_2^+ + A_2^-}{h^3},\ \zeta = \frac{z}{h},\ \eta = \frac{x}{l}, \\ s^{+/-} = \frac{\sigma_x^{+/-} b}{E_0 h},\ t^{+/-} = \frac{\tau_{zx}^{+/-} b}{E_0 h},\ u^* = \frac{ub}{l^2},\ w^* = \frac{wb}{l^2}, \end{aligned} \quad (83)$$

The two-dimensional solution for stress and displacement under pure bending, i.e., Equations (49) and (61), may be changed as, respectively,

$$s^+ = \frac{m}{a}\zeta e^{\alpha_1 \zeta},\ \text{for}\ 0 \le \zeta \le H_1;\ s^- = \frac{m}{a}\zeta e^{\alpha_2 \zeta},\ \text{for}\ -H_2 \le \zeta \le 0, \quad (84)$$

and

$$u^* = \frac{m}{a}\frac{h}{l}(\eta - 1)\zeta, w^* = -\frac{m}{2a}[\eta^2 + \mu(\frac{h}{l})^2 \zeta^2 - 2\eta + 1]. \quad (85)$$

The above dimensionless displacement is helpful for analyzing the approximation degree from a two-dimensional solution to a one-dimensional one. We note that there exists a common factor h/l in the expressions of u^* and w^*. If a typical shallow beam is considered here, the ratio of the beam height to span length will be much less than 1, i.e., $h/l \ll 1$; this makes the magnitude of the u^* value much less than the value of w^*. Thus, in one-dimensional beam theory the horizontal displacement u^* is generally neglected without much error. On the other hand, if $h/l \ll 1$, also the term $(h/l)^2 \ll 1$ and $0 < \mu < 0.5$ for common materials; thus, the second term $\mu(h/l)^2\zeta^2$ in w^* may be neglected comparing to other items. This yields

$$u^* = 0,\ w^* = -\frac{m}{2a}(\eta - 1)^2, \quad (86)$$

which is exactly the dimensionless one-dimensional solution for displacement.

Similarly, the two-dimensional solution for stress under lateral-force bending, i.e., Equation (77), may be changed as

$$s^+ = \frac{p}{a}\frac{l}{h}\eta\zeta e^{\alpha_1 \zeta},\ \text{for}\ 0 \le \zeta \le H_1;\ s^- = \frac{p}{a}\frac{l}{h}\eta\zeta e^{\alpha_2 \zeta},\ \text{for}\ -H_2 \le \zeta \le 0 \quad (87)$$

$$\begin{aligned} t^+ = \frac{p}{a}\left[\left(\frac{1}{\alpha_1^2} - \frac{\zeta}{\alpha_1}\right)e^{\alpha_1 \zeta} - \left(\frac{1}{\alpha_1^2} - \frac{H_1}{\alpha_1}\right)e^{\alpha_1 H_1}\right],\ \text{for}\ 0 \le \zeta \le H_1 \\ t^- = \frac{p}{a}\left[\left(\frac{1}{\alpha_2^2} - \frac{\zeta}{\alpha_2}\right)e^{\alpha_2 \zeta} - \left(\frac{1}{\alpha_2^2} + \frac{H_2}{\alpha_2}\right)e^{-\alpha_2 H_2}\right],\ \text{for}\ -H_2 \le \zeta \le 0 \end{aligned} \quad (88)$$

Considering the characteristics of the grade function $E^+(\zeta) = E_0 e^{\alpha_1 \zeta}$ where $0 \le \zeta \le H_1$ and $E^-(\zeta) = E_0 e^{\alpha_2 \zeta}$ where $-H_2 \le \zeta \le 0$, it is easily found from Figure 3 that if the grade indexes $\alpha_1 > 0$ and $\alpha_2 > 0$, $E^+(\zeta) > E^-(\zeta)$ holds; if $\alpha_1 < 0$ and $\alpha_2 < 0$, $E^+(\zeta) < E^-(\zeta)$ holds; obviously, $\alpha_1 = \alpha_2 = 0$ corresponds to the classical problem. Therefore, 13 representative examples concerning the taken values of α_1 and α_2 are selected, including ±0.5, ±1.0, and ±2.0. Some relative parameters, including H_1 and H_2 (from Equation (24)) and a (from Equation (83)), are computed and listed in Table 2.

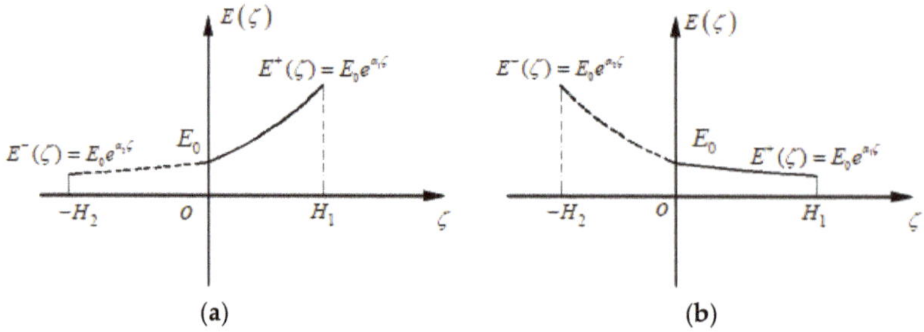

Figure 3. Variation of $E(\zeta)$ with the thickness direction. (**a**) $\alpha_1 > 0$ and $\alpha_2 > 0$; (**b**) $\alpha_1 < 0$ and $\alpha_2 < 0$.

From Table 2, it is easily found that for $E^+(\zeta) > E^-(\zeta)$, as the values of α_1 and α_2 increase, the tensile height decreases while the compressive height increases, which means that the neutral axis is tending downward (see Figures 1 and 2, z axis is down); for $E^+(\zeta) < E^-(\zeta)$, as the absolute values of α_1 and α_2 increase, the tensile height increases while the compressive height decreases, which means the neutral axis is tending upward. Besides this, we also note another interesting phenomenon, which is that due to the characteristic of an exponential function, the heights in tension and compression H_1 and H_2 are exactly reversed in some cases, including groups (a) and (g), (b) and (h), (c) and (i), and (e) and (k). For the values of a, they are the same as in the combinations above.

If the midspan displacement (i.e., $x = l/2$ or $\eta = 0.5$) of a beam under pure bending is considered, u^* in Equation (85) may be changed as

$$\frac{u^*}{m} = -\frac{\zeta}{20a}, \tag{89}$$

where h/l is taken as $1/10$. For the main three types of cases listed in Table 2, i.e., the representative groups (d), (f), and (j), the varying curves of u^*/m with $\zeta(= z/h)$ as well as the deflection curve of the neutral layer ($\zeta = 0$, see w^*/m in Equation (86)) with $\eta(= x/l)$ are plotted in Figures 4 and 5, respectively, in which the solid lines correspond to the case of $E^+(\zeta) > E^-(\zeta)$, the dashed lines correspond to the case of $E^+(\zeta) = E^-(\zeta)$, and the dotted lines correspond to the case of $E^+(\zeta) < E^-(\zeta)$.

Table 2. Numerical values of H_1, H_2, and a in different cases.

Cases	Groups	ff$_1$	ff$_2$	H_1	H_2	a
$E^+(\zeta) > E^-(\zeta)$	(a)	1.0	2.0	0.3725	0.6275	0.0560
	(b)	2.0	1.0	0.3859	0.6141	0.0836
	(c)	1.0	1.0	0.4180	0.5820	0.0762
	(d)	1.0	0.5	0.4399	0.5601	0.0872
	(e)	0.5	0.5	0.4585	0.5415	0.0815
$E^+(\zeta) = E^-(\zeta)$	(f)	0	0	1/2	1/2	1/12
$E^+(\zeta) < E^-(\zeta)$	(g)	−2.0	−1.0	0.6275	0.3725	0.0560
	(h)	−1.0	−2.0	0.6141	0.3859	0.0836
	(i)	−1.0	−1.0	0.5820	0.4180	0.0762
	(j)	−1.0	−0.5	0.5638	0.4362	0.0720
	(k)	−0.5	−0.5	0.5415	0.4585	0.0815

Similarly, we may use the midspan stress formulas ($\eta = 0.5$) of a beam under lateral-force bending to analyze the bimodular effect on the bending stress and shearing stress. Thus, Equation (87) is changed as

$$\frac{s^+}{p} = \frac{5}{a}\zeta e^{\alpha_1\zeta}, \text{ for } 0 \le \zeta \le H_1; \ \frac{s^-}{p} = \frac{5}{a}\zeta e^{\alpha_2\zeta}, \text{ for } -H_2 \le \zeta \le 0 \tag{90}$$

where $l/h = 10$. For the main three cases listed in Table 2, the variation of stresses with $\zeta(= z/h)$ are plotted in Figures 6 and 7, in which the shearing stress curve t/p is directly from Equation (88).

We should note such a fact that since the neutral layer is established on the x axis beforehand, the dividing line between tension and compression is always on $\zeta = 0$, which may be easily seen from Figures 4, 6 and 7. Figure 4 shows that the horizontal displacement varies in a linear relation along the direction of the beam thickness as indicated in Equation (89). The maximum horizontal displacement takes place at the edge of the compressive area for $E^+(\zeta) > E^-(\zeta)$ and at the edge of the tensile area for $E^+(\zeta) < E^-(\zeta)$, while the maximum displacement is equal for $E^+(\zeta) = E^-(\zeta)$. Figure 5 shows that, for any point on the neutral layer, the deflection value when $E^+(\zeta) > E^-(\zeta)$ is always less than the corresponding value when $E^+(\zeta) < E^-(\zeta)$.

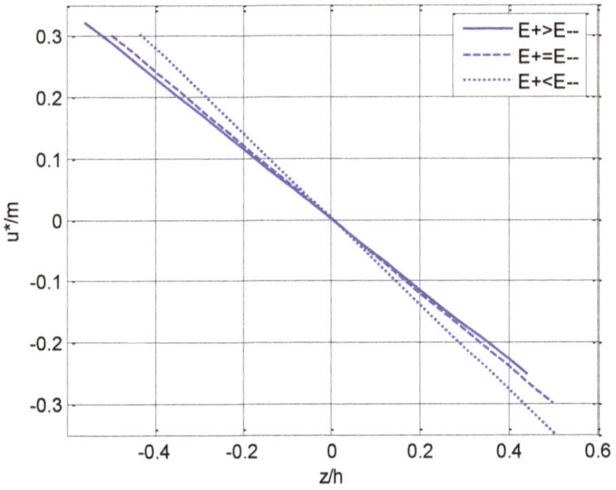

Figure 4. Variation of displacement u^* at midspan ($\eta = 0.5$) with $\zeta(= z/h)$ in three cases.

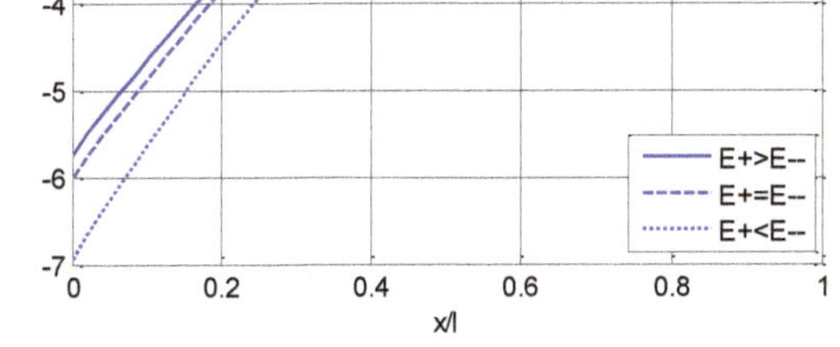

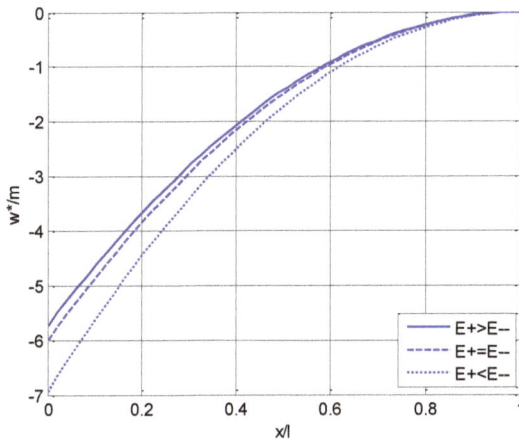

Figure 5. Variation of deflection w^* of the neutral layer ($\zeta = 0$) with $\eta(= x/l)$ in three cases.

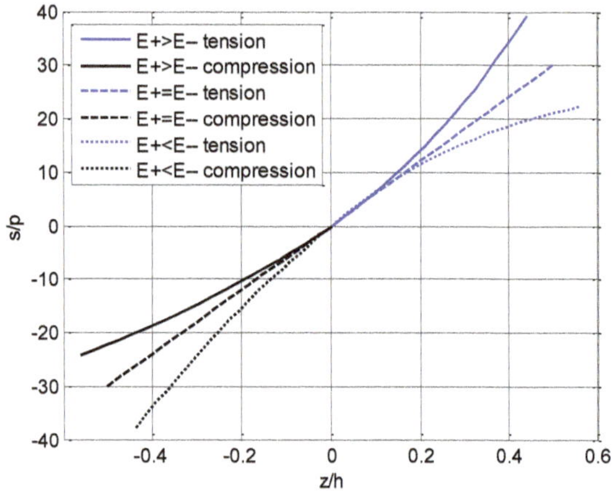

Figure 6. Variation of bending stress s at midspan ($\eta = 0.5$) with $\zeta(= z/h)$ in three cases.

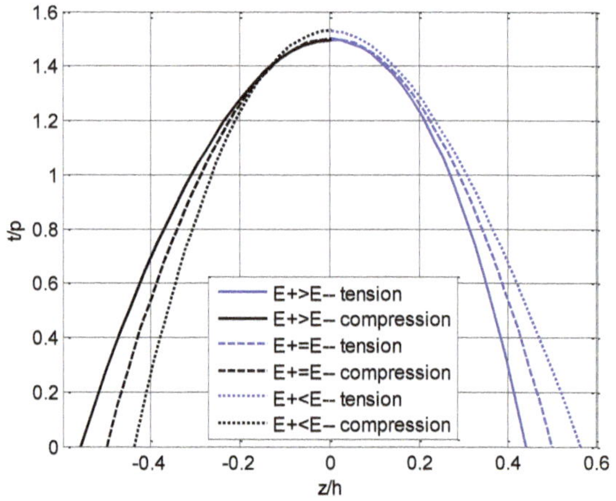

Figure 7. Variation of shearing stress t with $\zeta(= z/h)$ in three cases.

Figure 6 presents a typical exponent relation of bending stress varying along the direction of the beam thickness. Due to the variation of elastic modulus with the thickness direction, the location at which the maximum stress takes place may be changed. For $E^+(\zeta) > E^-(\zeta)$, the maximum tensile stress still takes place at the tensile edge of the beam while the maximum compressive stress will take place on a certain level between the compressive edge and the neutral layer; for $E^+(\zeta) < E^-(\zeta)$, the maximum compressive stress still takes place at the compressive edge of the beam while the maximum tensile stress will take place on a certain level between the tensile edge and the neutral layer; for $E^+(\zeta) = E^-(\zeta)$, the maximum tensile and compressive stress are equal and take place at the tensile and compressive edges of the beam, respectively, as we expected. This conclusion may be proved by the use of the extreme condition for an analytical solution of bending stress. We take the first-order derivative of bending stress with respect to the thickness direction, z, such that,

$$\frac{\partial \sigma_x^{+/-}}{\partial z} = \frac{\partial}{\partial z}\frac{M(x)}{b(A_2^+ + A_2^-)} z e^{\alpha_i z/h} = \frac{M(x)}{b(A_2^+ + A_2^-)}(e^{\alpha_i z/h} + z\frac{\alpha_i}{h}e^{\alpha_i z/h}), \tag{91}$$

where $M(x) = M$ for pure bending and $M(x) = Px$ for lateral bending. Via extreme conditions $\partial\sigma_x^{+/-}/\partial z = 0$, we have

$$e^{\alpha_i z/h}(1 + z\frac{\alpha_i}{h}) = 0, \quad (92)$$

$e^{\alpha_i z/h} > 0$ permanently holds true, we have

$$z = -\frac{h}{\alpha_i}, \quad (93)$$

which determines the location at which the maximum tensile or compression stress takes place. By referring to Figure 3, it is obvious that for $E^+(\zeta) > E^-(\zeta)$, the maximum compressive stress takes place at $z = -h/\alpha_2$; for $E^+(\zeta) < E^-(\zeta)$, the maximum tensile stress takes place at $z = -h/\alpha_1$. This phenomenon is quite different from the classical problem.

For the three cases of different moduli in tension and compression, Figure 7 uniformly indicates that the maximum shearing stress takes place at the neutral layer ($\zeta = 0$) and takes zero at the top and bottom of the beam. For $E^+(\zeta) = E^-(\zeta)$, the shearing stress in tension and compression is symmetrical with respect to $\zeta = 0$, while for the other two cases the rule does not hold. Moreover, the maximum shearing stress in the case of $E^+(\zeta) > E^-(\zeta)$ is less than the maximum stress in the case of $E^+(\zeta) < E^-(\zeta)$.

5. Concluding Remarks

In this study, one-dimensional and two-dimensional mechanical models for a functionally graded beam with different moduli in tension and compression were established. The corresponding analytical solutions under pure bending and lateral-force bending were obtained. The following three conclusions can be drawn.

(1) The mechanical models established on the one-dimensional and two-dimensional theory are consistent; the two-dimensional solution may regress to the corresponding one-dimensional solution.

(2) For pure bending problems, the plane section assumption still holds for a bimodular functionally graded beam; for lateral-force bending problems, the plane section assumption holds only in the case of a shallow beam.

(3) The introduction of the bimodular effect and functionally graded characteristic of materials will change the stress and deformation of the structure to some extent. Specifically, the maximum bending stress may take place at a certain level between the neutral layer and edge fibers of the beam, which should be given more attention in the analysis and design of similar structures.

The material considered in this study not only has a functionally graded characteristic but also exhibits different tensile and compressive moduli of elasticity, which further complicates the analysis of similar structures made from these materials. It will be worthwhile considering the plate model adopting classical plate theory for laminate (or higher order theory) to discretize the material properties along the direction of the plate thickness (or here along the beam height).

Moreover, since beams, plates, and shells can all be attributed to, from the point of view of loading and deformation, bending elements under external loads, this work may be extended to the static and dynamic responses of functionally graded beams [27], of functionally graded plates [28], as well as of functionally graded shells [29], in which the bimodular effect of the materials will be incorporated. At the same time, this work may also be extended to an investigation on the existing capabilities and limitations in numerical modeling of fracture problems in functionally graded materials by means of the well-known finite element code ABAQUS [30]. We will study these interesting issues in the future.

Author Contributions: X.H. and J.S. proposed the studied problem and the corresponding solving method; X.L. and J.D. conducted the theoretical derivation and the computation; X.H. and X.L. wrote the paper.

Funding: This project is supported by National Natural Science Foundation of China (Grant Nos. 11572061 and 11772072).

Conflicts of Interest: The authors declare no conflict of interest.

References

1. Jones, R.M. Stress-strain relations for materials with different moduli in tension and compression. *AIAA J.* **1977**, *15*, 16–23. [CrossRef]
2. Bertoldi, K.; Bigoni, D.; Drugan, W.J. Nacre: An orthotropic and bimodular elastic material. *Compos. Sci. Technol.* **2008**, *68*, 1363–1375. [CrossRef]
3. Tran, A.D.; Bert, C.W. Bending of thick beams of bimodulus materials. *Compos. Struct.* **1982**, *15*, 627–642. [CrossRef]
4. Reddy, J.N. Transient response of laminated, bimodular-material, composite rectangular plates. *J. Compos. Mater.* **1982**, *16*, 139–152. [CrossRef]
5. Bert, C.W.; Gordaninejad, F. Transverse shear effects in bimodular composite laminates. *J. Compos. Mater.* **1983**, *17*, 282–298. [CrossRef]
6. Ramana Murthy, P.V.; Rao, K.P. Analysis of curved laminated beams of bimodulus composite materials. *J. Compos. Mater.* **1983**, *17*, 435–438. [CrossRef]
7. Bruno, D.; Lato, S.; Zinno, R. Nonlinear analysis of doubly curved composite shells of bimodular material. *Compos. Part B* **1993**, *3*, 419–435. [CrossRef]
8. Zinno, R.; Greco, F. Damage evolution in bimodular laminated composites under cyclic loading. *Compos. Struct.* **2001**, *53*, 381–402. [CrossRef]
9. Ambartsumyan, S.A. Basic equations and relations in the theory of anisotropic bodies with different moduli in tension and compression. *Inzh. Zhur. MTT (Proc. Acad. Sci. USSR Eng. J. Mech. Solids)* **1969**, *3*, 51–61.
10. Liu, X.B.; Zhang, Y.Z. Modulus of elasticity in shear and accelerate convergence of different extension-compression elastic modulus finite element method. *J. Dalian Univ. Technol.* **2000**, *40*, 527–530.
11. Ye, Z.M.; Chen, T.; Yao, W.J. Progresses in elasticity theory with different modulus in tension and compression and related FEM. *Chin. J. Mech. Eng.* **2004**, *26*, 9–14.
12. Zhao, H.L.; Ye, Z.M. Analytic elasticity solution of bi-modulus beams under combined loads. *Appl. Math. Mech. (Engl. Ed.)* **2015**, *36*, 427–438. [CrossRef]
13. Du, Z.L.; Zhang, Y.P.; Zhang, W.S.; Guo, X. A new computational framework for materials with different mechanical responses in tension and compression and its applications. *Int. J. Solids Struct.* **2016**, *100–101*, 54–73. [CrossRef]
14. Sun, J.Y.; Zhu, H.Q.; Qin, S.H.; Yang, D.L.; He, X.T. A review on the research of mechanical problems with different moduli in tension and compression. *J. Mech. Sci. Technol.* **2010**, *24*, 1845–1854. [CrossRef]
15. Sankar, B.V. An elasticity solution for functionally graded beams. *Compos. Sci. Technol.* **2001**, *61*, 689–696. [CrossRef]
16. Sankar, B.V.; Tzeng, J.T. Thermal stresses in functionally graded beams. *AIAA J.* **2002**, *40*, 1228–1232. [CrossRef]
17. Venkataraman, S.; Sankar, B.V. Elasticity solution for stresses in a sandwich beam with functionally graded core. *AIAA J.* **2003**, *41*, 2501–2505. [CrossRef]
18. Zhu, H.; Sankar, B.V. A combined Fourier series-Galerkin method for the analysis of functionally graded beams. *ASME J. Appl. Mech.* **2004**, *71*, 421–424. [CrossRef]
19. Zhong, Z.; Yu, T. Analytical solution of a cantilever functionally graded beam. *Compos. Sci. Technol.* **2007**, *67*, 481–488. [CrossRef]
20. Nie, G.J.; Zhong, Z.; Chen, S. Analytical solution for a functionally graded beam with arbitrary graded material properties. *Compos. Part B* **2013**, *44*, 274–282. [CrossRef]
21. Daouadji, T.H.; Henni, A.H.; Tounsi, A.; Abbes, A.B. Elasticity solution of a cantilever functionally graded beam. *Appl. Compos. Mater.* **2013**, *20*, 1–15. [CrossRef]
22. Yao, W.J.; Ye, Z.M. Analytical solution for bending beam subject to lateral force with different modulus. *Appl. Math. Mech. (Engl. Ed.)* **2004**, *25*, 1107–1117.
23. He, X.T.; Chen, S.L.; Sun, J.Y. Elasticity solution of simple beams with different modulus under uniformly distributed load. *Chin. J. Eng. Mech.* **2007**, *24*, 51–56.
24. He, X.T.; Xu, P.; Sun, J.Y.; Zheng, Z.L. Analytical solutions for bending curved beams with different moduli in tension and compression. *Mech. Adv. Mater. Struct.* **2015**, *22*, 325–337. [CrossRef]
25. He, X.T.; Chen, Q.; Sun, J.Y.; Zheng, Z.L.; Chen, S.L. Application of the Kirchhoff hypothesis to bending thin plates with different moduli in tension and compression. *J. Mech. Mater. Struct.* **2010**, *5*, 755–769. [CrossRef]

26. He, X.T.; Li, W.M.; Sun, J.Y.; Wang, Z.X. An elasticity solution of functionally graded beams with different moduli in tension and compression. *J. Mech. Mater. Struct.* **2018**, *25*, 143–154. [CrossRef]
27. Wattanasakulpong, N.; Bui, T.Q. Vibration analysis of third-order shear deformable FGM beams with elastic support by Chebyshev collocation method. *Int. J. Struct. Stab. Dyn.* **2018**, *18*, 1850071. [CrossRef]
28. Fu, Y.; Yao, J.; Wan, Z.; Zhao, G. Free vibration analysis of moderately thick orthotropic functionally graded plates with general boundary restraints. *Materials* **2018**, *11*, 273. [CrossRef] [PubMed]
29. Nguyen Dinh, D.; Nguyen, P.D. The dynamic response and vibration of functionally graded carbon nanotube-reinforced composite (FG-CNTRC) truncated conical shells resting on elastic foundations. *Materials* **2017**, *10*, 1194. [CrossRef] [PubMed]
30. Martínez-Pañeda, E.; Gallego, R. Numerical analysis of quasi-static fracture in functionally graded materials. *Int. J. Mech. Mater. Des.* **2015**, *11*, 405–424. [CrossRef]

materials

MDPI

Article

An Electroelastic Solution for Functionally Graded Piezoelectric Circular Plates under the Action of Combined Mechanical Loads

Zhi-xin Yang [1], Xiao-ting He [1,2,*], Xue Li [1], Yong-sheng Lian [1] and Jun-yi Sun [1,2]

1 School of Civil Engineering, Chongqing University, Chongqing 400045, China; yangzhixin123@126.com (Z.-x.Y.); lixuecqu@126.com (X.L.); lianyongsheng@cqu.edu.cn (Y.-s.L.); sunjunyi@cqu.edu.cn (J.-y.S.)

2 Key Laboratory of New Technology for Construction of Cities in Mountain Area (Chongqing University), Ministry of Education, Chongqing 400045, China

* Correspondence: hexiaoting@cqu.edu.cn; Tel.: +86-(0)23-6512-0720

Received: 29 May 2018; Accepted: 3 July 2018; Published: 9 July 2018

Abstract: In this study, we obtained an electroelastic solution for functionally graded piezoelectric circular plates under the action of combined mechanical loads which include the uniformly distributed loads on the upper surface of the plate and the radial force and bending moment at the periphery of the plate. All electroelastic materials parameters are assumed to vary according to the same gradient function along the thickness direction. The influence of different functionally graded parameters on the elastic displacement and elastic stress, as well as the electric displacement and electric potential, was discussed by a numerical example. The solution presented in this study is not only applicable to the case of combined loads, but also to the case of a single mechanical load. In addition, this solution reflects the influence of the function gradient on the pure piezoelectric plate, which is helpful to the refined analysis and optimization design of similar structures.

Keywords: functionally graded piezoelectric materials; circular plate; combined mechanical loads; electroelastic solution

1. Introduction

The concept of functionally graded materials (FGMs) can be traced back to the eighties and nineties of last century, and at that time, to eliminate interface problems and relieve thermal stress concentrations in conventional laminated materials, a group of Japanese scientists suggested using this material as thermal barrier materials for aerospace structural applications and fusion reactors [1]. Generally, FGMs are a kind of inhomogeneous composite from the point of macroscopic view that are typically made from a mixture of two materials. This mixture can be obtained by gradually changing the composition of the constituent materials (along the thickness direction of components in most cases). The characteristics of FGMs vary gradually with the thickness direction within the structure, which eliminates interface problems, and thus the stress distributions are smooth. Moreover, FGMs possess many new properties that most traditional laminated materials do not have, which gives the use of FGMs many advantages in aerospace, automotive, and biomedical applications. During the past decades, FGMs have received a significant amount of attention from the academic community and engineering field, and many scholars have carried out research on functionally graded materials and structures [2–12].

On the other hand, piezoelectric materials have been used extensively in the design of sensors and actuators due to their high efficiency in electromechanical conversion [13–15]. Piezoelectric sensors are usually a laminated original made by ceramic slice. However, on this kind of laminated original, it is easy to cause stress concentration and promote the growth of interfacial microcracks which limit the application and development of the piezoelectric original. In order to solve this problem, functionally graded piezoelectric materials (FGPMs), whose material properties change continuously in one direction, were developed [16–19]. Because there is no obvious interface in this material, the damage caused by the stress concentration at the interface can be avoided.

With the increasing application of functionally graded piezoelectric materials, precise characterization of their mechanical properties is urgently needed. A great deal of research has been done on the mechanical properties of functionally graded piezoelectric materials. Dineva et al. [20] evaluated the stress and electric field concentrations around a circular hole in a functionally graded piezoelectric plane subjected to antiplane elastic SH-wave and in-plane, time-harmonic electric load. Chen and Ding [21] investigated the bending problem of a simply supported rectangular plate by introducing two displacement functions and stress functions and combining the state space method. Zhang et al. [22] studied the behavior of four parallel nonsymmetric permeable cracks with different lengths in a functionally graded piezoelectric material plane subjected to antiplane shear stress loading by the Schmidt method. Wu et al. [23] analyzed the electromechanical coupling effect for functionally graded piezoelectric plates. The coupled static analysis of thermal power and electricity for functionally graded piezoelectric rectangular plates was carried out by Zhong and Shang [24,25]. Based on the generalized Mindlin plate theory, Zhu et al. [26] derived the finite element equations of functionally graded material plates by using the variation principle and investigated and calculated the deflection and potential of a simply supported functionally graded piezoelectric square plate with linear gradient under uniformly distributed loads. Lu et al. [27,28] studied the bending problem of a simply supported functionally gradient piezoelectric plate and a cylindrical plate under mechanical load separately by using the similar Stroh equation. The exact solution of free vibration of functionally graded piezoelectric circular plates was studied by Zhang and Zhong [29]. Recently, Liu et al. [30] presented transient thermal dynamic analysis of stationary cracks in functionally graded piezoelectric materials based on the extended finite element method (X-FEM). Yu et al. [31] analyzed interfacial dynamic impermeable cracks in dissimilar piezoelectric materials under coupled electromechanical loading with the extended finite element method. Given that there are many studies in this field, here we do not review them in detail.

Among the studies above, we note that since the materials parameters vary with a certain direction and the electromechanical coupling effect exists, the obtainment of an analytical solution is relatively difficult. The basic equations of functionally graded piezoelectric structures are generally expressed in the form of partial differential equations except for the physical equations. The general practice is still the so-called separation of variables. According to the specific problem, for example, a spatial axisymmetric deformation problem in [32,33], the unknown stress or displacement function and the unknown electrical potential function are expressed as a polynomial with respect to two variables, i.e., $F(r,z) = \sum r^n f_n(z)$, in which r is the radial coordinate and z is the transverse coordinate along the thickness direction. By continuous substitution and integration, the partial differential equations are transformed into ordinary differential equations, and the integral constants may be determined by boundary conditions, thus obtaining the final solution. Besides, to the authors' knowledge, the existing work of functionally graded piezoelectric plates focused mostly on the problem of the plate subjected to a single load, and the problem under the action of combined mechanical loads seems to be relatively less.

In this study, we will analyze the axisymmetric deformation problem of functionally graded piezoelectric circular plates under the action of combined mechanical loads (i.e., uniformly distributed loads on the upper surface of the plate and radial force and bending moment at the periphery of the plate). The basic equations and their electroelastic solution are presented in Section 2. In Section 3, the influence of different functionally graded parameters on the elastic displacement and stress, as well as the electric displacement and electric potential, are discussed by a numerical example. Section 4 is the concluding remarks.

2. Basic Equations and Their Electroelastic Solution

Considering a simply supported functionally graded piezoelectric circular plate with radius a and thickness h, a uniformly distributed load q is applied on the upper surface of the plate and a radial force $\overline{N}$ and a bending moment $\overline{M}$ are applied at the periphery of the plate, as shown in Figure 1.

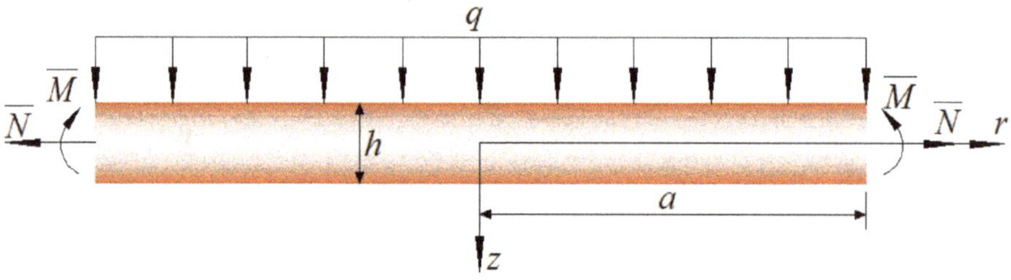

Figure 1. Sketch of a functionally graded piezoelectric circular plate.

Here, we introduce the cylindrical coordinate system (r, θ, z), where the upper and lower surfaces of the plate are $z = -h/2$ and $z = h/2$, respectively, the center of the plate is $r = 0$, and the periphery of the plate is $r = a$. The physical parameters of functionally graded piezoelectric materials are usually the functions of coordinates, and in many practical situations, the physical parameters change only in one direction. In this study, we assumed that the material parameters vary according to the same function along the thickness direction,

$$c_{ij} = c_{ij}^0 f(z), e_{ij} = e_{ij}^0 f(z), \lambda_{ij} = \lambda_{ij}^0 f(z), \tag{1}$$

in which $f(z) = e^{\alpha z/h}$ is the gradient function, α is the functional gradient parameter, c_{ij}, e_{ij}, λ_{ij} are elastic, piezoelectric, and dielectric parameters, respectively, and c_{ij}^0, e_{ij}^0, λ_{ij}^0 are the values of the corresponding material parameters at $z = 0$. Supposing that the polarization direction is the forward direction of the z axis, let us take a microelement in the circular plate, and from the balance of the force, we can obtain

$$\left.\begin{array}{c} \frac{\partial \sigma_r}{\partial r} + \frac{\sigma_r - \sigma_\theta}{r} + \frac{\partial \tau_{zr}}{\partial z} = 0 \\ \frac{\partial \tau_{rz}}{\partial r} + \frac{\partial \sigma_z}{\partial z} + \frac{\tau_{rz}}{r} = 0 \end{array}\right\}, \tag{2}$$

in which σ_r is the radial stress, σ_θ is the circumferential stress, σ_z is the stress in the thickness direction, and τ_{rz}, τ_{zr} are the tangential stress. The equation of Maxwell electric displacement conservation is

$$\frac{\partial D_r}{\partial r} + \frac{\partial D_z}{\partial z} + \frac{D_r}{r} = 0, \tag{3}$$

in which D_r and D_z are the electric displacement components, respectively. In the cylindrical coordinate system (r, θ, z), the physical equations of transversely isotropic, functionally graded piezoelectric materials with the z axis being normal to the plane of isotropy read

$$\begin{array}{c} \sigma_r = c_{11}\varepsilon_r + c_{12}\varepsilon_\theta + c_{13}\varepsilon_z - e_{31}E_z \\ \sigma_\theta = c_{12}\varepsilon_r + c_{11}\varepsilon_\theta + c_{13}\varepsilon_z - e_{31}E_z \\ \sigma_z = c_{13}\varepsilon_r + c_{13}\varepsilon_\theta + c_{33}\varepsilon_z - e_{33}E_z \\ \tau_{zr} = c_{44}\gamma_{zr} - e_{15}E_r \\ D_r = e_{15}\gamma_{zr} + \lambda_{11}E_r \\ D_z = e_{31}(\varepsilon_r + \varepsilon_\theta) + e_{33}\varepsilon_z + \lambda_{33}E_z \end{array}, \tag{4}$$

in which ε_r, ε_θ, ε_z, γ_{zr} are strain components, and E_r, E_z are the electric field in r and z directions, respectively. The geometric equations are

$$\left.\begin{array}{c} \varepsilon_r = \frac{\partial u_r}{\partial r}, \varepsilon_\theta = \frac{u_r}{r} \\ \varepsilon_z = \frac{\partial u_z}{\partial z}, \gamma_{rz} = \frac{\partial u_z}{\partial r} + \frac{\partial u_r}{\partial z} \end{array}\right\}, \tag{5}$$

in which u_r, u_z are the displacement in r and z directions, respectively. The relation of electric field and electric potential is

$$E_r = -\frac{\partial \phi}{\partial r}, E_z = -\frac{\partial \phi}{\partial z}, \tag{6}$$

in which ϕ is the electric potential. Those equations shown above are the basic equations of the problem presented here. The boundary conditions, which can be used for the solution of those basic equations, are shown as follows:

$$\sigma_z = -q, \tau_{rz} = 0, D_z = 0 \quad \text{at } z = -h/2, \tag{7a}$$

$$\sigma_z = 0, \tau_{rz} = 0, D_z = 0 \quad \text{at } z = h/2, \tag{7b}$$

$$N(r) = \overline{N}, M(r) = \overline{M}, u_z(r,0) = 0, \int_{-h/2}^{h/2} D_r dz = 0, \tau_{rz} = 0 \text{ at } r = a. \tag{7c}$$

Suppose that [32,33]

$$\left.\begin{array}{l} u_r(r,z) = r u_1(z) + r^3 u_3(z) \\ u_z(r,z) = w_0(z) + r^2 w_2(z) + r^4 w_4(z) \\ \phi(r,z) = \phi_0(z) + r^2\phi_2(z) + r^4\phi_4(z) \end{array}\right\}, \tag{8}$$

in which $u_i(z)$ and $w_i(z)$ are also looked at as the displacement functions, $\phi_i(z)$ is also looked at as the potential functions, and they depend only on z. The detailed reason for the assumption of Equation (8) is shown in the Appendix A, which includes some results from functionally graded piezoelectric beams [34,35]. Substituting Equation (8) into Equation (5), it gives

$$\left.\begin{array}{l} \varepsilon_r = u_1(z) + 3r^2 u_3(z),\ \varepsilon_\theta = u_1(z) + r^2 u_3(z) \\ \varepsilon_z = w_0'(z) + r^2 w_2'(z) + r^4 w_4'(z),\ \gamma_{rz} = 2r w_2(z) + 4r^3 w_4(z) + r u_1'(z) + r^3 u_3'(z) \end{array}\right\}. \tag{9}$$

Substituting Equations (6), (8), and (9) into Equation (4), we can obtain

$$\begin{array}{l} \sigma_r = [e_{31}\phi_4'(z) + c_{13}w_4'(z)]r^4 + [3c_{11}u_3(z) + c_{12}u_3(z) + c_{13}w_2'(z) + e_{31}\phi_2'(z)]r^2 \\ \quad + [c_{11}u_1(z) + c_{12}u_1(z) + c_{13}w_0'(z) + e_{31}\phi_0'(z)] \\ \sigma_\theta = [c_{13}w_4'(z) + e_{31}\phi_4'(z)]r^4 + [3c_{12}u_3(z) + c_{11}u_3(z) + c_{13}w_2'(z) + e_{31}\phi_2'(z)]r^2 \\ \quad + [c_{12}u_1(z) + c_{11}u_1(z) + c_{13}w_0'(z) + e_{31}\phi_0'(z)] \\ \sigma_z = [c_{33}w_4'(z) + e_{33}\phi_4'(z)]r^4 + [4c_{13}u_3(z) + c_{33}w_2'(z) + e_{33}\phi_2'(z)]r^2 \\ \quad + [2c_{13}u_1(z) + c_{33}w_0'(z) + e_{33}\phi_0'(z)] \\ \tau_{zr} = [2c_{44}w_2(z) + c_{44}u_1'(z) + 2e_{15}\phi_2(z)]r + [4c_{44}w_4(z) + c_{44}u_3'(z) + 4e_{15}\phi_4(z)]r^3 \\ D_r = [2e_{15}w_2(z) + e_{15}u_1'(z) - 2\lambda_{11}\phi_2(z)]r + [4e_{15}w_4(z) + e_{15}u_3'(z) - 4\lambda_{11}\phi_4(z)]r^3 \\ D_z = [2e_{31}u_1(z) + e_{33}w_0'(z) - \lambda_{33}\phi_0'(z)] + [4e_{31}u_3(z) + e_{33}w_2'(z) - \lambda_{33}\phi_2'(z)]r^2 \\ \quad + [e_{33}w_4'(z) - \lambda_{33}\phi_4'(z)]r^4 \end{array} \tag{10}$$

Then, substituting Equation (10) into Equations (2) and (3), respectively, we can also obtain

$$\begin{array}{l} \left\{[8c_{11}u_3(z) + 2e_{31}\phi'_2(z) + 2c_{13}w'_2(z)] + [2c_{44}w_2(z) + c_{44}u'_1(z) + 2e_{15}\phi_2(z)]_{,z}\right\}r \\ + \left\{[4c_{13}w'_4(z) + 4e_{31}\phi'_4(z)] + [4c_{44}w_4(z) + c_{44}u'_3(z) + 4e_{15}\phi_4(z)]_{,z}\right\}r^3 = 0 \end{array}, \tag{11}$$

$$\begin{aligned}&\left\{[4c_{44}w_2(z)+2c_{44}u'_1(z)+4e_{15}\phi_2(z)]+[2c_{13}u_1(z)+c_{33}w'_0(z)+e_{33}\phi'_0(z)]_{,z}\right\}\\&+\left\{[16c_{44}w_4(z)+4c_{44}u'_3(z)+16e_{15}\phi_4(z)]+[4c_{13}u_3(z)+c_{33}w'_2(z)+e_{33}\phi'_2(z)]_{,z}\right\}r^2\\&+[e_{33}\phi'_4(z)+c_{33}w'_4(z)]_{,z}r^4=0\end{aligned}, \tag{12}$$

$$\begin{aligned}&\left\{[4e_{15}w_2(z)+2e_{15}u'_1(z)-4\lambda_{11}\phi_2(z)]+[2e_{31}u_1(z)+e_{33}w'_0(z)-\lambda_{33}\phi'_0(z)]_{,z}\right\}\\&+\left\{[16e_{15}w_4(z)+4e_{15}u'_3(z)-16\lambda_{11}\phi_4(z)]+[4e_{31}u_3(z)+e_{33}w'_2(z)-\lambda_{33}\phi'_2(z)]_{,z}\right\}r^2\\&+[e_{33}w'_4(z)-\lambda_{33}\phi'_4(z)]_{,z}r^4=0\end{aligned}. \tag{13}$$

From Equations (11)–(13), we can obtain

$$[e_{33}\phi'_4(z)+c_{33}w'_4(z)]_{,z}=0, \tag{14}$$

$$[e_{33}w'_4(z)-\lambda_{33}\phi'_4(z)]_{,z}=0, \tag{15}$$

$$[4c_{13}w'_4(z)+4e_{31}\phi'_4(z)]+[4c_{44}w_4(z)+c_{44}u'_3(z)+4e_{15}\phi_4(z)]_{,z}=0, \tag{16}$$

$$[16c_{44}w_4(z)+4c_{44}u'_3(z)+16e_{15}\phi_4(z)]+[4c_{13}u_3(z)+c_{33}w'_2(z)+e_{33}\phi'_2(z)]_{,z}=0, \tag{17}$$

$$[16e_{15}w_4(z)+4e_{15}u'_3(z)-16\lambda_{11}\phi_4(z)]+[4e_{31}u_3(z)+e_{33}w'_2(z)-\lambda_{33}\phi'_2(z)]_{,z}=0, \tag{18}$$

$$[8c_{11}u_3(z)+2e_{31}\phi'_2(z)+2c_{13}w'_2(z)]+[2c_{44}w_2(z)+c_{44}u'_1(z)+2e_{15}\phi_2(z)]_{,z}=0, \tag{19}$$

$$[4c_{44}w_2(z)+2c_{44}u'_1(z)+4e_{15}\phi_2(z)]+[2c_{13}u_1(z)+c_{33}w'_0(z)+e_{33}\phi'_0(z)]_{,z}=0, \tag{20}$$

$$[4e_{15}w_2(z)+2e_{15}u'_1(z)-4\lambda_{11}\phi_2(z)]+[2e_{31}u_1(z)+e_{33}w'_0(z)-\lambda_{33}\phi'_0(z)]_{,z}=0. \tag{21}$$

Substituting Equation (10) into Equation (7a,b), respectively, we can obtain

$$\begin{aligned}&[c_{33}w'_4(z)+e_{33}\phi'_4(z)]\,|_{z=\pm h/2}=0\\&[e_{33}w'_4(z)-\lambda_{33}\phi'_4(z)]\,|_{z=\pm h/2}=0\\&[4c_{44}w_4(z)+c_{44}u'_3(z)+4e_{15}\phi_4(z)]\,|_{z=\pm h/2}=0\\&[4c_{13}u_3(z)+c_{33}w'_2(z)+e_{33}\phi'_2(z)]\,|_{z=\pm h/2}=0\\&[4e_{31}u_3(z)+e_{33}w'_2(z)-\lambda_{33}\phi'_2(z)]\,|_{z=\pm h/2}=0\\&[2c_{44}w_2(z)+c_{44}u'_1(z)+2e_{15}\phi_2(z)]\,|_{z=\pm h/2}=0\\&[2e_{31}u_1(z)+e_{33}w'_0(z)-\lambda_{33}\phi'_0(z)]\,|_{z=\pm h/2}=0\\&[2c_{13}u_1(z)+c_{33}w'_0(z)+e_{33}\phi'_0(z)]\,|_{z=-h/2}=-q\\&[2c_{13}u_1(z)+c_{33}w'_0(z)+e_{33}\phi'_0(z)]\,|_{z=h/2}=0\end{aligned} \tag{22}$$

We can obtain from the integration of Equations (14) and (15), respectively,

$$e_{33}\phi'_4(z)+c_{33}w'_4(z)=b_0, \tag{23a}$$

$$e_{33}w'_4(z)-\lambda_{33}\phi'_4(z)=b_1. \tag{23b}$$

Substituting Equation (23a,b) into the first and second ones of Equation (22), we can obtain

$$b_0=0, b_1=0. \tag{24}$$

From Equations (23a,b) and (24), we can obtain

$$(e_{33}^2+\lambda_{33}c_{33})w'_4(z)=0, \tag{25a}$$

$$(e_{33}^2+\lambda_{33}c_{33})\phi'_4(z)=0. \tag{25b}$$

As we all know, $(e_{33}^2 + \lambda_{33}c_{33}) \neq 0$, thus

$$w_4'(z) = 0, \tag{26a}$$

$$\phi_4'(z) = 0. \tag{26b}$$

We can obtain from the integration of Equation (26a,b), respectively,

$$w_4(z) = a_0, \tag{27a}$$

$$\phi_4(z) = a_1, \tag{27b}$$

in which, a_0, a_1 are integration constants. Substituting Equation (27a,b) into Equation (16), we can obtain

$$[4c_{44}a_0 + c_{44}u_3'(z) + 4e_{15}a_1]_{,z} = 0. \tag{28}$$

From the integration of Equation (28), one has

$$4c_{44}a_0 + c_{44}u_3'(z) + 4e_{15}a_1 = b_2. \tag{29}$$

Then, substituting Equation (29) into the third one of Equation (22), we can obtain

$$b_2 = 0. \tag{30}$$

Substituting Equation (30) into Equation (29) and integrating the two sides of Equation (29), it gives

$$u_3(z) = -(4a_0 + 4\frac{e_{15}}{c_{44}}a_1)z + a_2, \tag{31}$$

in which a_2 is an integration constant. Substituting Equations (27a,b) and (31) into Equations (17) and (18), respectively, we can obtain

$$[4c_{13}u_3(z) + c_{33}w_2'(z) + e_{33}\phi_2'(z)]_{,z} = 0, \tag{32}$$

$$[4e_{31}u_3(z) + e_{33}w_2'(z) - \lambda_{33}\phi_2'(z)]_{,z} = (16e_{15}\frac{e_{15}}{c_{44}} + 16\lambda_{11})a_1. \tag{33}$$

Integrating the two sides of Equations (32) and (33), we can obtain

$$4c_{13}u_3(z) + c_{33}w_2'(z) + e_{33}\phi_2'(z) = b_3, \tag{34}$$

$$4e_{31}u_3(z) + e_{33}w_2'(z) - \lambda_{33}\phi_2'(z) = (16\frac{(e_{15}^0)^2}{c_{44}^0} + 16\lambda_{11}^0)a_1\int_{-h/2}^{z} f(z)dz + b_4. \tag{35}$$

Then, substituting Equations (34) and (35) into the fourth and fifth ones of Equation (22), we can obtain

$$b_3 = 0,\ b_4 = 0,\ a_1 = 0. \tag{36}$$

Substituting Equation (36) into Equations (34) and (35), respectively, we can obtain

$$4c_{13}u_3(z) + c_{33}w_2'(z) + e_{33}\phi_2'(z) = 0, \tag{37}$$

$$4e_{31}u_3(z) + e_{33}w_2'(z) - \lambda_{33}\phi_2'(z) = 0. \tag{38}$$

From Equations (31), (37), and (38), we have

$$w_2'(z) = \frac{(4\lambda_{33}c_{13} + 4e_{33}e_{31})}{(\lambda_{33}c_{33} + e_{33}^2)}(4a_0z - a_2), \tag{39}$$

$$\phi_2'(z) = \frac{(4e_{33}c_{13} - 4c_{33}e_{31})}{(\lambda_{33}c_{33} + e_{33}^2)}(4a_0z - a_2). \tag{40}$$

Integrating the two sides of Equations (32) and (33), we can obtain

$$w_2(z) = \frac{(4\lambda_{33}c_{13} + 4e_{33}e_{31})}{(\lambda_{33}c_{33} + e_{33}^2)}(2a_0z^2 - a_2z) + a_3, \tag{41}$$

$$\phi_2(z) = \frac{(4e_{33}c_{13} - 4c_{33}e_{31})}{(\lambda_{33}c_{33} + e_{33}^2)}(2a_0z^2 - a_2z) + a_4, \tag{42}$$

in which a_3, a_4 are integration constants. From Equations (31), (41), and (42), Equation (19) gives

$$\begin{aligned} &[2c_{44}w_2(z) + c_{44}u_1'(z) + 2e_{15}\phi_2(z)]_{,z} = (8c_{11}\lambda_{33}c_{33} + 8c_{11}e_{33}^2 + 8c_{33}e_{31}^2 \\ &-8e_{31}e_{33}c_{13} - 8\lambda_{33}c_{13}^2 - 8c_{13}e_{33}e_{31})\tfrac{(4a_0z-a_2)}{\lambda_{33}c_{33}+e_{33}^2} \end{aligned}. \tag{43}$$

Integrating the two sides of Equation (43), we can obtain

$$[2c_{44}w_2(z) + c_{44}u_1'(z) + 2e_{15}\phi_2(z)] = 4a_0K_0F_1(z) - a_2K_0F_0(z) + b_5, \tag{44}$$

in which $K_0 = 8\frac{(c_{11}^0\lambda_{33}^0c_{33}^0 + c_{11}^0e_{33}^0e_{33}^0 + c_{33}^0e_{31}^0e_{31}^0 - e_{31}^0e_{33}^0c_{13}^0 - \lambda_{33}^0c_{13}^0c_{13}^0 - c_{13}^0e_{33}^0e_{31}^0)}{\lambda_{33}^0c_{33}^0 + e_{33}^0e_{33}^0}$, $F_0(z) = \int_{-h/2}^{z} f(z)dz$, $F_1(z) = \int_{-h/2}^{z} zf(z)dz$. Substituting Equation (44) into the sixth one of Equation (22), we can obtain

$$b_5 = 0,\ 4a_0F_1(h/2) - a_2F_0(h/2) = 0. \tag{45}$$

From the second one of Equation (45), we can obtain

$$a_2 = 4\frac{F_1(h/2)}{F_0(h/2)}a_0. \tag{46}$$

Substituting Equations (41) and (42) into Equation (44) and with the help of Equations (45) and (46), we get

$$u_1'(z) = 4a_0\frac{K_0}{c_{44}^0}\frac{F_1(z)}{f(z)} - a_2\frac{K_0}{c_{44}^0}\frac{F_0(z)}{f(z)} - K_1(2a_0z^2 - a_2z) - 2a_3 - 2\frac{e_{15}}{c_{44}}a_4, \tag{47}$$

in which $K_1 = \frac{(8c_{44}\lambda_{33}c_{13} + 8c_{44}e_{33}e_{31} + 8e_{15}e_{33}c_{13} - 8e_{15}c_{33}e_{31})}{c_{44}(\lambda_{33}c_{33} + e_{33}^2)}$. Integrating the two sides of Equation (47), one has

$$u_1(z) = 4a_0\frac{K_0}{c_{44}^0}H_1(z) - a_2\frac{K_0}{c_{44}^0}H_0(z) - \frac{2}{3}K_1a_0z^3 + K_1a_2\frac{z^2}{2} - (2a_3 + 2\frac{e_{15}}{c_{44}}a_4)z + a_5, \tag{48}$$

in which $H_0(z) = \int_{-h/2}^{z}\frac{F_0(z)}{f(z)}dz$, $H_1(z) = \int_{-h/2}^{z}\frac{F_1(z)}{f(z)}dz$. Substituting Equations (41), (42), and (48) into Equations (20) and (21), respectively, we can obtain

$$[2c_{13}u_1(z) + c_{33}w_0'(z) + e_{33}\phi_0'(z)]_{,z} = 2a_2K_0F_0(z) - 8a_0K_0F_1(z), \tag{49}$$

$$\begin{aligned} &[2e_{31}u_1(z) + e_{33}w_0'(z) - \lambda_{33}\phi_0'(z)]_{,z} = 2a_2K_0\tfrac{e_{15}}{c_{44}}F_0(z) - 8a_0K_0\tfrac{e_{15}}{c_{44}}F_1(z) \\ &-(4e_{15}\tfrac{e_{15}}{c_{44}} + 4\lambda_{11})(8K_2a_0z^2 - 4K_2a_2z - a_4) \end{aligned}, \tag{50}$$

in which $K_2 = \frac{c_{33}e_{31} - e_{33}c_{13}}{\lambda_{33}c_{33} + e_{33}^2}$. Integrating the two sides of Equations (49) and (50), respectively, we get

$$[2c_{13}u_1(z) + c_{33}w_0'(z) + e_{33}\phi_0'(z)] = 2a_2K_0G_0(z) - 8a_0K_0G_1(z) + b_6, \tag{51}$$

$$\begin{aligned}&[2e_{31}u_1(z)+e_{33}w_0'(z)-\lambda_{33}\phi_0'(z)]=2a_2K_0\tfrac{e_{15}}{c_{44}}G_0(z)-8a_0K_0\tfrac{e_{15}}{c_{44}}G_1(z)\\&-(4e^0_{15}\tfrac{e_{15}}{c_{44}}+4\lambda^0_{11})[8K_2a_0F_2(z)-4K_2a_2F_1(z)-a_4F_0(z)]+b_7\end{aligned}, \tag{52}$$

in which $G_0(z)=\int_{-h/2}^{z}F_0(z)dz$, $G_1(z)=\int_{-h/2}^{z}F_1(z)dz$, $F_2(z)=\int_{-h/2}^{z}z^2f(z)dz$. Substituting Equations (51) and (52) into the seventh, eighth, and ninth ones of Equation (22), respectively, we obtain the following

$$b_7=0, \tag{53}$$

$$a_4=K_3a_0, \tag{54}$$

$$b_6=-q, \tag{55}$$

$$a_0=K_4q, \tag{56}$$

in which

$$K_3=8K_2\frac{F_2(h/2)}{F_0(h/2)}-16K_2\frac{F_1^2(h/2)}{F_0^2(h/2)}-2\frac{e^0_{15}K_0G_0(h/2)F_1(h/2)}{(e^0_{15}e^0_{15}+c^0_{44}\lambda^0_{11})F_0^2(h/2)}+2\frac{e^0_{15}K_0G_1(h/2)}{(e^0_{15}e^0_{15}+c^0_{44}\lambda^0_{11})F_0(h/2)}$$

$$K_4=\frac{F_0(h/2)}{[8F_1(h/2)K_0G_0(h/2)-8F_0(h/2)K_0G_1(h/2)]}$$

Substituting Equation (48) into Equations (51) and (52), respectively, we get

$$\begin{aligned}&c_{33}w_0'(z)+e_{33}\phi_0'(z)=2a_2K_0G_0(z)-8a_0K_0G_1(z)-q-8c_{13}a_0\tfrac{K_0}{c^0_{44}}H_1(z)\\&+2c_{13}a_2\tfrac{K_0}{c^0_{44}}H_0(z)+\tfrac{4}{3}c_{13}K_1a_0z^3-c_{13}K_1a_2z^2+(4a_3c_{13}+4c_{13}\tfrac{e_{15}}{c_{44}}a_4)z-2c_{13}a_5\end{aligned}, \tag{57}$$

$$\begin{aligned}&e_{33}w_0'(z)-\lambda_{33}\phi_0'(z)=-(4e^0_{15}\tfrac{e_{15}}{c_{44}}+4\lambda^0_{11})[8K_2a_0F_2(z)-4K_2a_2F_1(z)-a_4F_0(z)]\\&+2a_2K_0\tfrac{e_{15}}{c_{44}}G_0(z)-8a_0K_0\tfrac{e_{15}}{c_{44}}G_1(z)-8e_{31}a_0\tfrac{K_0}{c^0_{44}}H_1(z)+2e_{31}a_2\tfrac{K_0}{c^0_{44}}H_0(z)\\&+\tfrac{4}{3}e_{31}K_1a_0z^3-e_{31}K_1a_2z^2+(4a_3e_{31}+4e_{31}\tfrac{e_{15}}{c_{44}}a_4)z-2e_{31}a_5\end{aligned}, \tag{58}$$

From Equations (57) and (58), we can obtain

$$w_0'(z)=J_0(z)a_0+J_1(z)a_2+J_2(z)a_3+J_3(z)a_4+J_4(z)a_5+J_5(z)q, \tag{59}$$

$$\phi_0'(z)=L_0(z)a_0+L_1(z)a_2+L_2(z)a_3+L_3(z)a_4+L_4(z)a_5+L_5(z)q, \tag{60}$$

in which

$$\begin{aligned}&J_0(z)=\tfrac{1}{(\lambda^0_{33}c^0_{33}+e^0_{33}e^0_{33})}[-8K_0\tfrac{G_1(z)}{f(z)}(\lambda^0_{33}+e^0_{33}\tfrac{e_{15}}{c_{44}})-8(\lambda^0_{33}c^0_{13}+e^0_{33}e^0_{31})\tfrac{K_0}{c^0_{44}}H_1(z)\\&+\tfrac{4}{3}(\lambda^0_{33}c^0_{13}+e^0_{33}e^0_{31})K_1z^3-32e^0_{33}(e^0_{15}\tfrac{e_{15}}{c_{44}}+\lambda^0_{11})K_2\tfrac{F_2(z)}{f(z)}]\end{aligned},$$

$$\begin{aligned}&J_1(z)=\tfrac{1}{(\lambda^0_{33}c^0_{33}+e^0_{33}e^0_{33})}[2(\lambda^0_{33}+e^0_{33}\tfrac{e_{15}}{c_{44}})K_0\tfrac{G_0(z)}{f(z)}+2(\lambda^0_{33}c^0_{13}+e^0_{33}e^0_{31})\tfrac{K_0}{c^0_{44}}H_0(z)\\&-(\lambda^0_{33}c^0_{13}+e^0_{33}e^0_{31})K_1z^2+16e^0_{33}(e^0_{15}\tfrac{e_{15}}{c_{44}}+\lambda^0_{11})K_2\tfrac{F_1(z)}{f(z)}]\end{aligned},$$

$$J_2(z)=4\frac{(\lambda^0_{33}c^0_{13}+e^0_{33}e^0_{31})}{(\lambda^0_{33}c^0_{33}+e^0_{33}e^0_{33})}z,$$

$$J_3(z)=\frac{1}{(\lambda^0_{33}c^0_{33}+e^0_{33}e^0_{33})}[4(\lambda^0_{33}c^0_{13}+e^0_{33}e^0_{31})\frac{e_{15}}{c_{44}}z+4e^0_{33}(e^0_{15}\frac{e_{15}}{c_{44}}+\lambda^0_{11})\frac{F_0(z)}{f(z)}],$$

$$J_4(z)=-2\frac{(\lambda^0_{33}c^0_{13}+e^0_{33}e^0_{31})}{(\lambda^0_{33}c^0_{33}+e^0_{33}e^0_{33})},$$

$$J_5(z) = -\frac{\lambda_{33}^0}{(\lambda_{33}^0 c_{33}^0 + e_{33}^0 e_{33}^0)} \frac{1}{f(z)},$$

$$L_0(z) = \frac{1}{(e_{33}^0 e_{33}^0 + \lambda_{33}^0 c_{33}^0)}[8(c_{33}^0 \frac{e_{15}}{c_{44}} - e_{33}^0) K_0 \frac{G_1(z)}{f(z)} + 8(c_{33}^0 e_{31}^0 - e_{33}^0 c_{13}^0) \frac{K_0}{c_{44}^0} H_1(z) + \frac{4}{3}(e_{33}^0 c_{13}^0 - c_{33}^0 e_{31}^0) K_1 z^3 + 32 c_{33}^0 (e_{15}^0 \frac{e_{15}}{c_{44}} + \lambda_{11}^0) K_2 \frac{F_2(z)}{f(z)}],$$

$$L_1(z) = \frac{1}{(e_{33}^0 e_{33}^0 + \lambda_{33}^0 c_{33}^0)}[2(e_{33}^0 - c_{33}^0 \frac{e_{15}}{c_{44}}) K_0 \frac{G_0(z)}{f(z)} + 2(e_{33}^0 c_{13}^0 - c_{33}^0 e_{31}^0) \frac{K_0}{c_{44}^0} H_0(z) + (c_{33}^0 e_{31}^0 - e_{33}^0 c_{13}^0) K_1 z^2 - 16 c_{33}^0 (e_{15}^0 \frac{e_{15}}{c_{44}} + \lambda_{11}^0) K_2 \frac{F_1(z)}{f(z)}],$$

$$L_2(z) = 4\frac{(e_{33}^0 c_{13}^0 - c_{33}^0 e_{31}^0)}{(e_{33}^0 e_{33}^0 + \lambda_{33}^0 c_{33}^0)} z,$$

$$L_3(z) = \frac{1}{(e_{33}^0 e_{33}^0 + \lambda_{33}^0 c_{33}^0)}[4(e_{33}^0 c_{13}^0 - c_{33}^0 e_{31}^0)\frac{e_{15}}{c_{44}} z - 4 c_{33}^0 (e_{15}^0 \frac{e_{15}}{c_{44}} + \lambda_{11}^0) \frac{F_0(z)}{f(z)}],$$

$$L_4(z) = 2\frac{(c_{33}^0 e_{31}^0 - e_{33}^0 c_{13}^0)}{(e_{33}^0 e_{33}^0 + \lambda_{33}^0 c_{33}^0)},$$

$$L_5(z) = -\frac{e_{33}^0}{(e_{33}^0 e_{33}^0 + \lambda_{33}^0 c_{33}^0)} \frac{1}{f(z)}.$$

Integrating the two sides of Equations (59) and (60), respectively, we can obtain

$$w_0(z) = j_0(z)a_0 + j_1(z)a_2 + j_2(z)a_3 + j_3(z)a_4 + j_4(z)a_5 + j_5(z)q + a_6, \tag{61}$$

$$\phi_0(z) = l_0(z)a_0 + l_1(z)a_2 + l_2(z)a_3 + l_3(z)a_4 + l_4(z)a_5 + l_5(z)q + a_7, \tag{62}$$

in which $j_i(z) = \int_{-h/2}^{z} J_i(z)dz$, $l_i(z) = \int_{-h/2}^{z} L_i(z)dz$, $(i = 0, 1, \ldots, 5)$.

From the above process, it can be seen that there are 8 integration constants $a_i (i = 0, 1, \ldots, 7)$ in total, in which a_0, a_1, a_2, a_4 have been determined and a_3, a_5, a_6, a_7 can be determined by the boundary conditions at $r = a$.

Substituting the displacement functions $u_i(z)$, $w_i(z)$, and the electric potential function $\phi(z)$ into Equation (10), the expressions of elastic stress and electric displacement components of the circular plate can be obtained

$$\begin{aligned}\sigma_r = {} & (c_{11} + c_{12})[4a_0 \tfrac{K_0}{c_{44}^0} H_1(z) - a_2 \tfrac{K_0}{c_{44}^0} H_0(z) - \tfrac{2}{3} K_1 a_0 z^3 + K_1 a_2 \tfrac{z^2}{2} - (2a_3 + 2\tfrac{e_{15}}{c_{44}} a_4) z + a_5] \\ & + K_5 f(z)(4a_0 z - a_2) r^2 + c_{13}[J_0(z)a_0 + J_1(z)a_2 + J_2(z)a_3 + J_3(z)a_4 + J_4(z)a_5 + J_5(z)q] \\ & + e_{31}[L_0(z)a_0 + L_1(z)a_2 + L_2(z)a_3 + L_3(z)a_4 + L_4(z)a_5 + L_5(z)q]\end{aligned}, \tag{63}$$

$$\begin{aligned}\sigma_\theta = {} & (c_{12} + c_{11})[4a_0 \tfrac{K_0}{c_{44}^0} H_1(z) - a_2 \tfrac{K_0}{c_{44}^0} H_0(z) - \tfrac{2}{3} K_1 a_0 z^3 + K_1 a_2 \tfrac{z^2}{2} - (2a_3 + 2\tfrac{e_{15}}{c_{44}} a_4) z + a_5] \\ & + K_6 f(z)(4a_0 z - a_2) r^2 + c_{13}[J_0(z)a_0 + J_1(z)a_2 + J_2(z)a_3 + J_3(z)a_4 + J_4(z)a_5 + J_5(z)q] \\ & + e_{31}[L_0(z)a_0 + L_1(z)a_2 + L_2(z)a_3 + L_3(z)a_4 + L_4(z)a_5 + L_5(z)q]\end{aligned}, \tag{64}$$

$$\begin{aligned}\sigma_z = {} & 2c_{13}[4a_0 \tfrac{K_0}{c_{44}^0} H_1(z) - a_2 \tfrac{K_0}{c_{44}^0} H_0(z) - \tfrac{2}{3} K_1 a_0 z^3 + K_1 a_2 \tfrac{z^2}{2} - (2a_3 + 2\tfrac{e_{15}}{c_{44}} a_4) z + a_5] \\ & + c_{33}[J_0(z)a_0 + J_1(z)a_2 + J_2(z)a_3 + J_3(z)a_4 + J_4(z)a_5 + J_5(z)q] \\ & + e_{33}[L_0(z)a_0 + L_1(z)a_2 + L_2(z)a_3 + L_3(z)a_4 + L_4(z)a_5 + L_5(z)q]\end{aligned}, \tag{65}$$

$$\tau_{zr} = [4a_0 K_0 F_1(z) - a_2 K_0 F_0(z)] r, \tag{66}$$

$$\begin{aligned}D_r = {} & [8\tfrac{(c_{44}\lambda_{11}c_{33}e_{31} - c_{44}\lambda_{11}e_{33}c_{13} + e_{15}e_{15}c_{33}e_{31} - e_{15}e_{15}e_{33}c_{13})}{c_{44}(\lambda_{33}c_{33} + e_{33}^2)}(2a_0 z^2 - a_2 z) \\ & + 4e_{15}a_0 \tfrac{K_0}{c_{44}^0} \tfrac{F_1(z)}{f(z)} - e_{15} a_2 \tfrac{K_0}{c_{44}^0} \tfrac{F_0(z)}{f(z)} - 2(\lambda_{11} + e_{15}\tfrac{e_{15}}{c_{44}}) a_4] r\end{aligned}, \tag{67}$$

$$\begin{aligned} D_z = & 8e_{31}a_0\tfrac{K_0}{c_{44}^0}H_1(z) - 2e_{31}a_2\tfrac{K_0}{c_{44}^0}H_0(z) - \tfrac{4}{3}e_{31}K_1a_0z^3 + e_{31}K_1a_2z^2 - 4e_{31}a_3z \\ & -4e_{31}\tfrac{e_{15}}{c_{44}}a_4z + 2e_{31}a_5 + e_{33}[J_0(z)a_0 + J_1(z)a_2 + J_2(z)a_3 + J_3(z)a_4 + J_4(z)a_5 \\ & +J_5(z)q] - \lambda_{33}[L_0(z)a_0 + L_1(z)a_2 + L_2(z)a_3 + L_3(z)a_4 + L_4(z)a_5 + L_5(z)q] \end{aligned}, \quad (68)$$

in which

$$K_5 = \frac{(4c_{13}\lambda_{33}c_{13} + 4c_{13}e_{33}e_{31} + 4e_{31}e_{33}c_{13} - 4e_{31}c_{33}e_{31} - 3c_{11}\lambda_{33}c_{33} - 3c_{11}e_{33}^2 - c_{12}\lambda_{33}c_{33} - c_{12}e_{33}^2)}{(\lambda_{33}c_{33} + e_{33}^2)},$$

$$K_6 = \frac{(4c_{13}\lambda_{33}c_{13} + 4c_{13}e_{33}e_{31} + 4e_{31}e_{33}c_{13} - 4e_{31}c_{33}e_{31})}{(\lambda_{33}c_{33} + e_{33}^2)} - (3c_{12} + c_{11}).$$

The expressions of the radial force and bending moment are

$$N(r) = \int_{-h/2}^{h/2} \sigma_r dz, \quad (69)$$

$$M(r) = \int_{-h/2}^{h/2} z\sigma_r dz, \quad (70)$$

and the expressions of the elastic displacement and electric potential are

$$\begin{aligned} u_r(r,z) = & [4a_0\tfrac{K_0}{c_{44}^0}H_1(z) - a_2\tfrac{K_0}{c_{44}^0}H_0(z) - \tfrac{2}{3}K_1a_0z^3 + K_1a_2\tfrac{z^2}{2} \\ & -(2a_3 + 2\tfrac{e_{15}}{c_{44}}a_4)z + a_5]r + (a_2 - 4a_0z)r^3 \end{aligned}, \quad (71)$$

$$\begin{aligned} u_z(r,z) = & j_0(z)a_0 + j_1(z)a_2 + j_2(z)a_3 + j_3(z)a_4 + j_4(z)a_5 + j_5(z)q + a_6 \\ & +[\tfrac{(4\lambda_{33}c_{13}+4e_{33}e_{31})}{(\lambda_{33}c_{33}+e_{33}^2)}(2a_0z^2 - a_2z) + a_3]r^2 + a_0r^4 \end{aligned}, \quad (72)$$

$$\begin{aligned} \phi(r,z) = & l_0(z)a_0 + l_1(z)a_2 + l_2(z)a_3 + l_3(z)a_4 + l_4(z)a_5 + l_5(z)q + a_7 \\ & +[\tfrac{(4e_{33}c_{13}-4c_{33}e_{31})}{(\lambda_{33}c_{33}+e_{33}^2)}(2a_0z^2 - a_2z) + a_4]r^2 \end{aligned}. \quad (73)$$

From Equation (7c), we can obtain

$$N(a) = \int_{-h/2}^{h/2} \sigma_r dz = \overline{N}, \quad (74)$$

$$M(a) = \int_{-h/2}^{h/2} z\sigma_r dz = \overline{M}. \quad (75)$$

There contain only two undetermined constants, a_3 and a_5, thus, from Equations (74) and (75), a_3 and a_5 can be determined. Then, from Equation (7c), one has

$$u_z(a,0) = 0. \quad (76)$$

With the help of determined a_3 and a_5, the undetermined constants a_6 can also be determined by Equation (76). Thus, we obtain the electroelastic solution of the axisymmetric deformation problem of simply supported functionally graded piezoelectric circular plates under the action of combined mechanical loads.

3. Comparisons and Discussions

3.1. Comparisions with Existing Result

Here, we use a numerical example to verify the results presented in this paper. Since there is no electroelastic solution for functionally graded piezoelectric circular plates under the action of combined mechanical loads, only the solution under a single load [32] is available, and we verify the correctness of the results presented in this paper according to the regression. That is, let the radial force and bending moment in this study be zero; the circular plate is now subjected to uniformly distributed loads only, thus the obtained result may be compared with the solution presented in [32] (subjected to uniformly distributed loads only). For this purpose, we consider a simply supported functionally graded piezoelectric circular plate with $a = 1$ m, $h = 0.1$ m and subjected to the action of uniformly-distributed loads $q = 1$ KPa on the upper surface of the plate, in which $\overline{N} = 0$ and $\overline{M}$ at the periphery of the plate. We here use two solutions, the solution presented in this study (denoted by I) and the solution presented in [32] (denoted by II), to conduct the numerical comparisons. In the comparisons, the functional gradient parameter α takes 2 and the material constants at z are listed in Table 1. The comparison results are shown in Figures 2–5, in which Figures 2 and 3 show the elastic displacement and stress, respectively; Figures 4 and 5 show the electric displacement and the electric potential, respectively. From Figures 2–5, it can be found that the solution presented in this study (I) and the solution presented in the previous study (II) are very close to each other, which demonstrates the validity of the results presented in this study.

Table 1. Material constants.

Property	Constants
Elastic(10^9N/m^2)	$c_{11}^0 = c_{22}^0 = 74.1$, $c_{33}^0 = 83.6$, $c_{12}^0 = 45.2$, $c_{13}^0 = c_{23}^0 = 39.3$, $c_{44}^0 = c_{55}^0 = 13.17$, $c_{66}^0 = 14.45$
Piezoelectric(C/m^2)	$e_{31}^0 = e_{32}^0 = -0.16$, $e_{33}^0 = 0.347$, $e_{15}^0 = -0.138$, $e_{24}^0 = 0$
Dielectric(F/m)	$\lambda_{11}^0 = \lambda_{22}^0 = 8.25 \times 10^{-11}$, $\lambda_{33}^0 = 9.02 \times 10^{-11}$

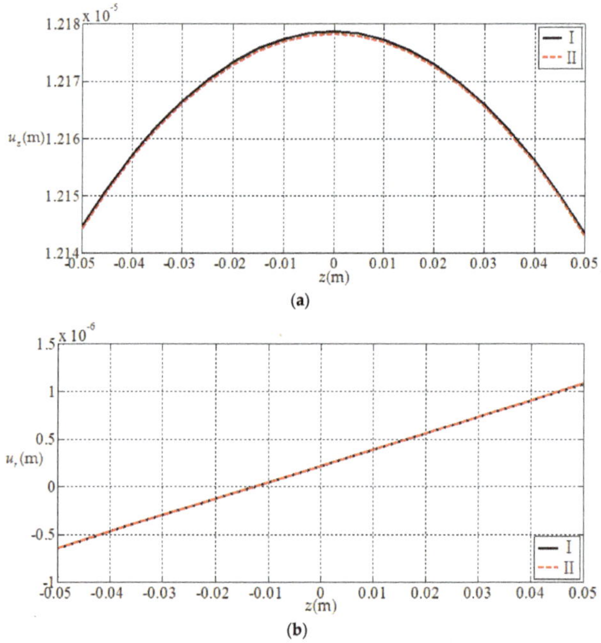

Figure 2. Variation of elastic displacements with coordinates z, where I denotes the solution presented in this study; II denotes the solution presented in [32]. (**a**) z-direction displacement at the center of plate $u_z(0, z)$; (**b**) radial displacement at the periphery of plate $u_r(1, z)$.

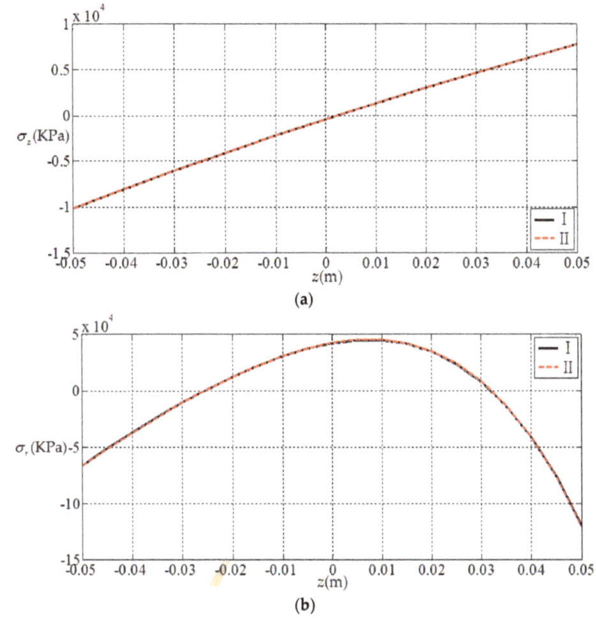

Figure 3. Variation of elastic stress with coordinates z, where I denotes the solution presented in this study; II denotes the solution presented in [32]. (**a**) z-direction stress at the periphery of plate $\sigma_z(1,z)$; (**b**) radial stress at the periphery of plate $\sigma_r(1,z)$.

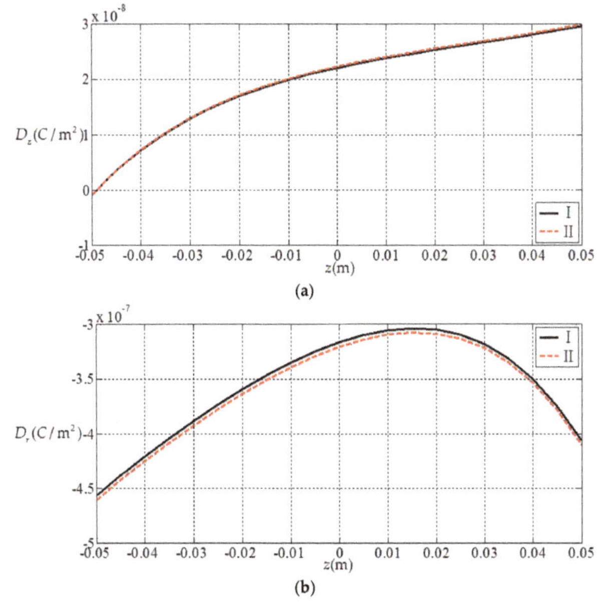

Figure 4. Variation of electric displacement with coordinates z, where I denotes the solution presented in this study; II denotes the solution presented in [32]. (**a**) Electric displacement at the periphery of plate $D_z(1,z)$; (**b**) electric displacement at the periphery of plate $D_r(1,z)$.

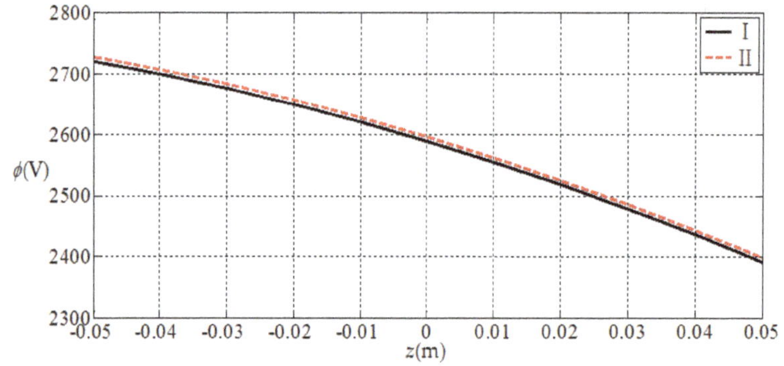

Figure 5. Variation of electric potential at the periphery of plate $\phi(1,z)$, where I denotes the solution presented in this study; II denotes the solution presented in [32].

3.2. *Influences of Functionally Graded Parameters*

Let us consider another numerical example of a simply supported functionally graded piezoelectric circular plate with $a = 1$ m, $h = 0.1$ m and subjected to the action of uniformly distributed loads $q = 1$ KPa on the upper surface of the plate and the action of the radial force $\overline{N} = 6$ kN/m and the bending moment $\overline{M} = 6$ kN at the periphery of the plate, to investigate the influence of different functionally graded parameters on the elastic displacement and elastic stress, as well as the electric displacement and electric potential of the circular plate. Suppose the functional gradient parameter α takes 0, 1, and 2, respectively. Besides, in the computation we still adopt the material constants at $z = 0$ in Table 1.

Figures 6–9, show the variation of the elastic displacement and stress, as well as the electric displacement and electric potential with the coordinate z. From Figures 6–9 it can be found that the variation curves of all physical quantities of the functionally graded piezoelectric circular plate ($\alpha \neq 0$) are deviated from the uniform piezoelectric plate ($\alpha = 0$), and the degree of deviation increases with the increase of functional gradient parameter α, in which the change of u_z, σ_z, D_r, D_z, ϕ are obvious, the change of σ_r is relatively small, and u_r has almost no change. For the functionally graded piezoelectric circular plate, u_r and σ_r change linearly along the thickness direction and u_z, σ_z, D_r, D_z and ϕ change nonlinearly along the thickness direction.

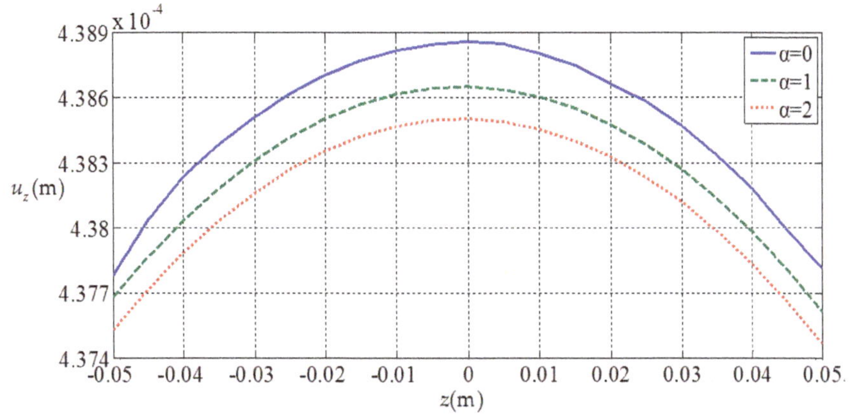

(**a**)

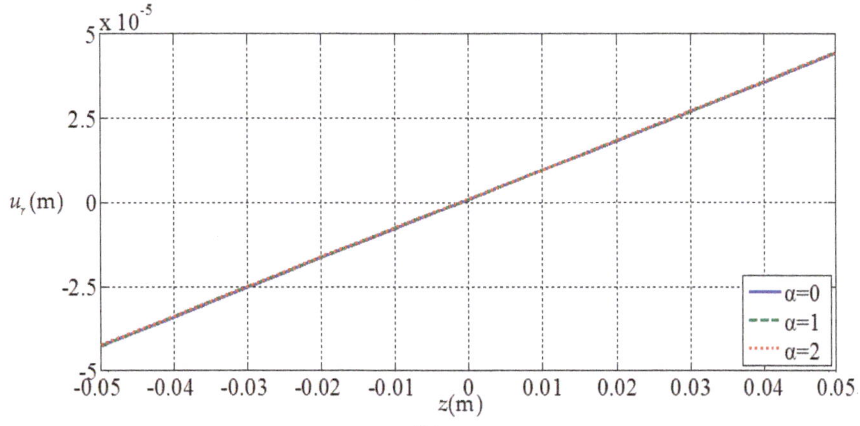

(**b**)

Figure 6. Variation of elastic displacements with coordinates z. (**a**) z-direction displacement at the center of plate $u_z(0,z)$; (**b**) radial displacement at the periphery of plate $u_r(1,z)$.

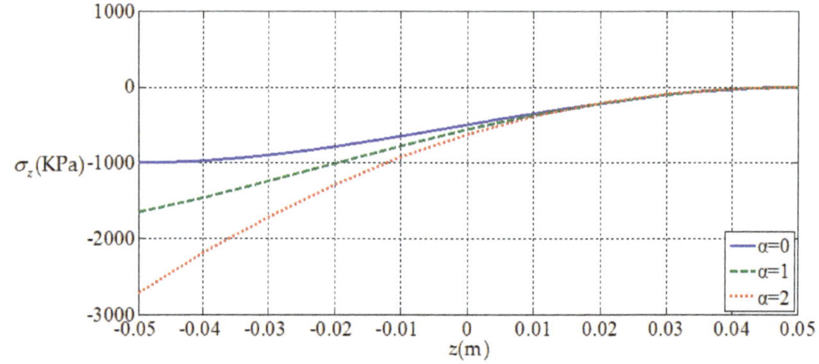

(**a**)

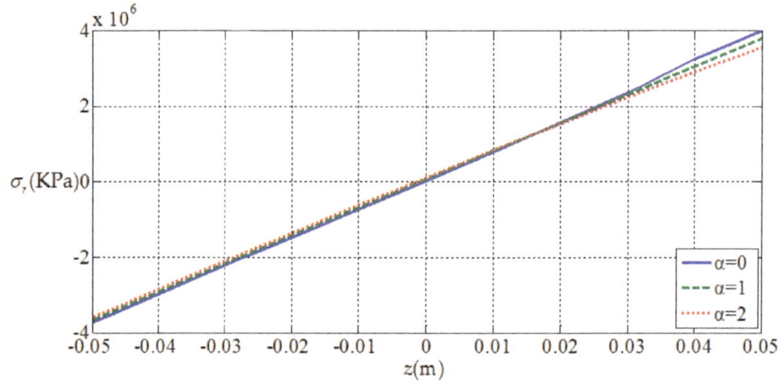

(**b**)

Figure 7. Variation of elastic stress with coordinates z. (**a**) z-direction stress at the periphery of plate $\sigma_z(1,z)$; (**b**) radial stress at the periphery of plate $\sigma_r(1,z)$.

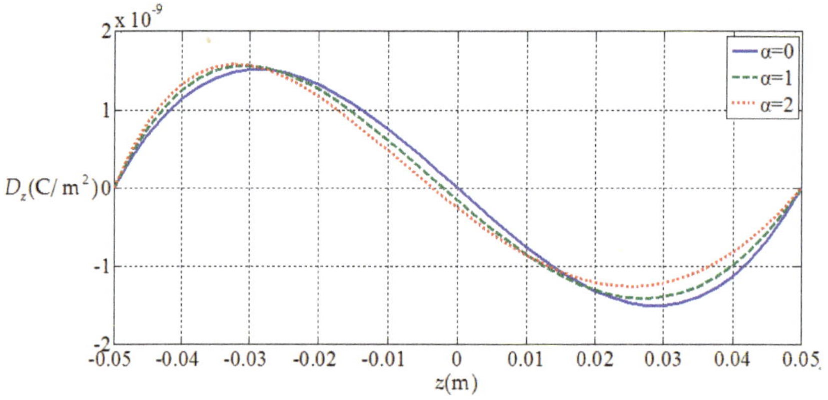

(**a**)

Figure 8. *Cont.*

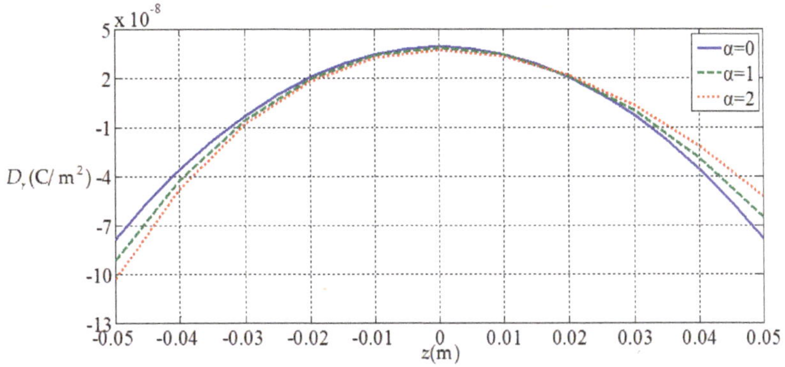

(b)

Figure 8. Variation of electric displacement with coordinates z. (**a**) Electric displacement at the periphery of plate $D_z(1,z)$; (**b**) electric displacement at the periphery of plate $D_r(1,z)$.

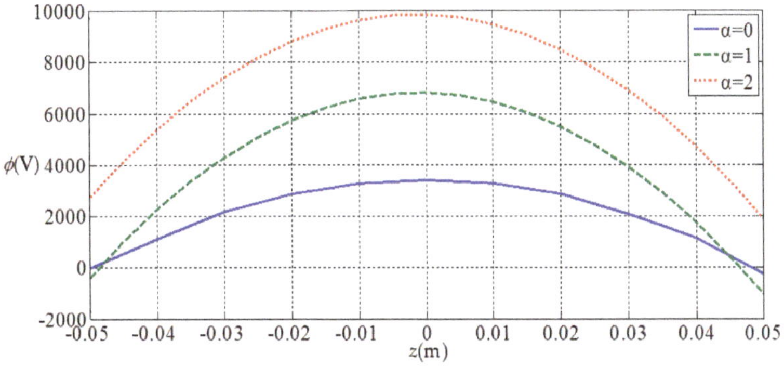

Figure 9. Variation of electric potential at the periphery of plate $\phi(1,z)$ with coordinates z.

Figures 10–13 show the variation of the elastic displacement and stress, as well as the electric displacement and electric potential with the coordinate r at $z = h/4$. From Figure 10, it can be found that the elastic displacements change linearly along r direction, and they have almost no change with the increases of functional gradient parameter α. From Figures 11–13, we can know that the variation curves of the elastic stress, electric displacement, and electric potential of the functionally graded piezoelectric circular plate ($\alpha \neq 0$) are deviated from the uniform piezoelectric plate ($\alpha = 0$) and the degree of deviation increases with the increase of functional gradient parameter α, in which D_r and ϕ increase from center to edge of the plate and σ_r decreases along the same direction while D_z and σ_z remain unchanged from center to edge of the plate. In addition, σ_z, D_r, and D_z change almost linearly along the r direction, and σ_r and ϕ change nonlinearly along the r direction. These characteristics can be used as a reference for the analysis and design of functionally gradient piezoelectric plates.

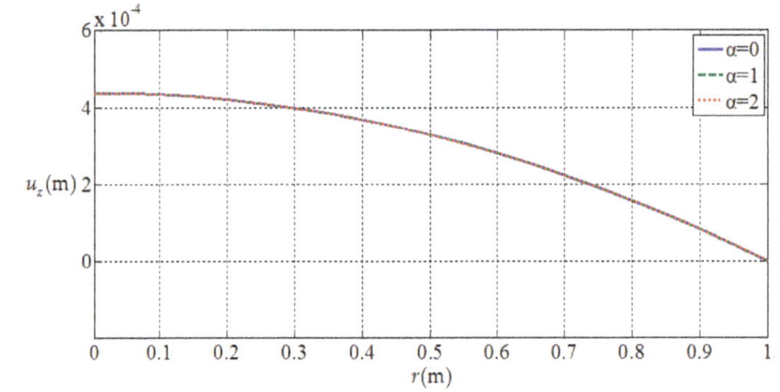

(a)

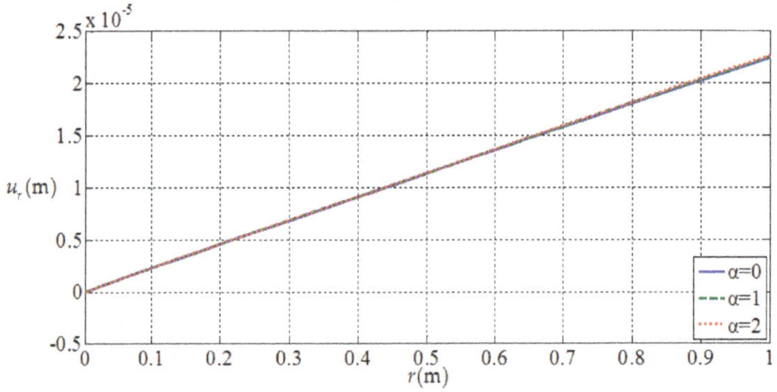

(b)

Figure 10. Variation of elastic displacement with coordinates r at $z = h/4$. (**a**) z-direction displacement $u_z(r, h/4)$ at $z = h/4$; (**b**) radial displacement $u_r(r, h/4)$ at $z = h/4$.

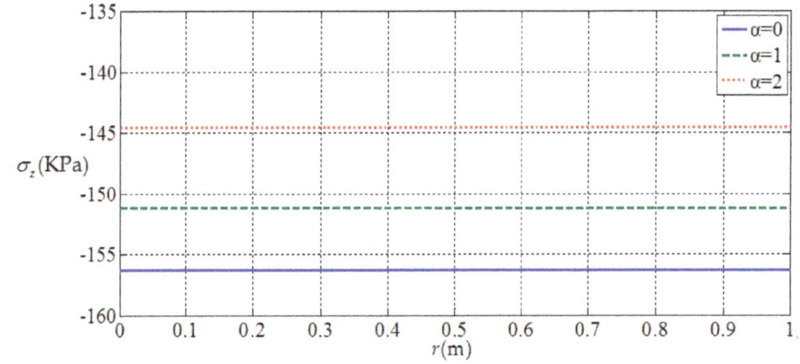

(a)

Figure 11. *Cont.*

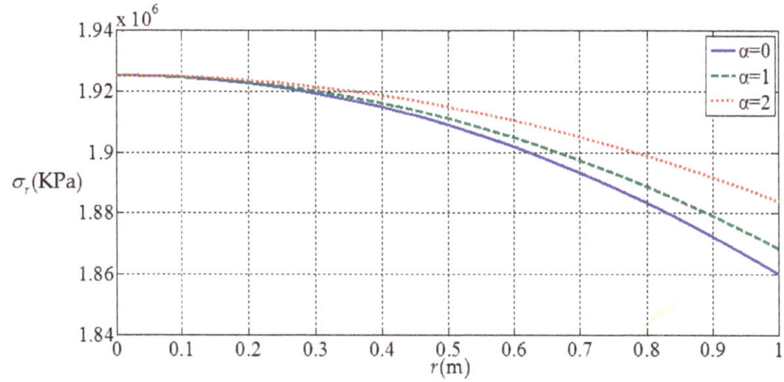

(b)

Figure 11. Variation of elastic stress with coordinates r at $z = h/4$. (**a**) z-direction stress $\sigma_z(r, h/4)$ at $z = h/4$; (**b**) radial stress $\sigma_r(r, h/4)$ at $z = h/4$.

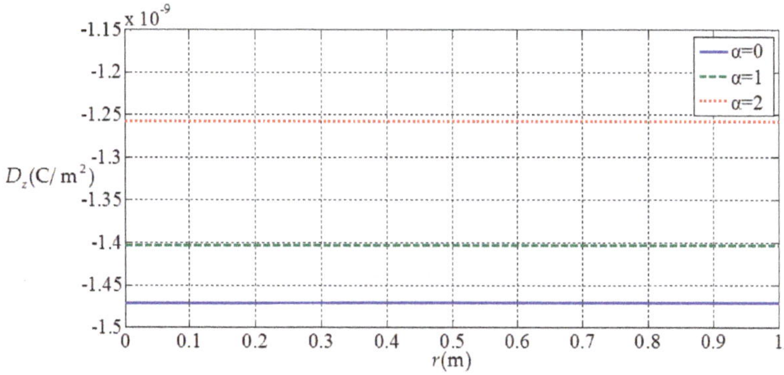

(a)

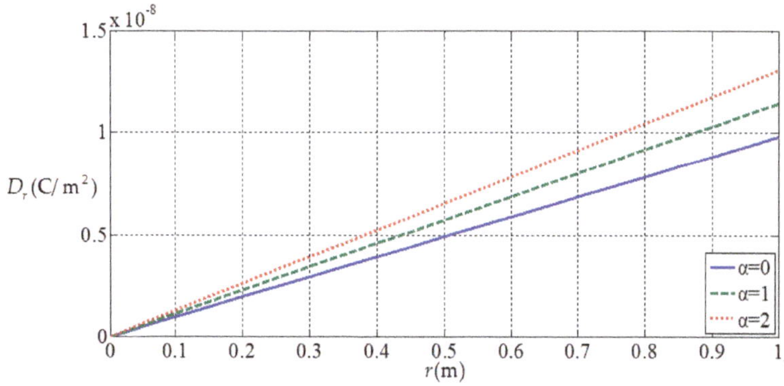

(b)

Figure 12. Variation of electric displacement with coordinates r at $z = h/4$. (**a**) Electric displacement $D_z(r, h/4)$ at $z = h/4$; (**b**) electric displacement $D_r(r, h/4)$ at $z = h/4$.

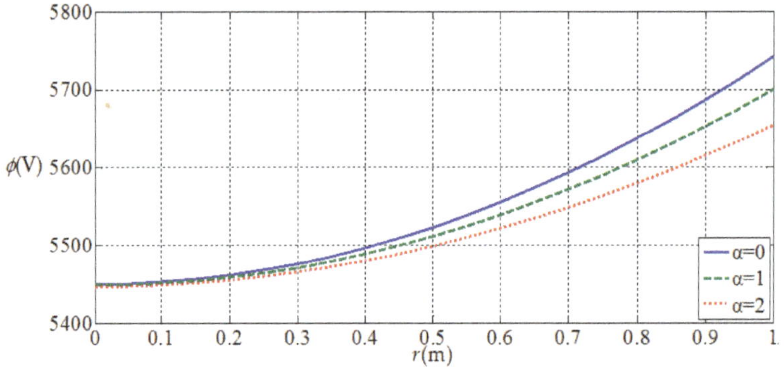

Figure 13. Variation of electric potential $\phi(r, h/4)$ with coordinates r at $z = h/4$.

4. Conclusions

In this study, the electroelastic solution of the axisymmetric deformation problem of functionally graded piezoelectric circular plates under the action of combined mechanical loads was derived by supposing the variable separation form of the displacement function and electrical potential function. Assuming that all the electroelastic materials parameters vary according to the same gradient function along the thickness direction, the electromechanical coupling effect of functionally graded piezoelectric circular plates under the combined mechanical loads was analyzed.

This work may be regarded as a theoretical reference for the analysis of functionally graded piezoelectric materials and structures. Specially, the solving method presented here can also be conveniently applied to other cases under the action of a single mechanical load or under different boundary conditions. Moreover, this work may be extended into the other problem under external electrical loads; in this case, the displacement function used for the solution needs to be modified to some extent. This work may also be extended to functionally graded beams and plates with different properties in tension and compression [36,37]. Obviously, the introduction of different moduli in tension and compression may bring some new issues, which will further complicate the solving of the problem. We will carry out these interesting works in the future.

Author Contributions: Conceptualization, Z.-x.Y., X.-t.H. and J.-y.S.; funding acquisition, X.-t.H.; methodology, Z.-x.Y., X.-t.H. and J.-y.S.; visualization, X.L. and Y.-s.L.; writing original draft, X.L. and Y.-s.L.

Funding: This work was funded by the National Natural Science Foundation of China (Grant No. 11572061 and 11772072).

Conflicts of Interest: The authors declare no conflict of interest.

Appendix A

Let us introduce two stress functions $F(r, z)$ and $\psi(r, z)$

$$\sigma_r = F_{,zz} + r^{-1}\psi_{,r},\ \sigma_\theta = F_{,zz} + \psi_{,rr},\ \sigma_z = r^{-1}(rF_{,r})_{,r},\ \tau_{rz} = -F_{,rz}. \tag{A1}$$

In this way, the Equation (2) is satisfied automatically. Then, suppose that the stress functions have the following form [32]

$$F(r, z) = \sum_{i=0}^{n} r^i F_i(z),\ \psi(r, z) = \sum_{i=0}^{n} r^i \psi_i(z),\ (n = 0, 1, 2 \ldots), \tag{A2}$$

in which $F_i(z)$ and $\psi_i(z)$ are undetermined functions. Substituting Equation (A2) into Equation (A1), we can obtain

$$\begin{aligned}&\sigma_z = \tfrac{F_1}{r} + \sum_{i=0}^{n} (i+2)^2 r^i F_{i+2},\ \sigma_r = \tfrac{\psi_1}{r} + \sum_{i=0}^{n} r^i [F_{i,zz} + (i+2)\psi_{i+2}],\\ &\sigma_\theta = \sum_{i=0}^{n} r^i [F_{i,zz} + (i+2)(i+1)\psi_{i+2}],\ \tau_{rz} = -\sum_{i=0}^{n} (i+1) r^i F_{i+1,z},\ (n = 0,1,2\ldots)\end{aligned}. \quad \text{(A3)}$$

The stress components σ_z and σ_r are limited values at $r = 0$, thus we have

$$F_1(z) = 0,\ \psi_1(z) = 0. \quad \text{(A4)}$$

From Equation (A4), we may infer that all the odd terms of r in Equation (A2) could be zero. From Equations (4) and (6) we can know the electric potential $\partial\phi/\partial z$ corresponds to the stresses σ_z, σ_θ, and σ_r. So, we can see that all the odd terms of r in the expression of electric potential ϕ are also equal to zero. Thus, the electric potential ϕ can be expressed as

$$\phi(r,z) = \sum_{i=0}^{n} r^i \phi_i(z),\ (n = 0,2,4\ldots). \quad \text{(A5)}$$

Moreover, the Equation (4) can be transformed into

$$\begin{aligned}\varepsilon_r &= s_{11}\sigma_r + s_{12}\sigma_\theta + s_{13}\sigma_z + d_{31}E_z\\ \varepsilon_\theta &= s_{12}\sigma_r + s_{11}\sigma_\theta + s_{13}\sigma_z + d_{31}E_z\ ,\\ \varepsilon_z &= s_{13}\sigma_r + s_{13}\sigma_\theta + s_{33}\sigma_z + d_{33}E_z\end{aligned} \quad \text{(A6)}$$

in which s_{ij} are the flexibility coefficients and d_{ij} are the piezoelectric constants. Substituting Equations (5), (6), and (A3) into (A6), we get

$$\begin{aligned}\tfrac{u_r}{r} &= s_{12}\sum_{i=0}^{n} r^i [F_{i,zz} + (i+2)\psi_{i+2}] + s_{11}\sum_{i=0}^{n} r^i [F_{i,zz} + (i+2)(i+1)\psi_{i+2}]\\ &+ s_{13}\sum_{i=0}^{n} (i+2)^2 r^i F_{i+2} - d_{31}\sum_{i=0}^{n} r^i \tfrac{\partial\phi_i(z)}{\partial z},\ (n = 0,2,4\ldots)\\ \tfrac{\partial u_z}{\partial z} &= s_{13}\sum_{i=0}^{n} r^i [F_{i,zz} + (i+2)\psi_{i+2}] + s_{13}\sum_{i=0}^{n} r^i [F_{i,zz} + (i+2)(i+1)\psi_{i+2}]\\ &+ s_{33}\sum_{i=0}^{n} (i+2)^2 r^i F_{i+2} - d_{33}\sum_{i=0}^{n} r^i \tfrac{\partial\phi_i(z)}{\partial z},\ (n = 0,2,4\ldots)\end{aligned}. \quad \text{(A7)}$$

From Equation (A7), we can obtain

$$\begin{aligned}&u_r = s_{12}[\sum_{i=0}^{n} r^{i+1} [F_{i,zz} + (i+2)\psi_{i+2}]] + s_{11}[\sum_{i=0}^{n} r^{i+1} [F_{i,zz} + (i+2)(i+1)\psi_{i+2}]]\\ &+ s_{13}[\sum_{i=0}^{n} (i+2)^2 r^{i+1} F_{i+2}] - d_{31}\sum_{i=0}^{n} r^{i+1} \tfrac{\partial\phi_i(z)}{\partial z},\ (n = 0,2,4\ldots)\\ &u_z = s_{13}[\sum_{i=0}^{n} r^i \int_0^z [F_{i,zz} + (i+2)\psi_{i+2}]dz] + s_{13}[\sum_{i=0}^{n} r^i \int_0^z [F_{i,zz} + (i+2)(i+1)\psi_{i+2}]dz]\\ &+ s_{33}[\sum_{i=0}^{n} (i+2)^2 r^i \int_0^z F_{i+2}dz] - d_{33}\sum_{i=0}^{n} r^i \phi_i(z),\ (n = 0,2,4\ldots)\end{aligned}. \quad \text{(A8)}$$

From Equation (A8) we can see that all the even items of r in the expression of u_r are zero and all the odd items of r in the expression of u_z are zero. Then, according to the boundary conditions of simply supported circular plates, we can finally get the forms of the displacement and the electric potential as follows (i.e., Equation (8); the more detailed derivation can be found in [32]):

$$\left.\begin{aligned} u_r(r,z) &= ru_1(z) + r^3u_3(z) \\ u_z(r,z) &= w_0(z) + r^2w_2(z) + r^4w_4(z) \\ \phi(r,z) &= \phi_0(z) + r^2\phi_2(z) + r^4\phi_4(z) \end{aligned}\right\}. \tag{A9}$$

In addition, similar expressions for displacement and potential function may be found in the analysis of functionally graded piezoelectric beams [34,35], in which the stress function U and the potential functions Φ were expressed in the form

$$\left.\begin{aligned} U(x,z) &= \tfrac{x^2}{2}f(z) + xf_1(z) + f_2(z) \\ \Phi(x,z) &= x^2f_3(z) + xf_4(z) + f_5(z) \end{aligned}\right\} \tag{A10}$$

in which x represents the longitudinal direction of the beam (similar to the radial direction r in the plate problem) and z stands for the thickness direction (similar to the transverse direction z in the plate problem). From the similarities of two sets of expression of beams and plates, we may find some consistencies in the analyses of beams and plates. For example, the derived analytical solutions both satisfy exactly the boundary conditions on the upper and lower surfaces of beams and plates (main boundary), while both satisfy approximately, in the Saint Venant sense, the end conditions for beams and the circumferential boundary conditions for plates (local boundary).

References

1. Koizumi, M. The concept of FGM, ceramic transactions. *Funct. Grad. Mater.* **1993**, *34*, 3–10.
2. Reddy, J.N.; Wang, C.M.; Kitipornchai, S. Axisymmetric bending of functionally graded circular and annular plates. *Eur. J. Mech. A/Solids* **1999**, *18*, 185–199. [CrossRef]
3. Ma, L.S.; Wang, T.J. Nonlinear bending and post-buckling of functionally graded circular plates under mechanical and thermal loadings. *Int. J. Solids Struct.* **2003**, *40*, 3311–3330. [CrossRef]
4. Chi, S.H.; Chung, Y.L. Mechanical behavior of functionally graded material plates under transverse load–Part I: Analysis. *Int. J. Solids Struct.* **2006**, *43*, 3657–3674. [CrossRef]
5. Li, X.Y.; Ding, H.J.; Chen, W.Q. Pure bending of simply supported circular plate of transversely isotropic functionally graded material. *J. Zhejiang Univ.* **2006**, *7*, 1324–1328. [CrossRef]
6. Naderi, A.; Saidi, A.R. On pre-buckling configuration of functionally graded Mindlin rectangular plates. *Mech. Res. Commun.* **2010**, *37*, 535–538. [CrossRef]
7. Martínez-Pañeda, E.; Gallego, R. Numerical analysis of quasi-static fracture in functionally graded materials. *Int. J. Mech. Mater. Des.* **2015**, *11*, 405–424. [CrossRef]
8. He, X.T.; Pei, X.X.; Sun, J.Y.; Zheng, Z.L. Simplified theory and analytical solution for functionally graded thin plates with different moduli in tension and compression. *Mech. Res. Commun.* **2016**, *74*, 72–80. [CrossRef]
9. Fu, Y.; Yao, J.; Wan, Z.; Zhao, G. Free vibration analysis of moderately thick orthotropic functionally graded plates with general boundary restraints. *Materials* **2018**, *11*, 273. [CrossRef] [PubMed]
10. Brischetto, S.; Torre, R. Effects of order of expansion for the exponential matrix and number of mathematical layers in the exact 3D static analysis of functionally graded plates and shells. *Appl. Sci.* **2018**, *8*, 110. [CrossRef]
11. Brischetto, S. A 3D layer-wise model for the correct imposition of transverse shear/normal load conditions in FGM shells. *Int. J. Mech. Sci.* **2018**, *136*, 50–66. [CrossRef]
12. Tang, Y.; Yang, T.Z. Post-buckling behavior and nonlinear vibration analysis of a fluid-conveying pipe composed of functionally graded material. *Compos. Struct.* **2018**, *185*, 393–400. [CrossRef]
13. Rao, S.S.; Sunar, M. Piezoelectricity and its use in disturbance sensing and control of flexible structures: A survey. *Appl. Mech. Rev.* **1994**, *47*, 113–123. [CrossRef]
14. Tani, J.; Takagi, T.; Qiu, J. Intelligent material systems: Application of functional materials. *Appl. Mech. Rev.* **1998**, *51*, 505–521. [CrossRef]
15. Pohanka, M. Overview of piezoelectric biosensors, immunosensors and DNA sensors and their applications. *Materials* **2018**, *11*, 448. [CrossRef] [PubMed]
16. Zhu, X.H.; Meng, Z.Y. Operational principle, fabrication and displacement characteristic of a functionally gradient piezoelectric ceramic actuator. *Sens. Actuators A Phys.* **1995**, *48*, 169–176. [CrossRef]

17. Wu, C.C.M.; Kahn, M.; Moy, W. Piezoelectric ceramics with functional gradients: A new application in material design. *J. Am. Ceram. Soc.* **1996**, *79*, 809–812. [CrossRef]
18. Shelley, W.F.; Wan, S.; Bowman, K.J. Functionally graded piezoelectric ceramics. *Mater. Sci. Forum* **1999**, *308–311*, 515–520. [CrossRef]
19. Taya, M.; Almajid, A.A.; Dunn, M.; Takahashi, H. Design of bimorph piezo-composite actuators with functionally graded microstructure. *Sens. Actuators A Phys.* **2003**, *107*, 248–260. [CrossRef]
20. Dineva, P.; Gross, D.; Müller, R.; Rangelov, T. Dynamic stress and electric field concentration in a functionally graded piezoelectric solid with a circular hole. *Z. Angew. Math. Mech.* **2011**, *91*, 110–124. [CrossRef]
21. Chen, W.Q.; Ding, H.J. Bending of functionally graded piezoelectric rectangular plates. *Acta Mech. Solida Sin.* **2000**, *13*, 312–319.
22. Zhang, P.W.; Zhou, Z.G.; Li, G.Q. Interaction of four parallel non-symmetric permeable mode-III cracks with different lengths in a functionally graded piezoelectric material plane. *Z. Angew. Math. Mech.* **2009**, *89*, 767–788. [CrossRef]
23. Wu, R.A.; Zhong, Z.; Jin, B. Three dimensional analysis of rectangular functionally graded piezoelectric plates. *Acta Mech. Solida Sin.* **2002**, *23*, 43–49.
24. Zhong, Z.; Shang, E.T. Three dimensional exact analysis of functionally gradient piezothermoelectrc material rectangular plate. *Acta Mech. Solida Sin.* **2003**, *35*, 533–552.
25. Zhong, Z.; Shang, E.T. Three-dimensional exact analysis of a simply supported functionally gradient piezoelectric plate. *Int. J. Solids Struct.* **2003**, *40*, 5335–5352. [CrossRef]
26. Zhu, H.W.; Li, Y.C.; Yang, C.J. Finite element solution of functionally graded piezoelectric plates. *Chin. Q. Mech.* **2005**, *26*, 567–571.
27. Lu, P.; Lee, H.P.; Lu, C. An exact solution for simply supported functionally graded piezoelectric laminates in cylindrical bending. *Int. J. Mech. Sci.* **2005**, *47*, 437–458. [CrossRef]
28. Lu, P.; Lee, H.P.; Lu, C. Exact solutions for simply supported functionally graded piezoelectric laminates by stroh-like formalism. *Compos. Struct.* **2006**, *72*, 352–363. [CrossRef]
29. Zhang, X.R.; Zhong, Z. Three dimensional exact solution for free vibration of functionally gradient piezoelectric circular plate. *Chin. Q. Mech.* **2005**, *26*, 81–86.
30. Liu, P.; Yu, T.T.; Bui, T.Q.; Zhang, C.Z.; Xu, Y.P.; Lim, C.W. Transient thermal shock fracture analysis of functionally graded piezoelectric materials by the extended finite element method. *Int. J. Solids Struct.* **2014**, *51*, 2167–2182. [CrossRef]
31. Yu, T.T.; Bui, T.Q.; Liu, P.; Zhang, C.Z.; Hirose, S. Interfacial dynamic impermeable cracks analysis in dissimilar piezoelectric materials under coupled electromechanical loading with the extended finite element method. *Int. J. Solids Struct.* **2015**, *67–68*, 205–218. [CrossRef]
32. Li, X.Y. Axisymmetric Problems of Functionally Graded Circular and Annular Plates with Transverse Isotropy. Ph.D. Thesis, Zhejiang University, Hangzhou, China, April 2007.
33. Li, X.Y.; Ding, H.J.; Chen, W.Q. Three-dimensional analytical solution for a transversely isotropic functionally graded piezoelectric circular plate subject to a uniform electric potential difference. *Sci. China Ser. G Phys. Mech. Astron.* **2008**, *51*, 1116–1125. [CrossRef]
34. Yu, T.; Zhong, Z. Bending analysis of a functionally graded piezoelectric cantilever beam. *Sci. China Ser. G Phys. Mech. Astron.* **2007**, *50*, 97–108. [CrossRef]
35. He, X.T.; Wang, Y.Z.; Shi, S.J.; Sun, J.Y. An electroelastic solution for functionally graded piezoelectric material beams with different moduli in tension and compression. *J. Intell. Mater. Syst. Struct.* **2018**, *29*, 1649–1669. [CrossRef]
36. Li, X.; Sun, J.Y.; Dong, J.; He, X.T. One-dimensional and two-dimensional analytical solutions for functionally graded beams with different moduli in tension and compression. *Materials* **2018**, *11*, 830. [CrossRef] [PubMed]
37. He, X.T.; Li, Y.H.; Liu, G.H.; Yang, Z.X.; Sun, J.Y. Non-Linear bending of functionally graded thin plates with different moduli in tension and compression and its general perturbation solution. *Appl. Sci.* **2018**, *8*, 731.

materials

MDPI

Article

Nonlinear Stability Analysis of Eccentrically Stiffened Functionally Graded Truncated Conical Sandwich Shells with Porosity

Duc-Kien Thai [1], Tran Minh Tu [2,*], Le Kha Hoa [3,4], Dang Xuan Hung [2] and Nguyen Ngọc Linh [2]

1 Department of Civil and Environmental Engineering, Sejong University, 98 Gunja-dong, Gwangjin-gu, Seoul 143-747, Korea; thaiduckien@sejong.ac.kr

2 Faculty of Industrial and Civil Engineering, University of Civil Engineering, Hanoi 100000, Vietnam; dangxuanhung@gmail.com (D.X.H.); nguyenngoclinh@gmail.com (N.N.L.)

3 Division of Computational Mathematics and Engineering, Institute for Computational Science, Ton Duc Thang University, Ho Chi Minh City 700000, Vietnam; lekhahoa@tdtu.edu.vn

4 Faculty of Civil Engineering, Ton Duc Thang University, Ho Chi Minh City 700000, Vietnam

* Correspondence: tpnt2002@yahoo.com; Tel.: +84-912-101-173

Received: 3 October 2018; Accepted: 2 November 2018; Published: 6 November 2018

Abstract: This paper analyzes the nonlinear buckling and post-buckling characteristics of the porous eccentrically stiffened functionally graded sandwich truncated conical shells resting on the Pasternak elastic foundation subjected to axial compressive loads. The core layer is made of a porous material (metal foam) characterized by a porosity coefficient which influences the physical properties of the shells in the form of a harmonic function in the shell's thickness direction. The physical properties of the functionally graded (FG) coatings and stiffeners depend on the volume fractions of the constituents which play the role of the exponent in the exponential function of the thickness direction coordinate axis. The classical shell theory and the smeared stiffeners technique are applied to derive the governing equations taking the von Kármán geometrical nonlinearity into account. Based on the displacement approach, the explicit expressions of the critical buckling load and the post-buckling load-deflection curves for the sandwich truncated conical shells with simply supported edge conditions are obtained by applying the Galerkin method. The effects of material properties, core layer thickness, number of stiffeners, dimensional parameters, semi vertex angle and elastic foundation on buckling and post-buckling behaviors of the shell are investigated. The obtained results are validated by comparing with those in the literature.

Keywords: porous materials; truncated conical sandwich shell; metal foam core layer; non-linear buckling analysis; orthogonal stiffener; elastic foundation

1. Introduction

Functionally graded (FG) materials are microscopically nonhomogeneous materials with smoothly and continuously varying mechanical properties in the preferred directions. The advantages of functionally graded material (FGM) include avoiding crack, avoiding delamination and eliminating residual stress. In micromechanics, FGM is considered to contain porosity during the production process, these porosities could be characterized to obtain the expected material properties such as the local density and to obtain the expected structural performance. Furthermore, porous materials such as metal foams have excellent energy-absorbing capability forming an important category of lightweight materials. As a result, porous materials have been considered in a wide range of application in practice for structures subjected to dynamic or impact loadings.

Truncated conical shells have been utilized in various engineering activities such as aerospace engineering, marine and ocean engineering structures, components of missiles and spacecrafts and nuclear reactors. Metallic sandwich structures are widely used in the aviation industry as well as in ship and railway engineering because of their low density, high specific strength, and effective energy absorption. The buckling and post-buckling behaviors of FG shells in cylindrical and conical forms under mechanical and thermal loads are prominent topics, drawing the considerable attention of many researchers. Huang and Han [1] used Donell shell theory to study the stability characteristics of functionally graded shells in cylindrical forms subjected to axially compressive loads employing the Ritz energy method. Naj et al. [2] analyze the instability of FG truncated conical shells under the coupling of thermal and mechanical loadings using the first-order shell theory. Sofiyev and his colleagues [3–10] published many studies on linear and nonlinear buckling of FG cylindrical and conical shells. By applying the Galerkin method and smeared stiffeners technique, Duc and his colleagues [11–17] investigated buckling and post-buckling behaviors of FG cylindrical and conical shells reinforced by eccentrically stiffeners (ES). Using the same approach, Bich et al. [18–20] examined the buckling behaviors and dynamic stability characteristics of eccentrically stiffened FG cylindrical shells and panels. Recently, Dung et al. [21,22] presented the theoretical solution for the buckling behaviors of FG truncated conical shells under different of mechanical loadings such as uniformly distributed loads and axially compressive loads. Dung and Chan [23] analyzed the orthogonally stiffened FG truncated conical shells in terms of the mechanical stability. Dung et al. [24] analyzed the nonlinear post-buckling behaviors of the eccentrically orthogonal stiffened FG truncated conical shells.

There are a few studies on the buckling of FG porous plates and beams in the available literature. Magnucki and Stasiewicz [25] examined the buckling features of beams with porosity considering the total potential energy using elastic formulations. Magnucka-Blandzi [26,27] mathematically modeled a porous sandwich plate to determine critical in-plane compressed loads. The work of Magnucka-Blandzi [28] focused on axis-symmetrical deflection and buckling of simply supported circular porous–cellular plates under lateral uniformly distributed pressures and compressive pressures in the radial direction uniform. Static buckling and bending features of FG beams with porosity taking the shear deformation into account are studied by Chen et al. in [29]. Kitipornchai et al. [30] studied elastic buckling and free vibration behaviors of closed-cell beams made of metal foam and reinforced by graphene platelets. Jabbari et al. [31] examined the buckling behaviors of an FG thin circle-shaped plate made of saturated porous materials. In another study, he also examined the buckling behaviors of a porous circular plate subjected to radial loadings employing the higher-order shear deformation theory [32]. To control the formation of porous structures, fabrication parameters need to be managed. In microelectromechanical systems (MEMS) and nanoelectromechanical systems (NEMS), we can improve the physical characteristic of micro/nano-scale structures by tailoring the architecture of porous materials. Examination and assessment of size-effects in NEMs structural problems, many researchers have been focused on size-dependent mechanical models [33–36]. Size effect plays important role in micron and sub-micron scales of metallic materials. Size effects in elastic-plastic functionally graded materials (FGMs) have been reported in work of Mathew et al. [37], Martínez-Paneda et al. [38,39].

From the above-mentioned literature context, it can be seen that there are very few studies focused on linear and non-linear stability of eccentrically stiffened FGM truncated conical shells. To the best of our knowledge, there are no publications on the nonlinear stability behaviors of the eccentrically stiffened functionally graded truncated conical sandwich shells with the porous core layer. The aim of the present paper is to meet this demand. The porous material core layer of the shell is made of metal foam. The outer and inner layers, eccentrically orthogonal stiffener systems are made of FGM. The shell is supported by Pasternak elastic foundation and subjected to the axial compressive load. The classical shell theory, the smeared stiffener technique, and the Galerkin method are applied to come up with explicit expressions of the critical buckling load and the post-buckling load-deflection curves for sandwich truncated conical shells with simply supported edge conditions. The effects of material properties, the number of stiffeners, geometry parameters, and elastic foundation on stability behaviors of the shell are also examined.

2. Model Configurations and Elastic Foundations

A porous eccentrically stiffened functionally graded truncated conical sandwich shells (PSTC) is considered with the geometry configurations and the coordinate system being shown in Figure 1. In which, α denotes the semi-vertex angle, R denotes the small base radius of the shell, L denotes the slant height and h denotes the shell thickness.

The shell consists of inner and outer layers (layers 1 and 3) made of FGM of the thickness h_{FG}, and the porous core layer (layer 2) of the thickness h_{core}. The PSTC is located in a curvilinear coordinate (x, θ, z) in which x and z axis share the origin at the vertex of the conical shell and together form a plane through the symmetry line of the shell. x axis exists along the shell slant and z axis is at right angles to the slant line. It is noted that the origin is located in the mid-surface of the shell and x_0 denotes the virtual slant height from the vertex to the adjacent base of the shell. Corresponding to x, θ and z axes, there are three displacements components u, v, and w of a point in the mid-surface, respectively. The displacement along the z axis (w) is also called the deflection of the PSTC which is also the primary variable of this work.

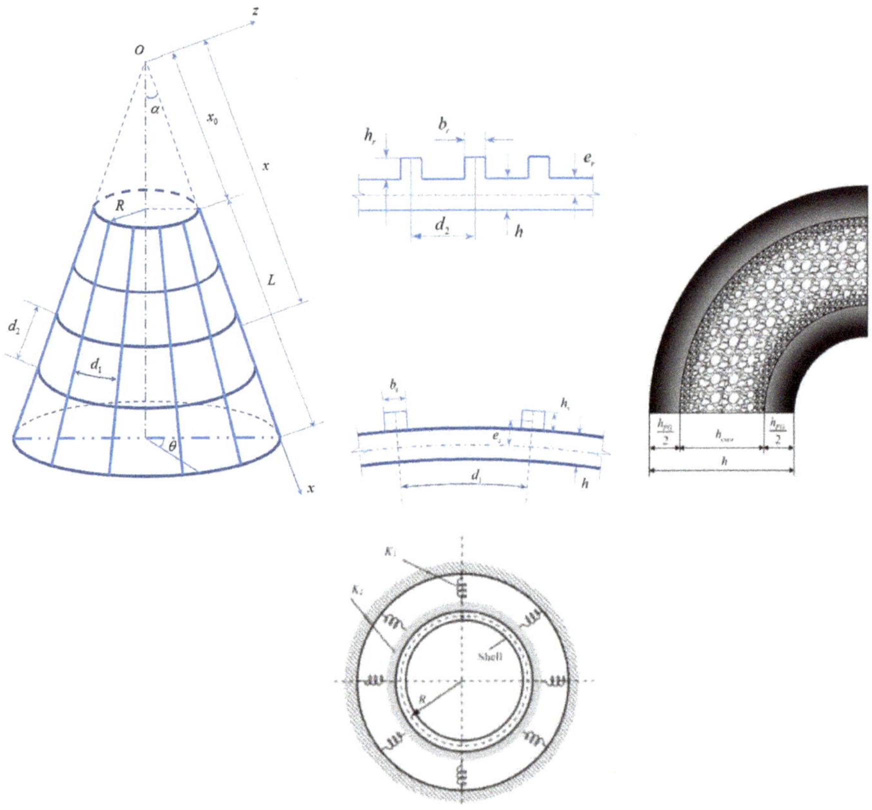

Figure 1. Geometry configurations and coordinates of the PSTC.

The space between FG stiffeners is assumed to be constant and closely spaced in the outer face of the PSTC. The Young moduli of FG cover layers and stiffeners vary according to a simple power distribution through the z direction with the exponent is the volume fraction of the constituents, and the Young moduli of the core follow a simple cosine rule of a symmetric distribution defined as follows:

$$E_{sh} = \begin{cases} E_c + E_{mc}\left(\frac{2z+h_{FG}+h_{core}}{h_{FG}}\right)^k & \text{at } -\frac{h}{2} \le z \le -\frac{h_{core}}{2} \\ E_m\left[1 - e_0\cos\left(\frac{\pi z}{h_{core}}\right)\right] & \text{at } -\frac{h_{core}}{2} \le z \le \frac{h_{core}}{2} \\ E_c + E_{mc}\left(\frac{-2z+h_{FG}+h_{core}}{h_{FG}}\right)^k & \text{at } \frac{h_{core}}{2} \le z \le \frac{h}{2} \end{cases} \quad (1a)$$

$$\begin{cases} h = h_{core} + h_{FG} \\ 0 < e_0 < 1 \end{cases} \quad (1b)$$

Reinforced stiffeners are considered in two following cases.

Case 1: Inside FGM stiffener

$$\begin{aligned} E_s &= E_c + E_{mc}\left(\frac{2z-h}{2h_s}\right)^{k_2} \quad \text{at } \frac{h}{2} \le z \le \frac{h}{2} + h_s \\ E_r &= E_c + E_{mc}\left(\frac{2z-h}{2h_r}\right)^{k_3} \quad \text{at } \frac{h}{2} \le z \le \frac{h}{2} + h_r \end{aligned} \quad (2a)$$

Case 2: Outside FGM stiffener

$$\begin{aligned} E_s &= E_c + E_{mc}\left(-\frac{2z+h}{2h_s}\right)^{k_2} \quad \text{at } -\frac{h}{2} - h_s \le z \le -\frac{h}{2} \\ E_r &= E_c + E_{mc}\left(-\frac{2z+h}{2h_r}\right)^{k_3} \quad \text{at } -\frac{h}{2} - h_r \le z \le -\frac{h}{2} \end{aligned} \quad (2b)$$

where:

$h_{FG}/2$ is the FG coating thickness,

$E_{mc} = E_m - E_c$, $E_{cm} = E_c - E_m$,
h_{core} is the core layer thickness,
h_s, h_r denote stringers and rings thickness respectively,
e_0 is the porosity coefficient of the core layer,
k, k_2, and k_3 are the shell, stringers, and rings volume fraction indexes respectively.
sh, m, c, r, and s denote shell, metal, ceramic, ring, and stringer respectively.
st denotes stiffeners in general, stiffeners are stringers and rings.
E_c, E_m are Young's moduli of ceramic and metal.
E_{sh}, E_s, and E_r are the Young moduli of shell, stringer, and ring of materials respectively.

The Poisson's ratios v of the shell and stiffeners materials are assumed to be independent of thickness coordinate [6].

It is noted from Equations (1) and (2) that the continuous variations of the material properties are satisfied between layers of the PSTC. From Equation (1), we can obtain equations for these different cases, namely the FG sigmoid sandwich shell with ($h_{core} = 0$), the metal foam sandwich shell with FG face sheets ($e_0 = 0$), or the full metal shell ($e_0 = k = 0$).

The reaction of the elastic foundation on the conical shell is described by using the Pasternak model. The shell-foundation interaction may be expressed as [40]

$$q_f = K_1 w - K_2\left(\frac{\partial^2 w}{\partial x^2} + \frac{1}{x}\frac{\partial w}{\partial x} + \frac{1}{x^2\sin^2\alpha}\frac{\partial^2 w}{\partial \theta^2}\right) \quad (3)$$

where K_1 (N/m^3) and K_2 (N/m) respectively are the Winkler foundation stiffness and the shear subgrade modulus of the foundation.

3. Theoretical Formulations

From the Donnell shell theory, at a distance z from the mid-surface of the shell, the normal and shear strains are given as follows [41]:

$$\varepsilon_x = \varepsilon_{xm} + zk_x,\ \varepsilon_\theta = \varepsilon_{\theta m} + zk_\theta,\ \gamma_{x\theta} = \gamma_{x\theta m} + 2zk_{x\theta} \quad (4)$$

in which ε_{xm} and $\varepsilon_{\theta m}$ are the normal strains $\gamma_{x\theta m}$ is the shear strain at a point on the shell mid-surface, and $k_x, k_\theta, k_{x\theta}$ are bending and twisting curvatures with respect to the $x-$axis, $\theta-$axis, and the plane (x,θ), respectively. Considering the von Karman geometrical nonlinearity, the strain–displacement relations are defined as [41]

$$\begin{aligned}
&\varepsilon_{xm} = \frac{\partial u}{\partial x} + \frac{1}{2}\left(\frac{\partial w}{\partial x}\right)^2,\ \varepsilon_{\theta m} = \frac{1}{x\sin\alpha}\frac{\partial v}{\partial \theta} + \frac{u}{x} + \frac{w}{x}\cot\alpha + \frac{1}{2x^2\sin^2\alpha}\left(\frac{\partial w}{\partial \theta}\right)^2,\\
&\gamma_{x\theta m} = \frac{1}{x\sin\alpha}\frac{\partial u}{\partial \theta} - \frac{v}{x} + \frac{\partial v}{\partial x} + \frac{1}{x\sin\alpha}\frac{\partial w}{\partial x}\frac{\partial w}{\partial \theta},\\
&k_x = -\frac{\partial^2 w}{\partial x^2},\ k_\theta = -\frac{1}{x^2\sin^2\alpha}\frac{\partial^2 w}{\partial \theta^2} - \frac{1}{x}\frac{\partial w}{\partial x},\ k_{x\theta} = -\frac{1}{x\sin\alpha}\frac{\partial^2 w}{\partial x\partial\theta} + \frac{1}{x^2\sin\alpha}\frac{\partial w}{\partial \theta}
\end{aligned} \quad (5)$$

The generalized Hooke law for the conical shell is presented as follows:

$$\sigma_x^{sh} = \frac{E(z)}{1-\nu^2}(\varepsilon_x + \nu\varepsilon_\theta),\quad \sigma_\theta^{sh} = \frac{E(z)}{1-\nu^2}(\varepsilon_\theta + \nu\varepsilon_x),\quad \sigma_{x\theta}^{sh} = \frac{E(z)}{2(1+\nu)}\gamma_{x\theta} \quad (6)$$

and for the stringer and ring stiffeners,

$$\sigma_x^{st} = E_s\varepsilon_x,\ \sigma_\theta^{st} = E_r\varepsilon_\theta \quad (7)$$

The material of the stiffeners is similar to the material of the FG coating at the outer surface. If the outside surface of the FG coating is ceramic-rich, the material of the stiffeners is ceramic, and vice versa.

Considering the change of stringers spacing, applying the Lekhnitskii smeared stiffener technique, and omitting the twisting effects of the stiffeners, we can define the force and moment resultants of the PSTC as follows:

$$\begin{aligned}
&N_x = \int_{-h/2}^{h/2} \sigma_x^{sh} dz + \frac{b_s}{d_1(x)}\int_{h/2}^{h/2+h_s} \sigma_x^s dz, \quad N_\theta = \int_{-h/2}^{h/2} \sigma_\theta^{sh} dz + \frac{b_r}{d_2}\int_{h/2}^{h/2+h_r} \sigma_\theta^s dz, \quad N_{x\theta} = \int_{-h/2}^{h/2} \sigma_{x\theta} dz\\
&M_x = \int_{-h/2}^{h/2} z\sigma_x^{sh} dz + \frac{b_s}{d_1(x)}\int_{h/2}^{h/2+h_s} z\sigma_x^s dz, \quad M_\theta = \int_{-h/2}^{h/2} z\sigma_\theta^{sh} dz + \frac{b_r}{d_2}\int_{h/2}^{h/2+h_r} z\sigma_\theta^s dz, \quad M_{x\theta} = \int_{-h/2}^{h/2} z\sigma_{x\theta} dz
\end{aligned} \quad (8)$$

Introducing Equations (6) and (7) into Equation (8) we obtain [22]

$$\begin{aligned}
&\begin{Bmatrix} N_x \\ N_\theta \\ N_{x\theta} \end{Bmatrix} = \begin{bmatrix} A_{11} + \frac{E_{1s}b_s}{d_1(x)} & A_{12} & 0 \\ A_{12} & A_{22} + \frac{E_{1r}b_r}{d_2} & 0 \\ 0 & 0 & A_{66} \end{bmatrix} \begin{Bmatrix} \varepsilon_{xm} \\ \varepsilon_{\theta m} \\ \gamma_{x\theta m} \end{Bmatrix} + \begin{bmatrix} B_{11} + C_1(x) & B_{12} & 0 \\ B_{12} & B_{22} + C_2 & 0 \\ 0 & 0 & 2B_{66} \end{bmatrix} \begin{Bmatrix} k_x \\ k_\theta \\ k_{x\theta} \end{Bmatrix}\\
&\begin{Bmatrix} M_x \\ M_\theta \\ M_{x\theta} \end{Bmatrix} = \begin{bmatrix} B_{11} + C(x) & B_{12} & 0 \\ B_{12} & B_{22} + C_2 & 0 \\ 0 & 0 & B_{66} \end{bmatrix} \begin{Bmatrix} \varepsilon_{xm} \\ \varepsilon_{\theta m} \\ \gamma_{x\theta m} \end{Bmatrix} + \begin{bmatrix} D_{11} + \frac{E_{3s}b_s}{d_1(x)} & D_{12} & 0 \\ D_{12} & D_{22} + \frac{E_{3r}b_r}{d_2} & 0 \\ 0 & 0 & 2D_{66} \end{bmatrix} \begin{Bmatrix} k_x \\ k_\theta \\ k_{x\theta} \end{Bmatrix}
\end{aligned} \quad (9)$$

in which the coefficients are presented in Appendix A.

The nonlinear equations of equilibrium of the PSTC resting on Pasternak foundation using the Donnell shell theory are given as follows [22]:

$$\begin{aligned}
&xN_{x,x} + \tfrac{1}{\sin\alpha} N_{x\theta,\theta} + N_x - N_\theta = 0 \\
&\tfrac{1}{\sin\alpha} N_{\theta,\theta} + xN_{x\theta,x} + 2N_{x\theta} = 0 \\
&xM_{x,xx} + 2M_{x,x} + \tfrac{2}{\sin\alpha}\left(M_{x\theta,x\theta} + \tfrac{1}{x} M_{x\theta,\theta}\right) + \tfrac{1}{x\sin^2\alpha} M_{\theta,\theta\theta} - M_{\theta,x} - N_\theta \cot\alpha \\
&+\left(xN_x w_{,x} + \tfrac{1}{\sin\alpha} N_{x\theta} w_{,\theta}\right)_{,x} + \tfrac{1}{\sin\alpha}\left(N_{x\theta} w_{,x} + \tfrac{1}{x\sin\alpha} N_\theta w_{,\theta}\right)_{,\theta} + (xN_x^o w_{,x})_{,x} \\
&-xK_1 w + xK_2\left(\tfrac{\partial^2 w}{\partial x^2} + \tfrac{1}{x}\tfrac{\partial w}{\partial x} + \tfrac{1}{x^2\sin^2\alpha}\tfrac{\partial^2 w}{\partial \theta^2}\right) = 0
\end{aligned} \tag{10}$$

where x, z and θ following the comma symbol $(,)$ indicates the partial derivative with respect to x, z and θ, respectively.

4. Prebuckling State Analysis

In this section, the PSTC is considered solely exposed to an axial compression P at the small base $x = x_0$. The equilibrium equations of the PSTC in the membrane-like form is derived from Equation (10) taking the symmetry of geometry and loading characteristics into account as follows:

$$x\frac{dN_x^0}{dx} + N_x^0 - N_\theta^0 = 0, \quad N_{x\theta}^0 = 0, \quad -N_\theta^0 \cot\alpha = 0 \tag{11}$$

Solving this system with condition

$$N_x^0 = -\frac{P}{\cos\alpha} \tag{12}$$

We obtain the prebuckling force resultants

$$N_x^0 = -\frac{px_o}{x\cos\alpha}, \; N_\theta^0 = 0, \; N_{x\theta}^0 = 0 \tag{13}$$

or in another form

$$N_x^o = -\frac{P}{\pi x \sin 2\alpha}, \text{ where } P = 2\pi p x_o \sin\alpha \tag{14}$$

5. Nonlinear Stability Formulations

Introducing Equation (4) into Equation (9) we obtain the force and moment resultants in term of displacements. The results are then substituted into Equation (10) in conjunction with Equation (14), and we have the stability equations as follows:

$$\Re_{11}(u) + \Re_{12}(v) + \Re_{13}(w) + G_{14} = 0 \tag{15}$$

$$\Re_{21}(u) + \Re_{22}(v) + \Re_{23}(w) + G_{24} = 0 \tag{16}$$

$$\Re_{31}(u) + \Re_{32}(v) + \Re_{33}(w) + P\Re_{34}(w) + G_{34} = 0 \tag{17}$$

where $\Re_{ij}$ with $i = (1-3)$ and $j = (1-4)$ are linear differential operators and G_{ij} with $i = (1-3)$ and $j = 4$ are nonlinear components, these values are listed in Appendix B. Equations (15)–(17) are employed to compute the critical buckling load and analyze post-buckling behavior of the PSTC. However, these equations are the coupling nonlinear partial differential equations whose difficulty would be overcome in the following section.

6. Buckling and Post-Buckling Analysis

The PSTC is considered simply supported at two bases such that

$$v = w = 0, \quad M_x = 0 \quad at \quad x = x_o, x_o + L \tag{18}$$

The solution approximately satisfying Equation (18) are chosen as [22,24]

$$\begin{array}{c} u = U\cos\frac{m\pi(x-x_0)}{L}\sin\frac{n\theta}{2} \\ v = V\sin\frac{m\pi(x-x_0)}{L}\cos\frac{n\theta}{2} \\ w = W\sin\frac{m\pi(x-x_0)}{L}\sin\frac{n\theta}{2} \end{array} \tag{19}$$

where n is the quantity of full-waves in the circumferential direction of the shell, and m is the number of half-waves along x axis. U, V and W are the corresponding displacement amplitudes which would be determined by then. In the integration domain given as $x_0 \leq x \leq x_0 + L$ and $0 \leq \theta \leq 2\pi$, Equations (15) and (16) are weighted by x and Equation (17) is weighted by x^2 before employing the Galerkin method to the obtained results. We have

$$\begin{array}{c} J_1 = \int\limits_{x_o}^{x_o+L}\int\limits_{0}^{2\pi}\Omega_1\sin\frac{n\theta}{2}\cos\frac{m\pi(x-x_0)}{L}\sin\alpha d\theta dx \\ J_2 = \int\limits_{x_o}^{x_o+L}\int\limits_{0}^{2\pi}\Omega_2\cos\frac{n\theta}{2}\sin\frac{m\pi(x-x_0)}{L}\sin\alpha d\theta dx \\ J_3 = \int\limits_{x_o}^{x_o+L}\int\limits_{0}^{2\pi}\Omega_3\sin\frac{n\theta}{2}\sin\frac{m\pi(x-x_0)}{L}\sin\alpha d\theta dx \end{array} \tag{20}$$

where

$$\begin{array}{c} \Omega_1 = x[\Re_{11}(u) + \Re_{12}(v) + \Re_{13}(w) + G_{14}] \\ \Omega_2 = x[\Re_{21}(u) + \Re_{22}(v) + \Re_{23}(w) + G_{24}] \\ \Omega_3 = x^2[\Re_{31}(u) + \Re_{32}(v) + \Re_{33}(w) + P\Re_{34}(w) + G_{34}] \end{array} \tag{21}$$

Introducing Equation (19) into Equation (21) and then the results into Equation (20), after integrations and other manipulations, we obtain

$$H_{11}U + H_{12}V + H_{13}W + L_{14}W^2 = 0 \tag{22}$$

$$H_{21}U + H_{22}V + H_{23}W + L_{24}W^2 = 0 \tag{23}$$

$$H_{31}U + H_{32}V + (H_{33} + H_{34}P)W + L_{34}W^2 + L_{35}VW + L_{36}UW + L_{37}W^3 = 0 \tag{24}$$

where H_{ij} and L_{ij} are given in Appendix C.

We obtain the expression for U and V from Equations (22) and (23) as follows:

$$\begin{array}{c} U = \frac{H_{13}H_{22}-H_{12}H_{23}}{H_{12}H_{21}-H_{11}H_{22}}W + \frac{L_{14}H_{22}-L_{24}H_{12}}{H_{12}H_{21}-H_{11}H_{22}}W^2 \\ V = \frac{H_{11}H_{23}-H_{13}H_{21}}{H_{12}H_{21}-H_{11}H_{22}}W + \frac{L_{24}H_{11}-L_{14}H_{21}}{H_{12}H_{21}-H_{11}H_{22}}W^2 \end{array}$$

Substituting U and V into Equation (24) we obtain the following equation.

$$\begin{array}{c} \left(\frac{L_{35}L_{24}H_{11}-L_{35}L_{14}H_{21}-L_{36}L_{24}H_{12}+L_{36}L_{14}H_{22}}{H_{12}H_{21}-H_{11}H_{22}} + L_{37}\right)W^3 \\ +\left(\begin{array}{c}\frac{-H_{31}L_{24}H_{12}+H_{31}L_{14}H_{22}+H_{32}L_{24}H_{11}-H_{32}L_{14}H_{21}}{H_{12}H_{21}-H_{11}H_{22}} + L_{34} \\ \frac{-L_{35}H_{13}H_{21}+L_{35}H_{11}H_{23}-L_{36}H_{12}H_{23}+L_{36}H_{13}H_{22}}{H_{12}H_{21}-H_{11}H_{22}}\end{array}\right)W^2 \\ +\left(\frac{H_{31}H_{13}H_{22}-H_{31}H_{12}H_{23}-H_{32}H_{13}H_{21}+H_{32}H_{11}H_{23}}{H_{12}H_{21}-H_{11}H_{22}} + H_{33}\right)W + H_{34}PW = 0 \end{array} \tag{25}$$

Solving the Equation (25), the analytical expression of P is obtained as follows:

$$\begin{aligned} P = & \frac{1}{H_{34}}\left(\frac{L_{35}L_{14}H_{21}-L_{35}L_{24}H_{11}+L_{36}L_{24}H_{12}-L_{36}L_{14}H_{22}}{H_{12}H_{21}-H_{11}H_{22}} - L_{37}\right)W^2 \\ & +\frac{1}{H_{34}}\left(\begin{array}{c}\frac{H_{31}L_{24}H_{12}-H_{31}L_{14}H_{22}+H_{32}L_{14}H_{21}-H_{32}L_{24}H_{11}}{H_{12}H_{21}-H_{11}H_{22}} - L_{34} \\ +\frac{L_{35}H_{13}H_{21}-L_{35}H_{11}H_{23}+L_{36}H_{12}H_{23}-L_{36}H_{13}H_{22}}{H_{12}H_{21}-H_{11}H_{22}}\end{array}\right)W \\ & +\frac{1}{H_{34}}\left(\frac{H_{31}H_{12}H_{23}-H_{31}H_{13}H_{22}+H_{32}H_{13}H_{21}-H_{32}H_{11}H_{23}}{H_{12}H_{21}-H_{11}H_{22}} - H_{33}\right) \end{aligned} \quad (26)$$

By then, the critical buckling load and the post-buckling load-deflection curve of the PSTC subjected to axial compressive loads could be obtained from Equation (26).

Setting $W \to 0$, Equation (26) yields the upper buckling compressive load as follows:

$$P = P_{upper} = \frac{1}{H_{34}}\left(\frac{H_{31}H_{12}H_{23} - H_{31}H_{13}H_{22} + H_{32}H_{13}H_{21} - H_{32}H_{11}H_{23}}{H_{12}H_{21} - H_{11}H_{22}} - H_{33}\right) \quad (27)$$

It is clear from Equation (26) and (27) that, the value of the buckling loads depends on m and n, as a result, it is worth considering the values of m and n in making these loads reaches the minimum values.

7. Numerical Results and Discussion

The geometric parameters of various model of truncated conical shell and stiffeners used in the present study are listed in Table 1.

Table 1. The geometric properties for the stiffened (un-stiffenedt) truncated conical shells.

Model	L/R	R/h	h (m)	α (°)	h_{core}/h_{FG}	b_r=b_s (m)	h_r=h_s (m)	n_r	n_s
M1	0.2; 0.5	100	0.01	1 to 80	-	-	-	-	-
M2	2	150	0.05	30	0 to 5	0.02	0.03	50	30
M3	2	150	0.01	45	0 to 8	-	-	-	-
M4	2	80	0.012	30	3	0.02	0.012	35	25

7.1. Verification Study

To verify the present study, firstly, the dimensionless buckling axial compressive loads P^* of single layer pure isotropic (Stainless steel—SUS304) un-stiffened truncated conical shell by setting ($h_{FG} = 0, e_0 = 0$) are compared with the results of Naj et al. [2] and Baruch et al. [42]. The results are presented in Table 2, and in this particular case, the circular cylindrical shell of model M1 without elastic foundation is considered. The material properties are $\nu = 0.3$, $E_m = 200$ GPa. We determine $P^* = P_{cr}/P_{cl}$ with $P_{cl} = \frac{2\pi E h^2 \cos^2\alpha}{\sqrt{3(1-\nu^2)}}$ [2] and is found from Equation (27).

Table 2. Dimensionless buckling axial compressive loads of un-stiffened isotropic truncated conical shells without elastic foundation.

α	L/R = 0.2			L/R = 0.5		
	Naj et al. [2]	Baruch et al. [42]	Present (P^*)	Naj et al. [2]	Baruch et al. [42]	Present (P^*)
1°	1.005 (7)	1.005 (7)	1.0002 (1,12) [a]	1.0017 (8)	1.002 (8)	1.0001 (2,17)
5°	1.006 (7)	1.006 (7)	1.0001 (1,12)	1.001 (8)	1.002 (8)	1.0002 (2,17)
10°	1.007 (7)	1.007 (7)	1.0002 (1,12)	1.000 (8)	1.002 (8)	1.0005 (2,17)
30°	1.0171 (5)	1.017 (5)	1.0017 (1,7)	0.987 (7)	1.001 (7)	1.0023 (2,15)
60°	1.148 (0)	1.144 (0)	1.1299 (1,1)	1.045 (7)	1.044 (7)	1.0150 (1,14)
80°	2.492 (0)	2.477 (0)	2.5091 (1,1)	1.004 (5)	1.015 (5)	1.0266 (1,4)

[a] Buckling mode (m,n).

The next verification is performed for stiffened FGM sandwich truncated conical shells with metal core ($e_0 = 0$), FG faces, and FG stiffeners (Model M2) resting on Pasternak's foundation. The obtained results are presented in Table 3 and are compared with the linear critical loads P_{cr} of Dung et al. [21]. In which, the Alumina has $E_c = 380$ GPa, Aluminum has $E_m = 70$ GPa, and $\nu = 0.3$ for both constituents. $k_2 = k_3 = k = 1$, $K_1 = 5 \times 10^5$ N/m^3, and $K_2 = 3 \times 10^4$ N/m. The expression P_{cr} is taken from Equation (27).

Table 3. Linear critical load of stiffened FG sandwich truncated conical shells.

P_{cr} (MN)	Case 1 (Outside Stiffeners)		Case 2 (Inside Stiffeners)	
h_{core}/h_{FG}	Dung et al. [21]	Present	Dung et al. [21]	Present
0	19.46667 (8,18)	19.4667 (8,18) [a]	19.14549 (7,21)	19.1455 (7,21)
0.5	16.12768 (8,16)	16.1277 (8,16)	15.79773 (6,22)	15.7977 (6,22)
1	14.09267 (8,16)	14.0927 (8,16)	13.76594 (6,22)	13.7659 (6,22)
2	11.74586 (8,15)	11.7459 (8,15)	11.42875 (6,22)	11.4288 (6,22)
3	10.43697 (8,16)	10.4370 (8,16)	10.12653 (6,22)	10.1265 (6,22)
4	9.60325 (8,16)	9.6033 (8,16)	9.29804 (6,22)	9.2980 (6,22)
5	9.02635 (8,16)	9.0264 (8,16)	8.72504 (6,22)	8.7250 (6,22)

[a] Buckling mode (*m*,*n*).

Finally, Table 4 compares the present results with those of Deniz [43] for un-stiffened three-layered FG/Metal/FG truncated conical shells (Model M3) subjected to an axial load without elastic foundation. The database is used in this example: $E_c = 348.43$ GPa; $E_m = 201.04$ GPa; $h = 0.01$ m; $\alpha = 45°$; $L/R = 2$; $R/h = 150$; $K_1 = K_2 = 0$; $e_0 = 0$. The author analyzed non-linear stability based on the Donnell shell theory with von Karman-type of kinematic non-linearity. Using stress approach and approximated solution with two terms may cause the considerable discrepancy between two results.

Table 4. Comparisons of nondimensional critical axial loads (calculated by Equation (27)) for un-stiffened three-layered FG/Metal/FG truncated conical shells with various ratio h_{core}/h_{FG}.

P_{cr} (GN)	$k = 1$			$k = 2$			$k = 5$		
	Deniz [43]	Present	Error	Deniz [43]	Present	Error	Deniz [43]	Present	Error
$h_{core}/h_{FG} = 0$	1.244	1.2914 (6,22) [a]	3.7%	1.314	1.3605 (6.22)	3.4%	1.390	1.4392 (6,22)	3.4%
$h_{core}/h_{FG} = 2$	1.190	1.1459 (6,22)	−3.8%	1.246	1.2021 (6,22)	−3.7%	1.297	1.2649 (6,22)	−3.8%
$h_{core}/h_{FG} = 4$	1.135	1.0915 (6,22)	−3.8%	1.178	1.1321 (6,22)	−3.5%	1.217	1.1713 (6,22)	−3.9%
$h_{core}/h_{FG} = 6$	1.105	1.0654 (6,23)	−3.6%	1.139	1.1086 (6,22)	−2.7%	1.171	1.1307 (6,22)	−3.6%
$h_{core}/h_{FG} = 8$	1.085	1.0502 (6,23)	−3.2%	1.113	1.0887 (6,22)	−2.2%	1.140	1.0968 (6,22)	−3.9%

[a] Buckling mode (*m*,*n*).

From above three verifications, we can conclude that the results of the present study agree well with the existing results in the available literature.

7.2. The PSTC on Pasternak Elastic Foundations

In the following subsections, the PSTC resting on Pasternak elastic foundations are considered. FG materials of the coatings are a blend of Si3N4 (Silicon nitride-ceramic) and SUS304 (Stainless steel-metal) with $E_c = 348.43$ GPa and with $E_m = 201.04$ GPa and the metal foam of the core layer has $E_m = 201.04$ GPa. The PSTC's model is M3 with volume fraction indices $k_2 = k_3 = k = 1$, and foundation parameters $K_1 = 6 \times 10^7$ N/m^3, $K_2 = 4 \times 10^5$ N/m.

7.2.1. Effect of Porosity Coefficients e_0 and Thickness of Core Layer h_{core}

Table 5 presents the critical buckling loads of the PSTC with different degrees of porosity, h_{core}/h_{FG} ratios, and the buckling mode parameters (*m*,*n*). Furthermore, two cases of stiffeners arrangement, namely outside and inside eccentrically FG stiffeners are considered. Figures 2 and 3 illustrate the ratio h_{core}/h_{FG} effect on the critical buckling loads and post-buckling load-deflection paths of the shell, respectively.

From the figures, it can be seen that when h_{core}/h_{FG} ratios increase, the buckling loads decrease for both cases of arranging stiffeners. Taking case 1, $e_0 = 0.5$ as an example, the critical load decreases by about 43% from $P_{cr} = 161.4554$ MN (with $h_{core}/h_{FG} = 0$) to $P_{cr} = 112.5450$ MN (with $h_{core}/h_{FG} = 20$). The stiffener arrangement has considerable influence on the critical buckling loads. Indeed, the P_{cr} value of the PSTC reinforced by inside stiffeners is always smaller than that by outside stiffeners.

Figure 4 depicts the influence of porosity coefficients on the behaviors of the PSTC in the post-buckling phase. From the figure, the loading capacity of the shell decreases when e_0 increases. Figure 5 examines the relation between the critical buckling loads of the PSTC and the porosity coefficients existed in the shell. It is found that with the increment of e_0, the critical buckling load P_{cr} of the PSTC decreases. Indeed, the porosity affects the Young modulus of porous shells significantly as can be seen from Equation (1).

Table 5. The critical buckling load P_{cr} of the PSTC for various ratios h_{core}/h_{FG}.

P_{cr} (MN)	Case 1: Outside Stiffener			Case 2: Inside Stiffener		
	e_0=0.2	e_0=0.5	e_0=0.8	e_0=0.2	e_0=0.5	e_0=0.8
$h_{core}/h_{FG} = 0$	161.4554 (7,1)	161.4554 (7,1)	161.4554 (7,1)	142.5447 (5,16)	142.5447 (5,16)	142.5447 (5,16)
$h_{core}/h_{FG} = 0.5$	152.0344 (7,1)	148.6324 (7,1)	145.2239 (7,1)	133.1968 (5,15)	129.6503 (5,15)	126.1000 (5,18)
$h_{core}/h_{FG} = 1$	146.3406 (7,1)	140.9428 (7,1)	135.5258 (7,1)	127.5050 (5,15)	121.9165 (5,15)	116.3167 (5,15)
$h_{core}/h_{FG} = 2$	139.6989 (7,1)	131.9538 (7,1)	124.1623 (7,1)	120.9373 (5,15)	112.9854 (5,15)	105.0065 (5,15)
$h_{core}/h_{FG} = 3$	135.9469 (7,1)	126.8605 (7,1)	117.7094 (7,1)	117.3071 (5,15)	108.0130 (5,15)	98.6459 (5,15)
$h_{core}/h_{FG} = 4$	133.5555 (7,1)	123.5999 (7,1)	113.5562 (7,1)	114.9159 (5,15)	104.7698 (5,15)	94.5725 (5,15)
$h_{core}/h_{FG} = 5$	131.8897 (7,1)	121.3272 (7,1)	110.6618 (7,1)	113.2905 (5,15)	102.5464 (5,15)	91.7425 (5,15)
$h_{core}/h_{FG} = 10$	127.8946 (7,1)	115.8633 (7,1)	103.6858 (7,1)	109.4029 (5,15)	97.2183 (5,15)	84.9488 (5,15)
$h_{core}/h_{FG} = 20$	125.4750 (7,1)	112.5450 (7,1)	99.4363 (7,1)	107.0552 (5,15)	93.9934 (5,15)	80.8281 (5,15)

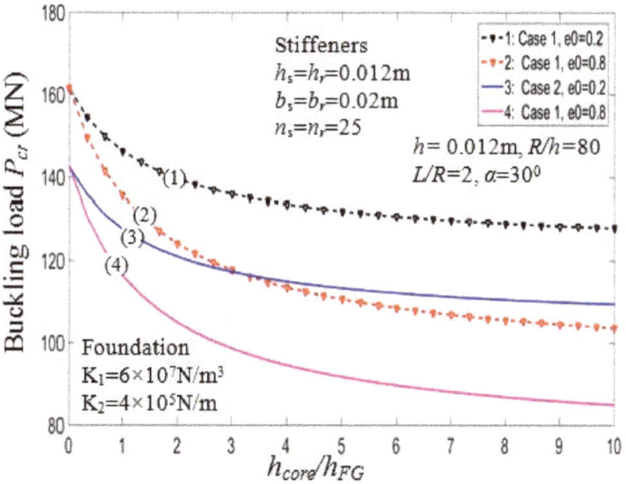

Figure 2. Effects of h_{core}/h_{FG} and e_0 on critical load P_{cr} ($k_2 = k_3 = k = 1$). Case 1: Outside stiffener; Case 2: Inside stiffener.

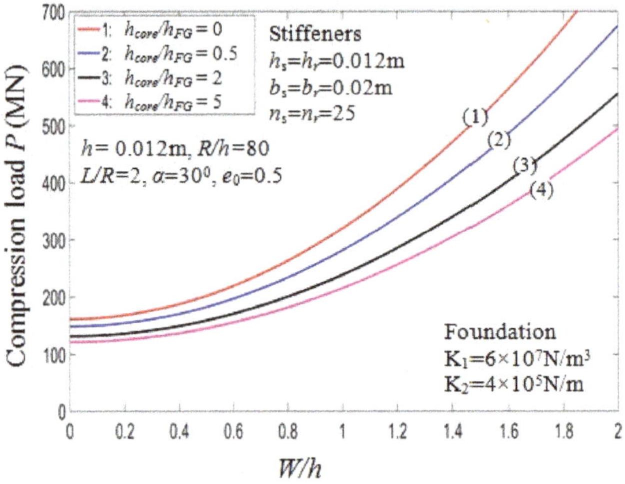

Figure 3. Effects of h_{core}/h_{FG} on postbuckling load—deflection curves (Case 1, $k_2 = k_3 = k = 1$).

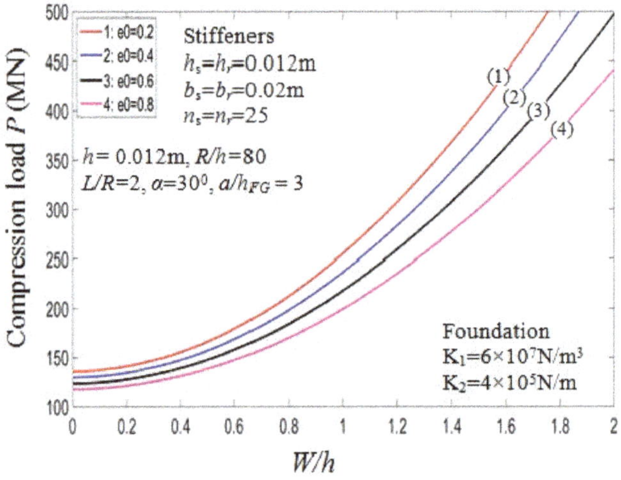

Figure 4. Effects of e_0 on postbuckling load—deflection curves (Outside stiffener, $k_2 = k_3 = k = 1$).

7.2.2. Effect of Semi-Vertex Angle α

The buckling loads of the PSTC in relation with the semi-vertex angle α are presented in Table 6. It could be noted from the table that when α increases, the critical buckling load of the PSTC decreases remarkably. Indeed, with $e_0 = 0.5$ in case 1, the value of P_{cr} experiences a reduction from 171.8857 MN to 10.9997 MN (93.6%) when the value varies from 50° to 80°. This observation has also been mentioned in Ref. [11,18]. The variation of critical axial compressive loads in relation with the semi-vertex angle is plotted in Figure 6 for various porosity coefficients and both cases of stiffener arrangements. Also, the influence of angle α on the equilibrium behaviors of the PSTC with outer stiffeners in the post-buckling phase is presented in Figure 7. The figure also shows that, when the value of angle α increases, P_{cr} decreases.

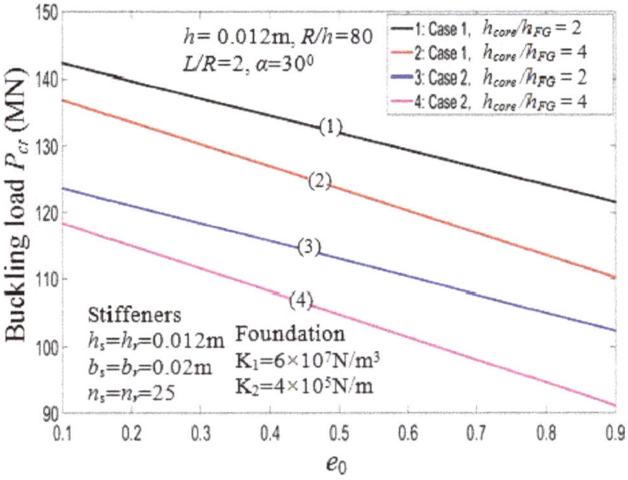

Figure 5. Effects of e_0 on critical load P_{cr} ($k_2 = k_3 = k = 1$).

Table 6. Critical compression load P_{cr} for various semi-vertex angles α.

P_{cr} (MN)	Case 1: Outside Stiffener		Case 2: Inside Stiffener	
	e_0=0.2	e_0=0.5	e_0=0.2	e_0=0.5
α = 5°	184.2470 (9,1)	171.8857 (9,1)	149.3844 (6,14)	136.8875 (6,14)
α = 10°	178.8700 (8,5)	166.0860 (8,3)	146.9110 (6,14)	133.9463 (5,14)
α = 20°	160.5859 (8,1)	150.1258 (8,1)	135.1141 (5,15)	123.7183 (5,15)
α = 30°	135.9469 (7,1)	126.8605 (7,1)	117.3071 (5,15)	108.0130 (5,15)
α = 45°	92.8172 (6,1)	86.9674 (6,1)	84.0426 (5,14)	78.3735 (5,14)
α = 60°	50.6289 (5,1)	47.8436 (5,1)	48.4738 (4,13)	45.6781 (4,12)
α = 70°	28.1649 (4,1)	26.8487 (4,1)	27.7523 (4,10)	26.5536 (4,9)
α = 80°	11.2994 (4,1)	10.9997 (4,1)	11.6098 (4,1)	11.3076 (4,2)

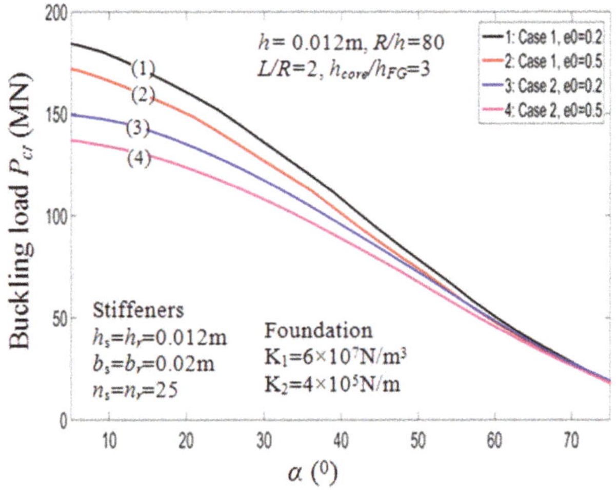

Figure 6. Effects of semi-vertex angle α on critical load P_{cr} ($k_2 = k_3 = k = 1$). Case 1: Outside stiffener; Case 2: Inside stiffener.

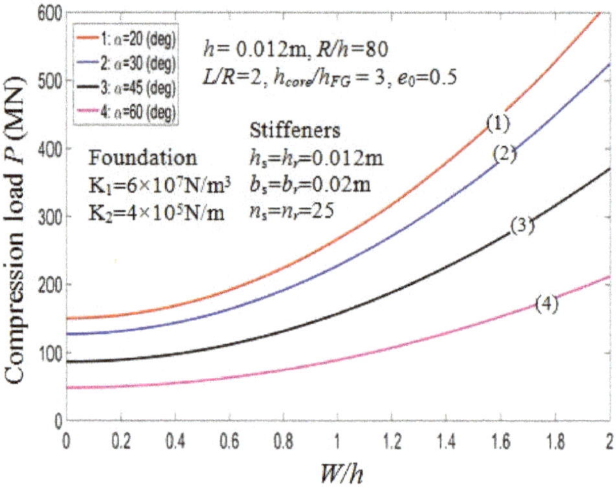

Figure 7. Effects of semi-vertex angle α on postbuckling load—deflection curves (Outside stiffener, mboxemphk$_2$ = k_3 = k = 1).

7.2.3. Effect of Geometrical Ratios

Effects of geometrical ratios L/R and R/h, on the buckling load P_{cr} of the PSTC are presented in Table 7 and graphically illustrated in Figure 8. When L/R and R/h ratios increase, P_{cr} decreases significantly. It is clear from the actual mechanical behavior of the structure that, in case of the shell structure, the thinner or the longer the shell, the smaller the value P_{cr}. Indeed, in Table 7, in the case of outside stiffeners, drawing the comparison between $P_{cr} = 684.7950$ MN (when $R/h = 60$, $L/R = 1$) and $P_{cr} = 197.9920$ MN (when $R/h = 60$, $L/R = 2$), the value of P_{cr} decreases by approximately 71.1%. This trend is also depicted in Figure 9 for the effect of R/h and L/R ratios on the post-buckling equilibrium paths of the PSTC in the case 1. Thus, the bearing capacity of the shell is quite sensitive to the variation of L/R and R/h ratios.

Table 7. Critical compression load P_{cr} for various values of L/R and R/h ratios.

P_{cr} (MN)	R/h = 60	R/h = 80	R/h = 100	R/h = 200	R/h = 300
			Case 1: Outside stiffeners		
$L/R = 1$	684.7950 (3,9)	398.8262 (4,1)	272.3611 (5,1)	93.1743 (6,1)	58.5103 (7,1)
$L/R = 1.5$	320.8777 (5,1)	197.8373 (6,1)	139.0107 (6,1)	58.4647 (8,1)	42.7500 (9,1)
$L/R = 2$	197.9920 (6,1)	126.8605 (7,1)	94.1463 (8,1)	47.0167 (9,1)	37.0240 (9,7)
$L/R = 3$	109.9757 (8,1)	77.4973 (8,6)	61.4766 (9,1)	37.9004 (9,10)	30.9346 (8,13)
			Case 2: Inside stiffeners		
$L/R = 1$	648.9722 (3,11)	379.1878 (3,14)	255.0781 (4,14)	82.7254 (5,20)	50.4788 (6,23)
$L/R = 1.5$	297.5119 (4,12)	177.0668 (4,15)	122.2705 (5,16)	47.5955 (6,20)	33.5999 (7,21)
$L/R = 2$	175.2790 (4,14)	108.0130 (5,15)	77.5981 (5,16)	35.7277 (6,18)	26.9945 (7,19)
$L/R = 3$	89.9204 (5,14)	60.0620 (6,15)	46.3523 (6,15)	26.2785 (7,16)	21.2033 (7,16)

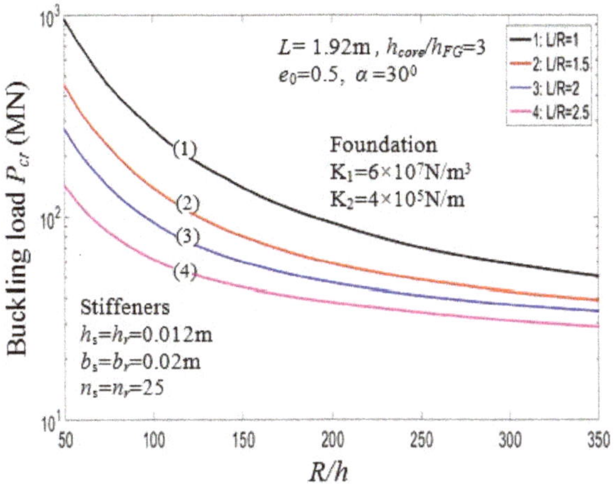

Figure 8. Effects of R/h and L/R on critical load P_{cr} (Case 1, $k_2 = k_3 = k = 1$).

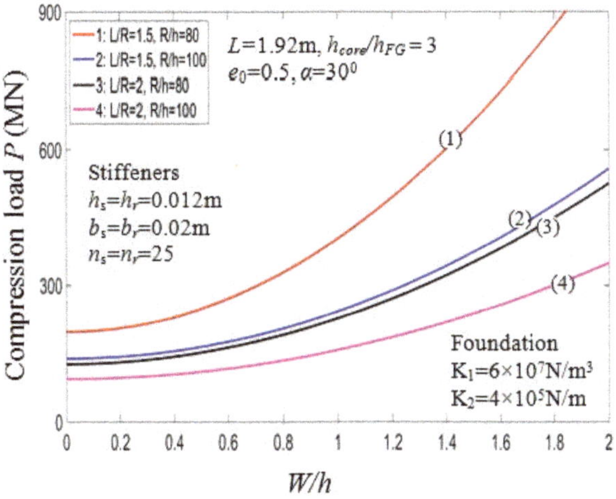

Figure 9. Effects of R/h and L/R on postbuckling load—deflection curves (Case 1, $k_2 = k_3 = k = 1$).

7.2.4. Effects of Volume Fraction Index

The critical buckling loads affected by the parameters k, k_2 and k_3 are shown in Table 8. The critical buckling loads vary according to the volume fraction index for two different values of the h_{core}/h_{FG} ratio depicted in Figure 10. From the figure, when the value of k increases, the critical loads P_{cr} increase. The reason is that the portion of the ceramic constituent in shell structure increase when the value of k increase. This is also confirmed by observing Figure 11, which depicts the load-deflection curves of the PSTC with outside stiffeners in relation to the volume fraction index in the post-buckling phase.

Table 8. Critical compression load P_{cr} for different values of volume fraction indexes.

P_{cr} (MN)	Case 1: Outside Stiffener ($k_2 = k_3 = 1/k$)		Case 2: Inside Stiffener ($k_2 = k_3 = k$)	
	e_0=0.2	e_0=0.5	e_0=0.2	e_0=0.5
$k = 0$	135.4442 (7,1)	126.2111 (7,1)	105.4436 (5,16)	96.2521 (5,15)
$k = 1$	135.9469 (7,1)	126.8605 (7,1)	117.3071 (5,15)	108.0130 (5,15)
$k = 5$	137.5855 (7,1)	128.5525 (7,1)	125.4939 (5,15)	116.1792 (5,15)
$k = 10$	137.9374 (7,1)	128.9140 (7,1)	127.4802 (5,15)	118.1596 (5,15)
$k = \infty$	138.3158 (7,1)	129.3029 (7,1)	130.0065 (5,15)	120.6781 (5,15)

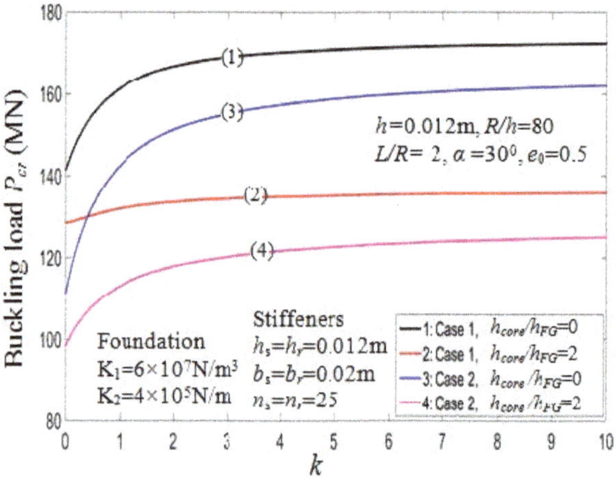

Figure 10. Effects of volume fraction indexes on critical load P_{cr} ($k_2, k_3 = 1/k$).

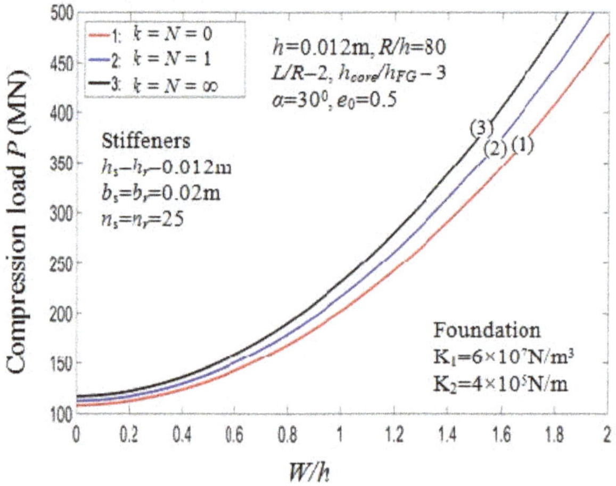

Figure 11. Effects of volume fraction indexes on postbuckling load—deflection curves (Case 1).

7.2.5. Effect of Stiffeners and Foundation

The effects of stiffeners and elastic foundations on the buckling loads P_{cr} of the PSTC are presented in Table 9. It is noted that the higher the number of stiffeners being used, the higher the buckling load. Indeed, for case 1 with $K_1 = 6 \times 10^7$ N/m^3, $K_2 = 4 \times 10^5$ N/m, drawing the comparison between $P_{cr} = 90.1237$ MN ($n_s = n_r = 0$) and $P_{cr} = 161.2914$ MN ($n_s = n_r = 50$), we could recognize the increment in the value of critical compressive load by about 79%. Furthermore, the critical compressive loads P_{cr} of the PSTC stiffened by rings are higher than that of the PSTC stiffened by stringers.

Table 9. Effects of stiffeners and foundation on buckling loads P_{cr}.

P_{cr} (MN)	K_1=0 K_2=0	K_1=3×10^7 N/m^3 K_2=2×10^5 N/m	K_1=6×10^7 N/m^3 K_2=4×10^5 N/m	K_1=9×10^7 N/m^3 K_2=6×10^5 N/m
Case 1: Conical Shell Reinforced by Outside Stiffener				
$n_s = 0, n_r = 0$	81.8418 (5,16)	86.4019 (7,4)	90.1237 (7,4)	93.8390 (7,3)
$n_s = 50, n_r = 0$	96.7739 (2,16)	110.7194 (4,17)	118.5695 (5,14)	124.6769 (5,14)
$n_s = 0, n_r = 50$	111.2085 (8,1)	114.4478 (8,1)	117.6871 (8,1)	120.9264 (8,1)
$n_s = 25, n_r = 25$	119.4573 (7,1)	123.1589 (7,1)	126.8605 (7,1)	130.5622 (7,1)
$n_s = 50, n_r = 50$	153.6013 (6,9)	157.5907 (7,1)	161.2924 (7,1)	164.9940 (7,1)
Case 2: Conical Shell Reinforced by Inside Stiffener				
$n_s = 0, n_r = 0$	81.8418 (5,16)	86.4019 (7,4)	90.1237 (7,4)	93.8390 (7,3)
$n_s = 50, n_r = 0$	85.6480 (2,16)	101.4708 (3,17)	111.2717 (4,17)	120.2202 (4,17)
$n_s = 0, n_r = 50$	84.7967 (6,15)	89.6205 (6,15)	94.4443 (6,15)	99.2681 (6,15)
$n_s = 25, n_r = 25$	94.3881 (4,15)	101.8290 (5,15)	108.0130 (5,15)	114.1970 (5,15)
$n_s = 50, n_r = 50$	103.4412 (4,14)	112.0060 (4,14)	120.2306 (5,15)	126.4147 (5,15)

It is also noted that the presence of elastic foundations enhances the buckling loads. The buckling load of the PSTC increases according to the increment of the foundation parameters. Indeed, for the PSTC with orthogonal stiffeners with ($n_s = n_r = 50$) the value of P_{cr} rises by about 9.3% from 119.4573MN with the absence of elastic foundation to 130.5622MN with the presence of elastic foundation: $K_1 = 9 \times 10^7$ N/m^3; $K_2 = 6 \times 10^5$ N/m).

Figure 12 depicts the effect of stiffeners quantity on the post-buckling equilibrium path $P - W/h$ of the PSTC. The value of the buckling loads is in a proportional relation with the quantity of the stiffeners. The curve for the stiffeners-free case and $n_s = n_r = 25$ case bottoms and tops the graph, respectively. The curves for $n_s = n_r = 15$ and $n_s = n_r = 10$ locate in the middle range. The effect of foundation parameters on the post-buckling equilibrium paths $P - W/h$ of the PSTC is also shown in Figure 13. It is observed that when the foundation parameters K_1, K_2 increases, the curves gradually rise, in other words, the post-buckling equilibrium loads increase. From the figure, the curve for $K_1 = 9 \times 10^7$ N/m^3; $K_2 = 6 \times 10^5$ N/m peaks, in other words, in this case, the buckling load at specific deflection value W/h is the highest among all the cases considered. The buckling load for the case with $K_1 = 6 \times 10^7$ N/m^3; $K_2 = 4 \times 10^5$ N/m is greater than that for the case with $K_1 = 3 \times 10^7$ N/m^3; $K_2 = 2 \times 10^5$ N/m in the post-buckling phase of the PSTC.

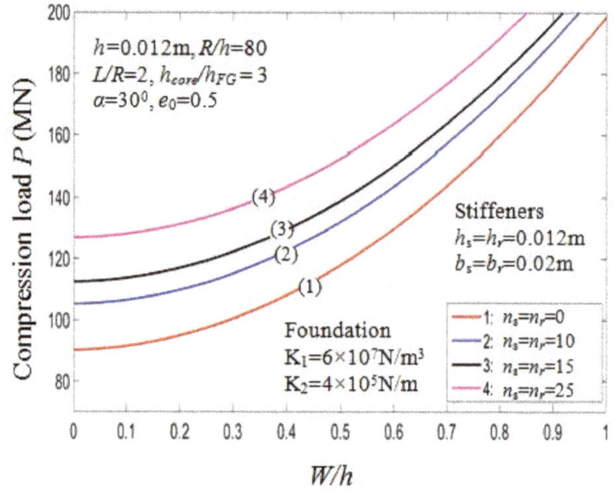

Figure 12. Effects of stiffeners on postbuckling load—deflection curves (Case 1, $k_2 = k_3 = k = 1$).

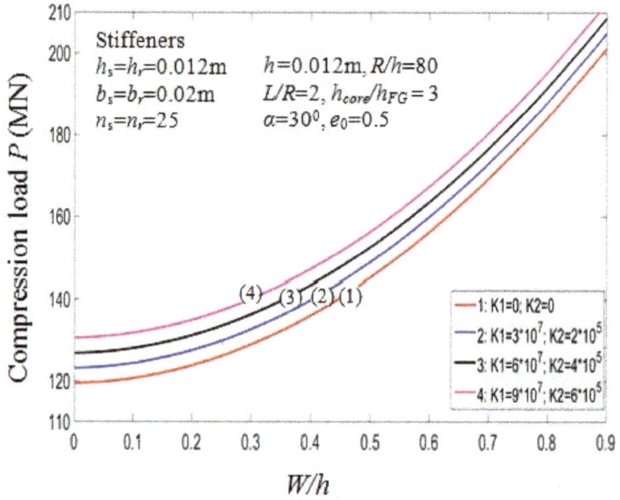

Figure 13. Effects of foundation on postbuckling load—deflection curves (Case 1, $k_2 = k_3 = k = 1$).

8. Conclusions

The paper produces an analytical procedure to analyze the nonlinear instability of the porous eccentrically stiffened functionally graded sandwich truncated conical shells surrounded by Pasternak elastic foundations using displacement approach. The core is made of a porous material (metal foam) with properties varying across its thickness according to a simple cosine law in term of a coefficient related to plate's porosity. The material properties of FG coatings and stiffeners are assumed to be graded through the thickness direction according to a simple power law distribution in terms of the volume fractions of the constituents. Two cases of stiffener arrangement: outside and inside stiffened are considered. The smeared stiffeners technique with von Karman geometrical nonlinearity and the classical shell theory are employed to bring about the governing equations. The Galerkin method is employed to obtain theoretical expressions of load-deflection curves or the post-buckling equilibrium paths. The numerical results show that the reinforced stiffeners, with volume fraction index k, the length-to-radius ratio L/R, the radius-to-thickness ratio R/h, and foundation parameters

K_1, K_2 significantly influence the buckling and post-buckling behaviors of the porous eccentrically stiffened functionally graded truncated conical sandwich shells. The study also shows the profound effects of the porosity coefficient e_0 and the core layer thickness on the critical buckling compressive loads and load-deflection curves in the post-buckling phase of the shell. Moreover, the stiffener arrangement has considerable influence on the critical buckling loads, the PSTC reinforced by outside stiffeners is always stiffer than that reinforced by inside stiffeners.

Author Contributions: Formal analysis, L.K.H.; Investigation, N.N.L.; Software, D.X.H.; Supervision—editing, T.M.T.; Writing—original draft, D.-K.T.

Funding: This research is funded by Vietnam National Foundation for Science and Technology Development (NAFOSTED) under grant number: 107.02-2018.17.

Acknowledgments: This research is funded by Vietnam National Foundation for Science and Technology Development (NAFOSTED) under grant number: 107.02-2018.17. The authors are grateful for this support.

Conflicts of Interest: The authors declared no potential conflicts of interest with respect to the research, authorship, and/or publication of this article.

Appendix A

$$
\begin{aligned}
E_1 &= \int_{-h/2}^{h/2} E_{sh}dz = E_c h_{FG} + E_{mc} h_{FG}\frac{1}{k+1} + E_m\left[h_{core} - e_0\frac{2h_{core}}{\pi}\right] \\
E_2 &= \int_{-h/2}^{h/2} zE_{sh}dz = 0 \\
E_3 &= \int_{-h/2}^{h/2} z^2E_{sh}dz = \frac{E_c}{12}\left[(h_{FG}+h_{core})^3 - h_{core}{}^3\right] \\
&+\frac{E_{mc}}{4}\left[\frac{h_{FG}^3}{k+3} - \frac{2h_{FG}^2(h_{FG}+h_{core})}{k+2} + \frac{h_{FG}(h_{FG}+h_{core})^2}{k+1}\right] + E_m\left[\frac{h_{core}{}^3}{12} - \frac{e_0 h_{core}{}^3(\pi^2-8)}{2\pi^3}\right]
\end{aligned}
$$

Case 1: Outside stiffener

$$
\begin{aligned}
E_{1s} &= E_c h_s + E_{mc}\frac{h_s}{k_2+1}, \\
E_{2s} &= E_c\frac{h_s^2+h_s(h_{FG}+h_{core})}{2} + E_{mc}\left(\frac{h_s^2}{k_2+2} + \frac{h_s(h_{FG}+h_{core})}{2k_2+2}\right), \\
E_{3s} &= E_c\frac{3(h_{Fg}+h_{core})^2h_s+6(h_{FG}+h_{core})h_s^2+4h_s^3}{12} + E_{mc}\left(\frac{h_s^3}{k_2+3} + \frac{h_s^2(h_{FG}+h_{core})}{k_2+2} + \frac{h_s(h_{FG}+h_{core})^2}{4k_2+4}\right),
\end{aligned}
$$

$$
\begin{aligned}
E_{1r} &= E_c h_r + E_{mc}\frac{h_r}{k_3+1}, \\
E_{2r} &= E_c\frac{h_r^2+h_r(h_{FG}+h_{core})}{2} + E_{mc}\left(\frac{h_r^2}{k_3+2} + \frac{h_r(h_{FG}+h_{core})}{2k_3+2}\right), \\
E_{3r} &= E_c\frac{3(h_{Fg}+h_{core})^2h_r+6(h_{FG}+h_{core})h_r^2+4h_r^3}{12} + E_{mc}\left(\frac{h_r^3}{k_3+3} + \frac{h_r^2(h_{FG}+h_{core})}{k_3+2} + \frac{h_r(h_{FG}+h_{core})^2}{4k_3+4}\right)
\end{aligned}
$$

Case 2: Inside stiffener

$$
\begin{aligned}
E_{1s} &= E_c h_s + E_{mc}\frac{h_s}{k_2+1}, \\
E_{2s} &= -E_c\frac{h_s(h_{FG}+h_{core})+h_s{}^2}{2} - E_{cm}\left(\frac{h_s{}^2}{k_2+2} + \frac{h_s(h_{FG}+h_{core})}{2k_2+2}\right), \\
E_{3s} &= E_c\frac{3h_s(h_{FG}+h_{core})^2+6h_s{}^2(h_{FG}+h_{core})+4h_s{}^3}{12} + E_{mc}\left(\frac{h_s{}^3}{k_2+3} + \frac{h_s{}^2(h_{FG}+h_{core})}{k_2+2} + \frac{h_s(h_{FG}+h_{core})^2}{4k_2+4}\right)
\end{aligned}
$$

$$E_{1r} = E_c h_r + E_{mc}\frac{h_r}{k_3+1},$$
$$E_{2r} = -E_c\frac{h_r(h_{FG}+h_{core})+h_r^2}{2} - E_{mc}\left(\frac{h_r^2}{k_3+2} + \frac{h_r(h_{FG}+h_{core})}{2k_3+2}\right),$$
$$E_{3r} = E_c\frac{3h_r(h_{FG}+h_{core})^2+6h_r^2(h_{FG}+h_{core})+4h_r^3}{12} + E_{mc}\left(\frac{h_r^3}{k_3+3} + \frac{h_r^2(h_{FG}+h_{core})}{k_3+2} + \frac{h_r(h_{FG}+h_{core})^2}{4k_3+4}\right)$$

In Equation (9), $A_{11} = A_{22} = \frac{E_1}{1-\nu^2}$, $A_{12} = \nu A_{11}$; $A_{66} = \frac{1-\nu}{2}A_{11}$; $B_{11} = B_{22} = \frac{E_2}{1-\nu^2}$, $B_{12} = \nu B_{11}$; $B_{66} = \frac{1-\nu}{2}B_{11}$; $D_{11} = D_{22} = \frac{E_3}{1-\nu^2}$, $D_{12} = \nu D_{11}$; $D_{66} = \frac{1-\nu}{2}D_{11}$; $d_1(x) = \lambda_0 x$, $d_2 = \frac{L}{n_r}$, $e_s = \frac{h+h_s}{2}$, $e_r = \frac{h+h_r}{2}$, $C_1(x) = \frac{C_1^0}{x}$, $C_1^0 = \frac{E_{2s}b_s}{\lambda_0}$, $C_2 = \frac{E_{2r}b_r}{d_2}$, $\lambda_0 = \frac{2\pi\sin\alpha}{n_s}$, in which n_s is the number of stringers, n_r is the number of rings; b_r are the width of rings, b_s is the width of stringers; $d_1 = d_1(x)$ is the span between stringers; d_2 is the span between rings as shown in Figure 2; e_s is the eccentricities of the stringers, e_r is the eccentricities of the rings to the mid-surface of the shell as shown in Figure 1.

Appendix B

In Equations (15)–(17)

$$E_1 = \int_{-h/2}^{h/2} E_{sh}dz = E_c h_{FG} + E_{mc}h_{FG}\frac{1}{k+1} + E_m\left[h_{core} - e_0\frac{2h_{core}}{\pi}\right]$$
$$E_2 = \int_{-h/2}^{h/2} zE_{sh}dz = 0$$
$$E_3 = \int_{-h/2}^{h/2} z^2E_{sh}dz = \frac{E_c}{12}\left[(h_{FG}+h_{core})^3 - h_{core}^3\right] + \frac{E_{mc}}{4}\left[\frac{h_{FG}^3}{k+3} - \frac{2h_{FG}^2(h_{FG}+h_{core})}{k+2} + \frac{h_{FG}(h_{FG}+h_{core})^2}{k+1}\right] + E_m\left[\frac{h_{core}^3}{12} - \frac{e_0 h_{core}^3(\pi^2-8)}{2\pi^3}\right]$$

Case 1: Outside stiffener

$$E_{1s} = E_c h_s + E_{mc}\frac{h_s}{k_2+1},$$
$$E_{2s} = E_c\frac{h_s^2+h_s(h_{FG}+h_{core})}{2} + E_{mc}\left(\frac{h_s^2}{k_2+2} + \frac{h_s(h_{FG}+h_{core})}{2k_2+2}\right),$$
$$E_{3s} = E_c\frac{3(h_{Fg}+h_{core})^2h_s+6(h_{FG}+h_{core})h_s^2+4h_s^3}{12} + E_{mc}\left(\frac{h_s^3}{k_2+3} + \frac{h_s^2(h_{FG}+h_{core})}{k_2+2} + \frac{h_s(h_{FG}+h_{core})^2}{4k_2+4}\right),$$

$$E_{1r} = E_c h_r + E_{mc}\frac{h_r}{k_3+1},$$
$$E_{2r} = E_c\frac{h_r^2+h_r(h_{FG}+h_{core})}{2} + E_{mc}\left(\frac{h_r^2}{k_3+2} + \frac{h_r(h_{FG}+h_{core})}{2k_3+2}\right),$$
$$E_{3r} = E_c\frac{3(h_{Fg}+h_{core})^2h_r+6(h_{FG}+h_{core})h_r^2+4h_r^3}{12} + E_{mc}\left(\frac{h_r^3}{k_3+3} + \frac{h_r^2(h_{FG}+h_{core})}{k_3+2} + \frac{h_r(h_{FG}+h_{core})^2}{4k_3+4}\right).$$

Case 2: Inside stiffener

$$E_{1s} = E_c h_s + E_{mc}\frac{h_s}{k_2+1},$$
$$E_{2s} = -E_c\frac{h_s(h_{FG}+h_{core})+h_s^2}{2} - E_{cm}\left(\frac{h_s^2}{k_2+2} + \frac{h_s(h_{FG}+h_{core})}{2k_2+2}\right),$$
$$E_{3s} = E_c\frac{3h_s(h_{FG}+h_{core})^2+6h_s^2(h_{FG}+h_{core})+4h_s^3}{12} + E_{mc}\left(\frac{h_s^3}{k_2+3} + \frac{h_s^2(h_{FG}+h_{core})}{k_2+2} + \frac{h_s(h_{FG}+h_{core})^2}{4k_2+4}\right)$$

$$E_{1r} = E_c h_r + E_{mc}\frac{h_r}{k_3+1},$$
$$E_{2r} = -E_c\frac{h_r(h_{FG}+h_{core})+h_r^2}{2} - E_{mc}\left(\frac{h_r^2}{k_3+2} + \frac{h_r(h_{FG}+h_{core})}{2k_3+2}\right),$$
$$E_{3r} = E_c\frac{3h_r(h_{FG}+h_{core})^2+6h_r^2(h_{FG}+h_{core})+4h_r^3}{12} + E_{mc}\left(\frac{h_r^3}{k_3+3} + \frac{h_r^2(h_{FG}+h_{core})}{k_3+2} + \frac{h_r(h_{FG}+h_{core})^2}{4k_3+4}\right)$$

In Equation (9), $A_{11} = A_{22} = \frac{E_1}{1-\nu^2}$, $A_{12} = \nu A_{11}$; $A_{66} = \frac{1-\nu}{2} A_{11}$; $B_{11} = B_{22} = \frac{E_2}{1-\nu^2}$, $B_{12} = \nu B_{11}$; $B_{66} = \frac{1-\nu}{2} B_{11}$; $D_{11} = D_{22} = \frac{E_3}{1-\nu^2}$, $D_{12} = \nu D_{11}$; $D_{66} = \frac{1-\nu}{2} D_{11}$; $d_1(x) = \lambda_0 x$, $d_2 = \frac{L}{n_r}$, $e_s = \frac{h+h_s}{2}$, $e_r = \frac{h+h_r}{2}$, $C_1(x) = \frac{C_1^0}{x}$, $C_1^0 = \frac{E_{2s} b_s}{\lambda_0}$, $C_2 = \frac{E_{2r} b_r}{d_2}$, $\lambda_0 = \frac{2\pi \sin\alpha}{n_s}$, in which n_s is the number of stringers, n_r is the number of rings; b_r are the width of rings, b_s is the width of stringers; $d_1 = d_1(x)$ is the span between stringers; d_2 is the span between rings as shown in Figure 2; e_s is the eccentricities of the stringers, e_r is the eccentricities of the rings to the mid-surface of the shell as shown in Figure 1.

In Equations (15)–(17)

$$F_{11} = \left[x A_{11} + \frac{E_{1s} b_s}{\lambda_o}\right] \frac{\partial^2}{\partial x^2} + A_{66} \frac{1}{x \sin^2\alpha} \frac{\partial^2}{\partial \theta^2} + A_{11} \frac{\partial}{\partial x} - \left[A_{22} + \frac{E_{1r} b_r}{d_2}\right] \frac{1}{x}$$

$$F_{12} = \frac{1}{\sin\alpha}(A_{12} + A_{66}) \frac{\partial^2}{\partial x \partial \theta} - \left[A_{22} + A_{66} + \frac{E_{1r} b_r}{d_2}\right] \frac{1}{x \sin\alpha} \frac{\partial}{\partial \theta}$$

$$\begin{aligned} F_{13} = - \quad & [x B_{11} + C_1^o] \frac{\partial^3}{\partial x^3} - \frac{1}{x \sin^2\alpha} [B_{12} + 2B_{66}] \frac{\partial^3}{\partial x \partial \theta^2} - B_{11} \frac{\partial^2}{\partial x^2} \\ & + \frac{1}{x^2 \sin^2\alpha} (B_{12} + 2B_{66} + B_{22} + C_2) \frac{\partial^2}{\partial \theta^2} + \left[A_{12} \cot\alpha + \frac{1}{x}(B_{22} + C_2)\right] \frac{\partial}{\partial x} - \frac{1}{x}\left(A_{22} + \frac{E_{1r} b_r}{d_2}\right) \cot\alpha \end{aligned}$$

$$F_{21} = \frac{1}{\sin\alpha} [A_{12} + A_{66}] \frac{\partial^2}{\partial x \partial \theta} + \frac{1}{x \sin\alpha} \left[A_{22} + A_{66} + \frac{E_{1r} b_r}{d_2}\right] \frac{\partial}{\partial \theta}$$

$$F_{22} = \left[A_{22} + \frac{E_{1r} b_r}{d_2}\right] \frac{1}{x \sin^2\alpha} \frac{\partial^2}{\partial \theta^2} + x A_{66} \frac{\partial^2}{\partial x^2} + A_{66} \frac{\partial}{\partial x} - A_{66} \frac{1}{x}$$

$$\begin{aligned} F_{23} = & -\frac{1}{\sin\alpha} [B_{12} + 2B_{66}] \frac{\partial^3}{\partial x^2 \partial \theta} - \frac{1}{x^2 \sin^3\alpha} [B_{22} + C_2] \frac{\partial^3}{\partial \theta^3} - \frac{1}{x \sin\alpha} [B_{22} + C_2] \frac{\partial^2}{\partial x \partial \theta} \\ & + \left[A_{22} + \frac{E_{1r} b_r}{d_2}\right] \frac{1}{x \sin\alpha} \frac{\partial}{\partial \theta} \cot\alpha \end{aligned}$$

$$\begin{aligned} F_{31} = & [x B_{11} + C_1^o] \frac{\partial^3}{\partial x^3} + \frac{1}{x \sin^2\alpha} [B_{12} + 2B_{66}] \frac{\partial^3}{\partial x \partial \theta^2} + 2B_{11} \frac{\partial^2}{\partial x^2} + \frac{1}{x^2 \sin^2\alpha} [B_{22} + C_2] \frac{\partial^2}{\partial \theta^2} \\ & - \left[\frac{1}{x}[B_{22} + C_2] + A_{12} \cot\alpha\right] \frac{\partial}{\partial x} + [B_{22} + C_2] \frac{1}{x^2} - \left[A_{22} + \frac{E_{1r} b_r}{d_2}\right] \frac{1}{x} \cot\alpha \end{aligned}$$

$$\begin{aligned} F_{32} = & \frac{1}{\sin\alpha} [B_{12} + 2B_{66}] \frac{\partial^3}{\partial x^2 \partial \theta} + \frac{1}{x^2 \sin^3\alpha} [B_{22} + C_2] \frac{\partial^3}{\partial \theta^3} - [B_{22} + C_2] \frac{1}{x \sin\alpha} \frac{\partial^2}{\partial x \partial \theta} + \\ & + \left[(B_{22} + C_2) \frac{1}{x^2 \sin\alpha} - \left(A_{22} + \frac{E_{1r} b_r}{d_2}\right) \frac{1}{x \sin\alpha} \cot\alpha\right] \frac{\partial}{\partial \theta} \end{aligned}$$

$$\begin{aligned} F_{33} = \quad & -\left[x D_{11} + \frac{E_{3s} b_s}{\lambda_o}\right] \frac{\partial^4}{\partial x^4} - \left[D_{22} + \frac{E_{3r} b_r}{d_2}\right] \frac{1}{x^3 \sin^4\alpha} \frac{\partial^4}{\partial \theta^4} - \frac{2}{x \sin^2\alpha} [D_{12} + 2D_{66}] \frac{\partial^4}{\partial x^2 \partial \theta^2} \\ & + \frac{2}{x^2 \sin^2\alpha} [D_{12} + 2D_{66}] \frac{\partial^3}{\partial x \partial \theta^2} - 2D_{11} \frac{\partial^3}{\partial x^3} + \left[\frac{1}{x}\left(D_{22} + \frac{E_{3r} b_r}{d_2}\right) + 2B_{12} \cot\alpha\right] \frac{\partial^2}{\partial x^2} \\ & + \frac{2}{x^2 \sin^2\alpha} \cot\alpha (B_{22} + C_2) \frac{\partial^2}{\partial \theta^2} - \frac{2}{x^3 \sin^2\alpha} \left[D_{12} + 2D_{66} + D_{22} + \frac{E_{3r} b_r}{d_2}\right] \frac{\partial^2}{\partial \theta^2} \\ & - \frac{1}{x^2} \left[D_{22} + \frac{E_{3r} b_r}{d_2}\right] \frac{\partial}{\partial x} + [B_{22} + C_2] \frac{1}{x^2} \cot\alpha - \frac{1}{x}\left[A_{22} + \frac{E_{1r} b_r}{d_2}\right] \cot^2\alpha \\ & - x K_1 + x K_2 \left(\frac{\partial^2}{\partial x^2} + \frac{1}{x} \frac{\partial}{\partial x} + \frac{1}{x^2 \sin^2\alpha} \frac{\partial^2}{\partial \theta^2}\right) \end{aligned}$$

$$F_{34} = -\frac{P}{\pi \sin 2\alpha} \frac{\partial^2}{\partial x^2}$$

$$\begin{aligned} G_{14} = & \left[x A_{11} + \frac{E_{1s} b_s}{\lambda_o}\right] \frac{\partial w}{\partial x} \frac{\partial^2 w}{\partial x^2} + \frac{1}{2}[A_{11} - A_{12}] \left(\frac{\partial w}{\partial x}\right)^2 + \frac{A_{12}}{2x \sin^2\alpha} \frac{\partial}{\partial x} \left(\frac{\partial w}{\partial \theta}\right)^2 \\ & - \frac{1}{2x^2 \sin^2\alpha} \left(A_{12} + A_{22} + \frac{E_{1r} b_r}{d_2}\right) \left(\frac{\partial w}{\partial \theta}\right)^2 + \frac{1}{x \sin^2\alpha} A_{66} \frac{\partial^2 w}{\partial x \partial \theta} \frac{\partial w}{\partial \theta} + \frac{1}{x \sin^2\alpha} A_{66} \frac{\partial w}{\partial x} \frac{\partial^2 w}{\partial \theta^2} \end{aligned}$$

$$\begin{aligned} G_{24} = & \frac{1}{2 \sin\alpha} A_{12} \frac{\partial}{\partial \theta} \left(\frac{\partial w}{\partial x}\right)^2 + \frac{1}{x^2 \sin^3\alpha} \left[A_{22} + \frac{E_{1r} b_r}{d_2}\right] \frac{\partial w}{\partial \theta} \frac{\partial^2 w}{\partial \theta^2} + A_{66} \frac{1}{\sin\alpha} \frac{\partial^2 w}{\partial x^2} \frac{\partial w}{\partial \theta} \\ & + A_{66} \frac{1}{\sin\alpha} \frac{\partial w}{\partial x} \frac{\partial^2 w}{\partial x \partial \theta} + A_{66} \frac{1}{x \sin\alpha} \frac{\partial w}{\partial x} \frac{\partial w}{\partial \theta} \end{aligned}$$

$$
\begin{aligned}
G_{34} = & B_{11}\frac{\partial w}{\partial x}\frac{\partial^2 w}{\partial x^2} + \frac{1}{x^3\sin^2\alpha}B_{12}\left(\frac{\partial w}{\partial \theta}\right)^2 - \frac{2}{x^2\sin^2\alpha}B_{12}\frac{\partial^2 w}{\partial x\partial\theta}\frac{\partial w}{\partial\theta} + \frac{2}{x\sin^2\alpha}B_{12}\left(\frac{\partial^2 w}{\partial x\partial\theta}\right)^2 \\
& +2B_{66}\frac{1}{x\sin^2\alpha}\frac{\partial^2 w}{\partial x^2}\frac{\partial^2 w}{\partial\theta^2} - 3B_{12}\frac{\partial w}{\partial x}\frac{\partial^2 w}{\partial x^2} + [B_{22}+C_2]\frac{1}{x^3\sin^2\alpha}\left(\frac{\partial w}{\partial\theta}\right)^2 \\
& -[B_{22}+C_2]\frac{2}{x^2\sin^2\alpha}\frac{\partial w}{\partial\theta}\frac{\partial^2 w}{\partial x\partial\theta} + \frac{1}{2}A_{12}\left(\frac{\partial w}{\partial x}\right)^2\cot\alpha + \left[A_{22}+\frac{E_{1r}b_r}{d_2}\right]\frac{1}{2x^2\sin^2\alpha}\left(\frac{\partial w}{\partial\theta}\right)^2\cot\alpha \\
& +\left[xA_{11}+\frac{E_{1s}b_s}{\lambda_o}\right]\frac{\partial^2 u}{\partial x^2}\frac{\partial w}{\partial x} + \left[xA_{11}+\frac{E_{1s}b_s}{\lambda_o}\right]\frac{\partial u}{\partial x}\frac{\partial^2 w}{\partial x^2} + A_{11}\frac{\partial u}{\partial x}\frac{\partial w}{\partial x} + A_{12}\frac{\partial u}{\partial x}\frac{\partial w}{\partial x} + A_{12}u\frac{\partial^2 w}{\partial x^2} \\
& +A_{12}\frac{1}{\sin\alpha}\frac{\partial^2 v}{\partial x\partial\theta}\frac{\partial w}{\partial x} + A_{12}\frac{1}{\sin\alpha}\frac{\partial v}{\partial\theta}\frac{\partial^2 w}{\partial x^2} + \frac{3}{2}\left[xA_{11}+\frac{E_{1s}b_s}{\lambda_o}\right]\left(\frac{\partial w}{\partial x}\right)^2\frac{\partial^2 w}{\partial x^2} + \frac{1}{2}A_{11}\left(\frac{\partial w}{\partial x}\right)^3 \\
& +A_{12}w\frac{\partial^2 w}{\partial x^2}\cot\alpha - A_{12}\frac{1}{2x^2\sin^2\alpha}\frac{\partial w}{\partial x}\left(\frac{\partial w}{\partial\theta}\right)^2 + A_{12}\frac{1}{2x\sin^2\alpha}\frac{\partial^2 w}{\partial x^2}\left(\frac{\partial w}{\partial\theta}\right)^2 + A_{12}\frac{2}{x\sin^2\alpha}\frac{\partial w}{\partial x}\frac{\partial w}{\partial\theta}\frac{\partial^2 w}{\partial x\partial\theta} \\
& +B_{12}\frac{1}{x^2\sin^2\alpha}\frac{\partial w}{\partial x}\frac{\partial^2 w}{\partial\theta^2} - B_{12}\frac{2}{x\sin^2\alpha}\frac{\partial^2 w}{\partial x^2}\frac{\partial^2 w}{\partial\theta^2} - A_{66}\frac{1}{x^2\sin^2\alpha}\frac{\partial u}{\partial\theta}\frac{\partial w}{\partial\theta} + A_{66}\frac{1}{x\sin^2\alpha}\frac{\partial^2 u}{\partial x\partial\theta}\frac{\partial w}{\partial\theta} \\
& +A_{66}\frac{2}{x\sin^2\alpha}\frac{\partial u}{\partial\theta}\frac{\partial^2 w}{\partial x\partial\theta} + A_{66}\frac{1}{\sin\alpha}\frac{\partial^2 v}{\partial x^2}\frac{\partial w}{\partial\theta} + A_{66}\frac{2}{\sin\alpha}\frac{\partial v}{\partial x}\frac{\partial^2 w}{\partial x\partial\theta} - A_{66}\frac{1}{x\sin\alpha}\frac{\partial v}{\partial x}\frac{\partial w}{\partial\theta} \\
& +A_{66}\frac{1}{x^2\sin\alpha}v\frac{\partial w}{\partial\theta} \\
& -A_{66}\frac{2}{x\sin\alpha}v\frac{\partial^2 w}{\partial x\partial\theta} - A_{66}\frac{2}{x^2\sin^2\alpha}\frac{\partial w}{\partial\theta}\frac{\partial^2 w}{\partial x\partial\theta} + A_{66}\frac{1}{x\sin^2\alpha}\frac{\partial^2 w}{\partial x^2}\left(\frac{\partial w}{\partial\theta}\right)^2 \\
& +A_{66}\frac{4}{x\sin^2\alpha}\frac{\partial w}{\partial x}\frac{\partial w}{\partial\theta}\frac{\partial^2 w}{\partial x\partial\theta} + B_{66}\frac{8}{x^2\sin^2\alpha}\frac{\partial^2 w}{\partial x\partial\theta}\frac{\partial w}{\partial\theta} - B_{66}\frac{4}{x^3\sin^2\alpha}\left(\frac{\partial w}{\partial\theta}\right)^2 + A_{66}\frac{1}{x\sin^2\alpha}\frac{\partial^2 u}{\partial\theta^2}\frac{\partial w}{\partial x} \\
& +A_{66}\frac{1}{\sin\alpha}\frac{\partial^2 v}{\partial x\partial\theta}\frac{\partial w}{\partial x} - A_{66}\frac{1}{x\sin\alpha}\frac{\partial v}{\partial\theta}\frac{\partial w}{\partial x} + A_{66}\frac{1}{x\sin^2\alpha}\left(\frac{\partial w}{\partial x}\right)^2\frac{\partial^2 w}{\partial\theta^2} - 2B_{66}\frac{1}{x\sin^2\alpha}\left(\frac{\partial^2 w}{\partial x\partial\theta}\right)^2 \\
& +2B_{66}\frac{1}{x^2\sin^2\alpha}\frac{\partial^2 w}{\partial\theta^2}\frac{\partial w}{\partial x} + A_{12}\frac{1}{x\sin^2\alpha}\frac{\partial^2 u}{\partial x\partial\theta}\frac{\partial w}{\partial\theta} + A_{12}\frac{1}{x\sin^2\alpha}\frac{\partial u}{\partial x}\frac{\partial^2 w}{\partial\theta^2} \\
& +\left[A_{22}+\frac{E_{1r}b_r}{d_2}\right]\frac{1}{x^2\sin^2\alpha}\frac{\partial u}{\partial\theta}\frac{\partial w}{\partial\theta} \\
& +\left[A_{22}+\frac{E_{1r}b_r}{d_2}\right]\frac{1}{x^2\sin^2\alpha}u\frac{\partial^2 w}{\partial\theta^2} + \left[A_{22}+\frac{E_{1r}b_r}{d_2}\right]\frac{1}{x^2\sin^3\alpha}\frac{\partial^2 v}{\partial\theta^2}\frac{\partial w}{\partial\theta} \\
& +\left[A_{22}+\frac{E_{1r}b_r}{d_2}\right]\frac{1}{x^2\sin^3\alpha}\frac{\partial v}{\partial\theta}\frac{\partial^2 w}{\partial\theta^2} + \frac{1}{2}A_{12}\frac{1}{x\sin^2\alpha}\left(\frac{\partial w}{\partial x}\right)^2\frac{\partial^2 w}{\partial\theta^2} + \left[A_{22}+\frac{E_{1r}b_r}{d_2}\right]\frac{1}{x^2\sin^2\alpha}w\frac{\partial^2 w}{\partial\theta^2}\cot\alpha \\
& +\left[A_{22}+\frac{E_{1r}b_r}{d_2}\right]\frac{3}{2x^3\sin^4\alpha}\left(\frac{\partial w}{\partial\theta}\right)^2\frac{\partial^2 w}{\partial\theta^2} \\
& -[B_{22}+C_2]\frac{1}{x^2\sin^2\alpha}\frac{\partial w}{\partial x}\frac{\partial^2 w}{\partial\theta^2}
\end{aligned}
$$

Appendix C

In Equations (22)–(24)

$$
\begin{aligned}
H_{11} &= -\left(J_{11}A_{11}+J_{12}\tfrac{E_{1s}b_s}{\lambda_o}\right)\tfrac{m^2\pi^2}{L^2}\sin\alpha - J_{13}A_{66}\tfrac{1}{\sin\alpha}\tfrac{n^2}{4} - J_{13}\left(A_{22}+\tfrac{E_{1r}b_r}{d_2}\right)\sin\alpha - J_{14}A_{11}\tfrac{m\pi}{L}\sin\alpha \\
H_{12} &= -J_{12}(A_{12}+A_{66})\tfrac{nm\pi}{2L} + J_{15}\left(A_{22}+A_{66}+\tfrac{E_{1r}b_r}{d_2}\right)\tfrac{n}{2} \\
H_{13} &= \left(J_{11}B_{11}+J_{12}C_1^o\right)\tfrac{m^3\pi^3}{L^3}\sin\alpha + J_{13}\tfrac{1}{\sin\alpha}(B_{12}+2B_{66})\tfrac{n^2m\pi}{4L} + J_{12}A_{12}\tfrac{m\pi}{L}\sin\alpha\cot\alpha \\
&+J_{13}(B_{22}+C_2)\tfrac{m\pi}{L}\sin\alpha - J_{15}\left(A_{22}+\tfrac{E_{1r}b_r}{d_2}\right)\sin\alpha\cot\alpha + J_{14}B_{11}\tfrac{m^2\pi^2}{L^2}\sin\alpha
\end{aligned}
$$

$$
L_{14} = -\left(J_{18}A_{11}+J_{19}\tfrac{E_{1s}b_s}{\lambda_o}\right)\tfrac{m^3\pi^3}{L^3}\sin\alpha + (J_{112}A_{12}+J_{112}A_{66}-J_{110}A_{66})\tfrac{n^2m\pi}{4L\sin\alpha} + J_{111}(A_{11}-A_{12})\tfrac{m^2\pi^2}{2L^2}\sin\alpha
$$

$$
\begin{aligned}
H_{21} &= -J_{22}(A_{12}+A_{66})\tfrac{nm\pi}{2L} + J_{26}\left(A_{22}+A_{66}+\tfrac{E_{1r}b_r}{d_2}\right)\tfrac{n}{2} \\
H_{22} &= -J_{23}\left(A_{22}+\tfrac{E_{1r}b_r}{d_2}\right)\tfrac{1}{\sin\alpha}\tfrac{n^2}{4} - J_{21}A_{66}\tfrac{m^2\pi^2}{L^2}\sin\alpha - J_{23}A_{66}\sin\alpha + J_{25}A_{66}\tfrac{m\pi}{L}\sin\alpha \\
H_{23} &= J_{23}\left(A_{22}+\tfrac{E_{1r}b_r}{d_2}\right)\tfrac{n}{2}\cot\alpha + J_{22}(B_{12}+2B_{66})\tfrac{nm^2\pi^2}{2L^2} + \tfrac{J_{24}}{\sin^2\alpha}(B_{22}+C_2)\tfrac{n^3}{8} - J_{26}(B_{22}+C_2)\tfrac{nm\pi}{2L}
\end{aligned}
$$

$$L_{24} = J_{27}(A_{66} + A_{12})\frac{nm^2\pi^2}{2L^2} - \frac{J_{29}}{\sin^2\alpha}\left(A_{22} + \frac{E_{1r}b_r}{d_2}\right)\frac{n^3}{8} - J_{28}A_{66}\frac{nm^2\pi^2}{2L^2} + J_{210}A_{66}\frac{nm\pi}{2L}$$

$$\begin{aligned} H_{31} = {} & (J_{32}B_{11} + J_{33}C_1^o)\tfrac{m^3\pi^3}{L^3}\sin\alpha + \tfrac{J_{34}}{\sin\alpha}(B_{12} + 2B_{66})\tfrac{n^2m\pi}{4L} + J_{34}(B_{22} + C_2)\tfrac{m\pi}{L}\sin\alpha + J_{33}A_{12}\tfrac{m\pi}{L}\sin\alpha\cot\alpha \\ & -2J_{38}B_{11}\tfrac{m^2\pi^2}{L^2}\sin\alpha - \tfrac{J_{310}}{\sin\alpha}(B_{22} + C_2)\tfrac{n^2}{4} + J_{310}(B_{22} + C_2)\sin\alpha - J_{39}\left(A_{22} + \tfrac{E_{1r}b_r}{d_2}\right)\sin\alpha\cot\alpha \end{aligned}$$

$$\begin{aligned} H_{32} = {} & J_{33}(B_{12} + 2B_{66})\tfrac{nm^2\pi^2}{2L^2} + \tfrac{J_{35}}{\sin^2\alpha}(B_{22} + C_2)\tfrac{n^3}{8} - J_{35}(B_{22} + C_2)\tfrac{n}{2} \\ & +J_{34}\left(A_{22} + \tfrac{E_{1r}b_r}{d_2}\right)\tfrac{n}{2}\cot\alpha + J_{39}(B_{22} + C_2)\tfrac{nm\pi}{2L} \end{aligned}$$

$$\begin{aligned} H_{33} = {} & -\left(J_{32}D_{11} + J_{33}\tfrac{E_{3s}b_s}{\lambda_o}\right)\tfrac{m^4\pi^4}{L^4}\sin\alpha - J_{36}\left(D_{22} + \tfrac{E_{3r}b_r}{d_2}\right)\tfrac{1}{\sin^3\alpha}\tfrac{n^4}{16} - \tfrac{2J_{34}}{\sin\alpha}(D_{12} + 2D_{66})\tfrac{n^2m^2\pi^2}{4L^2} \\ & -J_{34}\left(D_{22} + \tfrac{E_{3r}b_r}{d_2}\right)\tfrac{m^2\pi^2}{L^2}\sin\alpha - 2J_{33}B_{12}\tfrac{m^2\pi^2}{L^2}\sin\alpha\cot\alpha - \tfrac{2J_{35}}{\sin\alpha}\cot\alpha(B_{22} + C_2)\tfrac{n^2}{4} \\ & +\tfrac{2J_{36}}{\sin\alpha}\left(D_{12} + 2D_{66} + D_{22} + \tfrac{E_{3r}b_r}{d_2}\right)\tfrac{n^2}{4} + J_{35}(B_{22} + C_2)\sin\alpha\cot\alpha - J_{34}\left(A_{22} + \tfrac{E_{1r}b_r}{d_2}\right)\sin\alpha\cot^2\alpha \\ & -J_{32}K_1\sin\alpha - K_2J_{32}\tfrac{m^2\pi^2}{L^2}\sin\alpha - K_2\tfrac{J_{34}}{\sin\alpha}\tfrac{n^2}{4} - \tfrac{2J_{310}}{\sin\alpha}(D_{12} + 2D_{66})\tfrac{n^2m\pi}{4L} \\ & +2J_{38}D_{11}\tfrac{m^3\pi^3}{L^3}\sin\alpha - J_{310}\left(D_{22} + \tfrac{E_{3r}b_r}{d_2}\right)\tfrac{m\pi}{L}\sin\alpha + J_{38}K_2\tfrac{m\pi}{L}\sin\alpha \end{aligned}$$

$$H_{34} = \frac{J_{33}m^2\pi}{2L^2\cos\alpha}$$

$$\begin{aligned} L_{34} = {} & J_{314}\tfrac{1}{\sin\alpha}B_{12}\tfrac{n^2}{4} + J_{314}(B_{22} + C_2)\tfrac{1}{\sin\alpha}\tfrac{n^2}{4} + J_{313}\left(A_{22} + \tfrac{E_{1r}b_r}{d_2}\right)\tfrac{1}{2\sin\alpha}\tfrac{n^2}{4}\cot\alpha - J_{314}B_{66}\tfrac{1}{\sin\alpha}n^2 \\ & +\tfrac{1}{2}J_{316}A_{12}\tfrac{m^2\pi^2}{L^2}\sin\alpha\cot\alpha - J_{319}B_{11}\tfrac{m^3\pi^3}{L^3}\sin\alpha + 3J_{319}B_{12}\tfrac{m^3\pi^3}{L^3}\sin\alpha - J_{320}B_{12}\tfrac{1}{\sin\alpha}\tfrac{n^2m\pi}{4L} \\ & -2J_{320}B_{66}\tfrac{1}{\sin\alpha}\tfrac{n^2m\pi}{4L} + J_{320}(B_{22} + C_2)\tfrac{1}{\sin\alpha}\tfrac{n^2m\pi}{4L} + 2J_{322}B_{66}\tfrac{1}{\sin\alpha}\tfrac{n^2m^2\pi^2}{4L^2} - J_{322}A_{12}\tfrac{m^2\pi^2}{L^2}\sin\alpha\cot\alpha \\ & -J_{324}\left(A_{22} + \tfrac{E_{1r}b_r}{d_2}\right)\tfrac{\cot\alpha}{\sin\alpha}\tfrac{n^2}{4} - 2J_{323}B_{12}\tfrac{1}{\sin\alpha}\tfrac{n^2m^2\pi^2}{4L^2} - \tfrac{J_{333}}{\sin\alpha}B_{12}\tfrac{n^2m\pi}{2L} - (B_{22} + C_2)\tfrac{J_{333}}{\sin\alpha}\tfrac{n^2m\pi}{2L} \\ & -J_{333}A_{66}\tfrac{1}{\sin\alpha}\tfrac{n^2m\pi}{2L} + 2B_{66}\tfrac{J_{333}}{\sin\alpha}\tfrac{n^2m\pi}{L} + \tfrac{J_{336}}{\sin\alpha}B_{12}\tfrac{n^2m^2\pi^2}{2L^2} - \tfrac{J_{336}}{\sin\alpha}B_{66}\tfrac{n^2m^2\pi^2}{2L^2} \end{aligned}$$

$$\begin{aligned} L_{35} = {} & -A_{66}J_{311}\tfrac{nm^2\pi^2}{2L^2} + J_{313}A_{66}\tfrac{n}{2} - J_{313}\left(A_{22} + \tfrac{E_{1r}b_r}{d_2}\right)\tfrac{1}{\sin^2\alpha}\tfrac{n^3}{8} - J_{316}A_{12}\tfrac{nm^2\pi^2}{2L^2} - J_{316}A_{66}\tfrac{nm^2\pi^2}{2L^2} \\ & +J_{319}A_{66}\tfrac{nm\pi}{2L} + J_{322}A_{12}\tfrac{nm^2\pi^2}{2L^2} + J_{324}\left(A_{22} + \tfrac{E_{1r}b_r}{d_2}\right)\tfrac{1}{\sin^2\alpha}\tfrac{n^3}{8} - \tfrac{3}{2}J_{332}A_{66}\tfrac{nm\pi}{L} + J_{335}A_{66}\tfrac{nm^2\pi^2}{L^2} \end{aligned}$$

$$\begin{aligned} L_{36} = {} & -J_{312}A_{66}\tfrac{1}{\sin\alpha}\tfrac{n^2m\pi}{4L} - J_{312}A_{12}\tfrac{1}{\sin\alpha}\tfrac{n^2m\pi}{4L} - \left(J_{315}A_{11} + J_{316}\tfrac{E_{1s}b_s}{\lambda_o}\right)\tfrac{m^3\pi^3}{L^3}\sin\alpha \\ & -J_{317}A_{66}\tfrac{1}{\sin\alpha}\tfrac{n^2m\pi}{4L} - J_{319}A_{11}\tfrac{m^2\pi^2}{L^2}\sin\alpha - 2J_{318}A_{12}\tfrac{m^2\pi^2}{L^2}\sin\alpha - J_{320}\left(A_{22} + \tfrac{E_{1r}b_r}{d_2}\right)\tfrac{1}{\sin\alpha}\tfrac{n^2}{4} \\ & +\left(J_{321}A_{11} + J_{322}\tfrac{E_{1s}b_s}{\lambda_o}\right)\tfrac{m^3\pi^3}{L^3}\sin\alpha + J_{323}A_{12}\tfrac{1}{\sin\alpha}\tfrac{n^2m\pi}{4L} - A_{66}\tfrac{J_{333}}{\sin\alpha}\tfrac{n^2}{4} + \tfrac{J_{333}}{\sin\alpha}\left(A_{22} + \tfrac{E_{1r}b_r}{d_2}\right)\tfrac{n^2}{4} \\ & +J_{336}A_{66}\tfrac{1}{\sin\alpha}\tfrac{n^2m\pi}{2L} \end{aligned}$$

$$\begin{aligned} L_{37} = {} & -\tfrac{3}{2}\left(J_{325}A_{11} + J_{326}\tfrac{E_{1s}b_s}{\lambda_0}\right)\tfrac{m^4\pi^4}{L^4}\sin\alpha - A_{66}\tfrac{J_{326}}{\sin\alpha}\tfrac{n^2m^2\pi^2}{4L^2} - A_{12}\tfrac{J_{327}}{2\sin\alpha}\tfrac{n^2m^2\pi^2}{4L^2} \\ & -A_{12}\tfrac{J_{328}}{2\sin\alpha}\tfrac{n^2m\pi}{4L} + \tfrac{1}{2}J_{329}A_{11}\tfrac{m^3\pi^3}{L^3}\sin\alpha - \tfrac{J_{330}}{\sin\alpha}A_{66}\tfrac{n^2m^2\pi^2}{4L} - \tfrac{J_{330}}{2\sin\alpha}A_{12}\tfrac{n^2m^2\pi^2}{4L^2} \\ & -J_{331}\left(A_{22} + \tfrac{E_{1r}b_r}{d_2}\right)\tfrac{3}{2\sin^3\alpha}\tfrac{n^4}{16} + A_{12}\tfrac{J_{334}}{\sin\alpha}\tfrac{n^2m^3\pi^3}{2L^3} + J_{334}A_{66}\tfrac{1}{\sin\alpha}\tfrac{n^2m^3\pi^3}{L^3} \end{aligned}$$

$$J_{11} = \pi\left[\tfrac{(L+x_o)^4}{8} - \tfrac{x_o^4}{8} - \tfrac{3L^2x_o^2}{8\pi^2m^2} + \tfrac{3L^2(L+x_o)^2}{8\pi^2m^2}\right]$$

$$J_{12} = \pi\left[\frac{(L+x_o)^3}{6} - \frac{x_o^3}{6} + \frac{L^3}{4\pi^2m^2}\right],$$

$$J_{13} = \pi\left(\frac{L^2}{4} + \frac{Lx_o}{2}\right)$$

$$J_{14} = -\frac{L^2(L+2x_o)}{4m},$$

$$J_{15} = -\frac{L^2}{4m}$$

$$J_{18} = \frac{4-4(-1)^n}{3n}\left[-\frac{L(L+x_o)^3(-1)^m}{3\pi m} + \frac{Lx_o^3}{3\pi m} - \frac{14L^3x_o}{9\pi^3m^3} + \frac{14L^3(L+x_o)(-1)^m}{9\pi^3m^3}\right]$$

$$J_{19} = \frac{4-4(-1)^n}{3n}\left[-\frac{14L^3}{27\pi^3 m^3} + \frac{Lx_o^2}{3\pi m} - \frac{L(L+x_o)^2(-1)^m}{3\pi m} + \frac{14L^3(-1)^m}{27\pi^3 m^3}\right]$$

$$J_{110} = \frac{4-4(-1)^n}{3n}\left[-\frac{L(L+x_o)(-1)^m}{3\pi m} + \frac{Lx_o}{3\pi m}\right]$$

$$J_{111} = -\frac{56(-1)^n - 56}{27\pi^2 m^2 n}\left[L^3(-1)^m - L^2 x_o + L^2 x_o(-1)^m\right]$$

$$J_{112} = \frac{2}{9\pi mn}\left[L^2(-1)^{m+n} - L^2(-1)^m + Lx_o(-1)^{m+n} - x_o L(-1)^m - x_o L(-1)^n + x_o L\right]$$

$$J_{21} = \pi\left[\frac{(L+x_o)^4}{8} - \frac{x_o^4}{8} + \frac{3L^2x_o^2}{8\pi^2 m^2} - \frac{3L^2(L+x_o)^2}{8\pi^2 m^2}\right],$$

$$J_{22} = \pi\left[\ \frac{(L+x_o)^3}{6} - \frac{x_o^3}{6} - \frac{L^3}{4\pi^2 m^2}\ \right]$$

$$J_{23} = \pi\left(\frac{L^2}{4} + \frac{Lx_o}{2}\right),$$

$$J_{24} = \frac{L\pi}{2},$$

$$J_{25} = -\frac{L^2(L+2x_o)}{4m},$$

$$J_{26} = -\frac{L^2}{4m}$$

$$J_{27} = \frac{1}{81\pi^3 m^3 n}\left[\begin{array}{l} 18\pi^2 L^3 m^2 \cos^3(\pi m)\cos^3(\pi n)\ -18\pi^2 L^3 m^2\cos^3(\pi m) - 4L^3\cos^3(\pi m)\cos^3(\pi n) \\ +4L^3\cos^3(\pi m) - 72L^3\cos(\pi m)\cos(\pi n) + 72L^3\cos(\pi m) \\ +48L^3\cos^3(\pi n)\sin^2\left(\frac{\pi m}{2}\right) + 4L^3\cos^3(\pi n) - 144L^3\cos(\pi n)\sin^2\left(\frac{\pi m}{2}\right) \\ +72L^3\cos(\pi n) + 96L^3\sin^2\left(\frac{\pi m}{2}\right) - 76L^3 + 36\pi^2L^2m^2x_o\cos^3(\pi m)\cos^3(\pi n) \\ -36\pi^2L^2m^2x_o\cos^3(\pi m) + 18\pi^2Lm^2x_o^2\cos^3(\pi m)\cos^3(\pi n) \\ -18\pi^2Lm^2x_o^2\cos^3(\pi m) - 18\pi^2Lm^2x_o^2\cos^3(\pi n) + 18\pi^2Lm^2x_o^2 \end{array}\right]$$

$$J_{28} = \frac{2-2(-1)^n}{3n}\left[-\frac{40L^3}{27\pi^3 m^3} + \frac{2Lx_o{}^2}{3\pi m} - \frac{2L(L+x_o)^2(-1)^m}{3\pi m} + \frac{40L^3(-1)^m}{27\pi^3 m^3}\right]$$

$$J_{29} = \frac{4L}{9\pi mn}\left[-(-1)^n + 1 + (-1)^{m+n} - (-1)^m\right]$$

$$J_{210} = \frac{4L^2}{27\pi^2 m^2 n}\left[(-1)^m - 1 + (-1)^n - (-1)^{m+n}\right]$$

$$J_{31} = \pi\left[\frac{(L+x_o)^6 - x_o^6}{12} + \frac{5L^2\left(x_o^4 - (L+x_o)^4\right)}{8\pi^2 m^2} + \frac{15L^4\left((L+x_o)^2 - x_o^2\right)}{8\pi^4 m^4}\right]$$

$$J_{32} = \pi\left[\frac{(L+x_o)^5 - x_o^5}{10} + \frac{3L^5}{4\pi^4 m^4} + \frac{L^2\left(x_o^3 - (L+x_o)^3\right)}{2\pi^2 m^2}\right]$$

$$J_{33} = \pi\left[\frac{(L+x_o)^4 - x_o^4}{8} - \frac{3L^3(L+2x_o)}{8\pi^2 m^2}\right],$$

$$J_{34} = J_{22},$$

$$J_{35} = J_{23},$$

$$J_{36} = J_{24}$$

$$J_{37} = \pi\left[\frac{L\left(x_o^4-(L+x_o)^4\right)}{4\pi m}+\frac{3L^4(L+2x_o)}{4\pi^3m^3}\right],$$

$$J_{38} = \pi\left[\frac{L\left(x_o^3-(L+x_o)^3\right)}{4\pi m}+\frac{3L^4}{8\pi^3m^3}\right]$$

$$J_{39} = \frac{-L^2(L+2x_o)}{4m},$$

$$J_{310} = -\frac{L^2}{4m}$$

$$J_{311} = \frac{2-2(-1)^n}{3n}\left[-\frac{2L(L+x_o)^3(-1)^m}{3\pi m}+\frac{40L^3(L+x_o)(-1)^m}{9\pi^3m^3}+\frac{2Lx_o^3}{3m\pi}-\frac{40L^3x_o}{9\pi^3m^3}\right],$$

$$J_{312} = J_{28}$$

$$J_{313} = \frac{2-2(-1)^n}{3n}\left[\frac{2Lx_o}{3\pi m}-\frac{2L(L+x_o)(-1)^m}{3\pi m}\right],$$

$$J_{314} = \frac{4L}{9\pi mn}\left[-(-1)^n+1+(-1)^{m+n}-(-1)^m\right]$$

$$J_{315} = \frac{4(-1)^n-4}{3n}\left\{\frac{L(L+x_o)^4(-1)^m-Lx_o^4}{3\pi m}+\frac{488L^5\left[(-1)^m-1\right]}{81\pi^5m^5}+\frac{28L^3x_o^2-28L^3(L+x_o)^2(-1)^m}{9\pi^3m^3}\right\}$$

$$J_{316} = \frac{4-4(-1)^n}{3n}\left[-\frac{L(L+x_o)^3(-1)^m}{3\pi m}+\frac{14L^3(L+x_o)(-1)^m}{9\pi^3m^3}+\frac{Lx_o^3}{3\pi m}-\frac{14L^3x_o}{9\pi^3m^3}\right]$$

$$J_{317} = \frac{4-4(-1)^n}{3n}\left[-\frac{L(L+x_o)^2(-1)^m}{3\pi m}+\frac{14L^3(-1)^m}{27\pi^3m^3}+\frac{Lx_o^2}{3\pi m}-\frac{14L^3}{27\pi^3m^3}\right]$$

$$J_{318} = \frac{1-(-1)^n}{3n}\left[\frac{160L^4-160L^4(-1)^m}{27\pi^4m^4}+\frac{8L^2(L+x_o)^2(-1)^m}{3\pi^2m^2}-\frac{8L^2x_o^2}{3\pi^2m^2}\right]$$

$$J_{319} = \frac{1-(-1)^n}{3n}\left(\frac{16L^2(L+x_o)(-1)^m}{9\pi^2m^2}-\frac{16L^2x_o}{9\pi^2m^2}\right)$$

$$J_{320} = \frac{8L^2}{27\pi^2m^2n}\left[-1+(-1)^n+(-1)^m-(-1)^{m+n}\right]$$

$$J_{321} = \frac{4-4(-1)^n}{3n}\left\{\frac{2L\left[x_o^4-(L+x_o)^4(-1)^m\right]}{3\pi m}+\frac{1456L^5\left[1-(-1)^m\right]}{81\pi^5m^5}+\frac{80L^3\left[(L+x_o)^2(-1)^m-x_o^2\right]}{9\pi^3m^3}\right\}$$

$$J_{322} = \frac{4-4(-1)^n}{3n}\left[-\frac{2L(L+x_o)^3(-1)^m}{3\pi m}+\frac{40L^3(L+x_o)(-1)^m}{9\pi^3m^3}+\frac{2Lx_o^3}{3\pi m}-\frac{40L^3x_o}{9\pi^3m^3}\right]$$

$$J_{323} = \frac{4-4(-1)^n}{3n}\left[-\frac{2L(L+x_o)^2(-1)^m}{3\pi m}+\frac{40L^3(-1)^m}{27\pi^3m^3}+\frac{2Lx_o^2}{3\pi m}-\frac{40L^3}{27\pi^3m^3}\right]$$

$$J_{324} = \frac{4-4(-1)^n}{3n}\left[\frac{2Lx_o}{3\pi m}-\frac{2L(L+x_o)(-1)^m}{3\pi m}\right]$$

$$J_{325} = -\frac{3\pi}{4}\left[\frac{x_o^5-(L+x_o)^5}{40}-\frac{3L^5}{256\pi^4m^4}+\frac{L^2\left((L+x_o)^3-x_o^3\right)}{32\pi^2m^2}\right]$$

$$J_{326} = -\frac{3\pi}{4}\left[\frac{(L+x_o)^4}{32}-\frac{x_o^4}{32}-\frac{3L^3(L+2x_o)}{128\pi^2m^2}\right],$$

$$J_{327} = -\frac{3\pi}{4}\left[\frac{(L+x_o)^3}{24} - \frac{x_o^3}{24} - \frac{L^3}{64\pi^2 m^2}\right]$$

$$J_{328} = \frac{\pi}{4}\left[\frac{3Lx_o}{32\pi m} - \frac{3L(L+x_o)}{32\pi m}\right],$$

$$J_{329} = -\frac{3\pi}{4}\left[-\frac{51L^4}{256\pi^3 m^3} + \frac{5L\left((L+x_o)^3 - x_o^3\right)}{32\pi m} \right]$$

$$J_{330} = \frac{\pi}{4}\left(\frac{(L+x_o)^3}{8} - \frac{x_o^3}{8} - \frac{15L^3}{64\pi^2 m^2}\right),$$

$$J_{331} = \frac{3L\pi}{32\pi m}$$

$$J_{332} = \frac{16 - 16(-1)^n}{51\pi^2 m^2 n}\left[L^3(-1)^m - L^2 x_o + L^2 x_o(-1)^m\right]$$

$$J_{333} = \frac{4}{27\pi^2 m^2 n}\left[L^2(-1)^m - L^2 + L^2(-1)^n - L^2(-1)^{m+n}\right],$$

$$J_{334} = \frac{\pi}{4}\left[\frac{(L+x_o)^3}{24} - \frac{x_o^3}{24} + \frac{L^2 x_o}{64\pi^2 m^2}\right]$$

$$J_{335} = \frac{-1}{81\pi^3 m^3 n}\left\{\begin{array}{l} 84L^3 x_o\left[1 - (-1)^n - (-1)^m + (-1)^{m+n}\right] + 84L^4\left[(-1)^{m+n} - (-1)^m\right] \\ +18\pi^3 m^3 L^4(-1)^m + 18\pi^2 m^2 L x_o^3\left[-1 - (-1)^{m+n} + (-1)^m + (-1)^n\right] \\ -18\pi^2 m^2 L^4(-1)^{m+n} + 54\pi^2 m^2\left[L^2 x_o^2(-1)^m - L^2 x_o^2(-1)^{m+n}\right] \\ +54\pi^2 m^2\left[L^3 x_o(-1)^m - L^3 x_o(-1)^{m+n}\right] \end{array}\right\}$$

$$J_{336} = \frac{1}{81\pi^3 m^3 n}\left\{\begin{array}{l} 18\pi^2 m^2 L^3\left[(-1)^{m+n} - (-1)^m\right] + 76L^3\left[-1 - (-1)^{m+n} + (-1)^m + (-1)^n\right] \\ +18\pi^2 m^2 L x_o^2\left[(-1)^{m+n} - (-1)^m - (-1)^n + 1\right] \\ -96L^3(-1)^n \sin^2\left(\frac{\pi m}{2}\right) + 96L^3 \sin^2\left(\frac{\pi m}{2}\right) + 36\pi^2 m^2 L^2 x_o\left[(-1)^{m+n} - (-1)^m\right] \end{array}\right\}$$

References

1. Huang, H.; Han, Q. Nonlinear elastic buckling and postbuckling of axially compressed functionally graded cylindrical shells. *Int. J. Mech. Sci.* **2009**, *51*, 500–507. [CrossRef]
2. Naj, R.; Boroujerdy, M.S.; Eslami, M. Thermal and mechanical instability of functionally graded truncated conical shells. *Thin-Walled Struct.* **2008**, *46*, 65–78.
3. Sofiyev, A. Influences of shear stresses on the dynamic instability of exponentially graded sandwich cylindrical shells. *Compos. Part B Eng.* **2015**, *77*, 349–362.
4. Sofiyev, A.; Kuruoglu, N. Non-linear buckling of an FGM truncated conical shell surrounded by an elastic medium. *Int. J. Press. Vessels Pip.* **2013**, *107*, 38–49.
5. Sofiyev, A. Buckling analysis of freely-supported functionally graded truncated conical shells under external pressures. *Compos. Struct.* **2015**, *132*, 746–758. [CrossRef]
6. Sofiyev, A.; Kuruoglu, N. On the solution of the buckling problem of functionally graded truncated conical shells with mixed boundary conditions. *Compos. Struct.* **2015**, *123*, 282–291.
7. Sofiyev, A.H.; Avcar, M. The stability of cylindrical shells containing an FGM layer subjected to axial load on the Pasternak foundation. *Engineering* **2010**, *2*, 228–236. [CrossRef]
8. Sofiyev, A. Influence of the initial imperfection on the non-linear buckling response of FGM truncated conical shells. *Int. J. Mech. Sci.* **2011**, *53*, 753–761. [CrossRef]
9. Sofiyev, A. The buckling of FGM truncated conical shells subjected to combined axial tension and hydrostatic pressure. *Compos. Struct.* **2010**, *92*, 488–498.
10. Sofiyev, A. Non-linear buckling behavior of FGM truncated conical shells subjected to axial load. *Int. J. Non-Linear Mech.* **2011**, *46*, 711–719.

11. Duc, N.D.; Thang, P.T. Nonlinear response of imperfect eccentrically stiffened ceramic–metal–ceramic FGM thin circular cylindrical shells surrounded on elastic foundations and subjected to axial compression. *Compos. Struct.* **2014**, *110*, 200–206. [CrossRef]
12. Duc, N.D.; Thang, P.T. Nonlinear buckling of imperfect eccentrically stiffened metal–ceramic–metal S-FGM thin circular cylindrical shells with temperature-dependent properties in thermal environments. *Int. J. Mech. Sci.* **2014**, *81*, 17–25. [CrossRef]
13. Duc, N.D.; Van Tung, H. Nonlinear analysis of stability for functionally graded cylindrical panels under axial compression. *Comput. Mater. Sci.* **2010**, *49*, S313–S316. [CrossRef]
14. Duc, N.D.; Thang, P.T.; Dao, N.T.; Tac, H.V. Nonlinear buckling of higher deformable S-FGM thick circular cylindrical shells with metal–ceramic–metal layers surrounded on elastic foundations in thermal environment. *Compos. Struct.* **2015**, *121*, 134–141. [CrossRef]
15. Duc, N.D.; Cong, P.H.; Anh, V.M.; Quang, V.D.; Tran, P.; Tuan, N.D.; Thinh, N.H. Mechanical and thermal stability of eccentrically stiffened functionally graded conical shell panels resting on elastic foundations and in thermal environment. *Compos. Struct.* **2015**, *132*, 597–609. [CrossRef]
16. Duc, N.D.; Quan, T.Q. Nonlinear response of imperfect eccentrically stiffened FGM cylindrical panels on elastic foundation subjected to mechanical loads. *Eur. J. Mech. A Solids* **2014**, *46*, 60–71. [CrossRef]
17. Duc, N.D.; Tuan, N.D.; Quan, T.Q.; Quyen, N.V.; Anh, T.V. Nonlinear mechanical, thermal and thermo-mechanical postbuckling of imperfect eccentrically stiffened thin FGM cylindrical panels on elastic foundations. *Thin-Walled Struct.* **2015**, *96*, 155–168. [CrossRef]
18. Phuong, N.T.; Bich, D.H. Buckling analysis of eccentrically stiffened functionally graded circular cylindrical thin shells under mechanical load. *VNU J. Sci. Math. Phys.* **2013**, *29*, 55–72.
19. Ninh, D.G.; Bich, D.H. Nonlinear torsional buckling and post-buckling of eccentrically stiffened ceramic functionally graded material metal layer cylindrical shell surrounded by elastic foundation subjected to thermo-mechanical load. *J. Sandw. Struct. Mater.* **2016**, *18*, 712–738. [CrossRef]
20. Bich, D.H.; van Dung, D.; Nam, V.H. Nonlinear dynamical analysis of eccentrically stiffened functionally graded cylindrical panels. *Compos. Struct.* **2012**, *94*, 2465–2473. [CrossRef]
21. Dung, D.V.; Hoa, L.K.; Thuyet, B.T.; Nga, N.T. Buckling analysis of functionally graded material (FGM) sandwich truncated conical shells reinforced by FGM stiffeners filled inside by elastic foundations. *Appl. Math. Mech.* **2016**, *37*, 879–902. [CrossRef]
22. Van Dung, D.; Nga, N.T. Instability of eccentrically stiffened functionally graded truncated conical shells under mechanical loads. *Compos. Struct.* **2013**, *106*, 104–113. [CrossRef]
23. Van Dung, D.; Chan, D.Q. Analytical investigation on mechanical buckling of FGM truncated conical shells reinforced by orthogonal stiffeners based on FSDT. *Compos. Struct.* **2017**, *159*, 827–841. [CrossRef]
24. Van Dung, D.; Hoai, B.T.T. Postbuckling nonlinear analysis of FGM truncated conical shells reinforced by orthogonal stiffeners resting on elastic foundations. *Acta Mech.* **2017**, *228*, 1457–1479. [CrossRef]
25. Magnucki, K.; Stasiewicz, P. Elastic buckling of a porous beam. *J. Theor. Appl. Mech.* **2004**, *42*, 859–868.
26. Magnucka-Blandzi, E. Mathematical Modeling of a Rectangular Sandwich Plate with a Non-Homogeneous Core. *AIP Conf. Proc.* **2007**, *936*. [CrossRef]
27. Magnucka-Blandzi, E. Mathematical modelling of a rectangular sandwich plate with a metal foam core. *J. Theor. Appl. Mech.* **2011**, *49*, 439–455.
28. Magnucka-Blandzi, E. Axi-symmetrical deflection and buckling of circular porous-cellular plate. *Thin-Walled Struct.* **2008**, *46*, 333–337. [CrossRef]
29. Chen, D.; Yang, J.; Kitipornchai, S. Elastic buckling and static bending of shear deformable functionally graded porous beam. *Compos. Struct.* **2015**, *133*, 54–61. [CrossRef]
30. Kitipornchai, S.; Chen, D.; Yang, J. Free vibration and elastic buckling of functionally graded porous beams reinforced by graphene platelets. *Mater. Des.* **2017**, *116*, 656–665. [CrossRef]
31. Jabbari, M.; Mojahedin, A.; Khorshidvand, A.R.; Eslami, M.R. Buckling analysis of a functionally graded thin circular plate made of saturated porous materials. *J. Eng. Mech.* **2013**, *140*, 287–295. [CrossRef]
32. Mojahedin, A.; Jabbari, M.; Khorshidvand, A.R.; Eslami, M.R. Buckling analysis of functionally graded circular plates made of saturated porous materials based on higher order shear deformation theory. *Thin-Walled Struct.* **2016**, *99*, 83–90. [CrossRef]

33. Barretta, R.; Faghidian, S.A.; Luciano, R.; Medaglia, C.M.; Penna, R. Free vibrations of FG elastic Timoshenko nano-beams by strain gradient and stress-driven nonlocal models. *Compos. Part B Eng.* **2018**, *154*, 20–32. [CrossRef]
34. Mahmoudpour, E.; Hosseini-Hashemi, S.; Faghidian, S. Nonlinear vibration analysis of FG nano-beams resting on elastic foundation in thermal environment using stress-driven nonlocal integral model. *Appl. Math. Model.* **2018**, *57*, 302–315. [CrossRef]
35. Tang, H.; Li, L.; Hu, Y. Coupling effect of thickness and shear deformation on size-dependent bending of micro/nano-scale porous beams. *Appl. Math. Model.* **2018**, *66*, 527–547. [CrossRef]
36. Barretta, R.; Luciano, R.; de Sciarra, F.M. Stress-driven nonlocal integral model for Timoshenko elastic nano-beams. *Eur. J. Mech. A Solids* **2018**, *72*, 275–286. [CrossRef]
37. Mathew, T.V.; Natarajan, S.; Martínez-Pañeda, E. Size effects in elastic-plastic functionally graded materials. *Compos. Struct.* **2018**, *204*, 43–51. [CrossRef]
38. Martínez-Pañeda, E.; Niordson, C.F. On fracture in finite strain gradient plasticity. *Int. J. Plast.* **2016**, *80*, 154–167. [CrossRef]
39. Martínez-Pañeda, E.; Niordson, C.F.; Bardella, L. A finite element framework for distortion gradient plasticity with applications to bending of thin foils. *Int. J. Solids Struct.* **2016**, *96*, 288–299. [CrossRef]
40. Sofiyev, A. The buckling of FGM truncated conical shells subjected to axial compressive load and resting on Winkler–Pasternak foundations. *Int. J. Press. Vessels Pip.* **2010**, *87*, 753–761. [CrossRef]
41. Brush, D.O.; Almroth, B.O. *Buckling of Bars, Plates, and Shells*; McGraw-Hill: New York, NY, USA, 1975; Volume 6.
42. Baruch, M.; Harari, O.; Singer, J. Low buckling loads of axially compressed conical shells. *J. Appl. Mech.* **1970**, *37*, 384–392. [CrossRef]
43. Deniz, A. Non-linear stability analysis of truncated conical shell with functionally graded composite coatings in the finite deflection. *Compos. Part B Eng.* **2013**, *51*, 318–326. [CrossRef]

materials

MDPI

Article

New Shape Function for the Bending Analysis of Functionally Graded Plate

Dragan Čukanović [1], Aleksandar Radaković [2,*], Gordana Bogdanović [3], Milivoje Milanović [2], Halit Redžović [2] and Danilo Dragović [2]

1. Faculty of Technical Sciences, University of Priština, 38220 Kosovska Mitrovica, Serbia; dragan.cukanovic@pr.ac.rs
2. Department of Technical Sciences, State University of Novi Pazar, 36300 Novi Pazar, Serbia; milanovicnp@gmail.com (M.M.); halit.redzovic@gmail.com (H.R.); danilo.dragovic@yahoo.com (D.D.)
3. Faculty of Engineering, University of Kragujevac, 34000 Kragujevac, Serbia; gocab@kg.ac.rs

* Correspondence: aradakovic@np.ac.rs; Tel.: +381-606157757

Received: 2 November 2018; Accepted: 22 November 2018; Published: 26 November 2018

Abstract: The bending analysis of thick and moderately thick functionally graded square and rectangular plates as well as plates on Winkler–Pasternak elastic foundation subjected to sinusoidal transverse load is presented in this paper. The plates are assumed to have isotropic, two-constituent material distribution through the thickness, and the modulus of elasticity of the plate is assumed to vary according to a power-law distribution in terms of the volume fractions of the constituents. This paper presents the methodology of the application of the high order shear deformation theory based on the shape functions. A new shape function has been developed and the obtained results are compared to the results obtained with 13 different shape functions presented in the literature. Also, the validity and accuracy of the developed theory was verified by comparing those results with the results obtained using the third order shear deformation theory and 3D theories. In order to determine the procedure for the analysis and the prediction of behavior of functionally graded plates, the new program code in the software package MATLAB has been developed based on the theories studied in this paper. The effects of transversal shear deformation, side-to-thickness ratio, and volume fraction distributions are studied and appropriate conclusions are given.

Keywords: functionally graded plate; power-law distribution; high order shear deformation theory; elastic foundation

---

1. Introduction

Failure and delamination at the border between two layers are the biggest and the most frequently studied problem of the conventional composite laminates. Delamination of layers due to high local inter-laminar stresses causes a reduction of stiffness and a loss of structural integrity of a construction. In order to eliminate these problems, improved materials such as functionally graded materials (FGM), which are getting more and more popular, are used for innovative engineering constructions.

FGM is a composite material consisting of two or more constituents with the continuous change of properties in a certain direction. In other words, these materials can also be defined as materials which possess a gradient change of properties due to material heterogeneity. A gradient property can go in one or more directions and it can also be continuous or discontinuous from one surface to another depending on the production technique [1–3]. One of the most common uses of FGM materials is found in thermal barriers, one surface of which is in contact with high temperatures and is made of ceramic which can provide adequate thermal stability, low thermal conductivity, and fine antioxidant properties. The low-temperature side of the barrier is made of metal, which is superior in terms of mechanical strength, toughness, and high thermal conductivity. Functionally graded materials, which contain metal and ceramic constituents, improve thermo-mechanical properties between layers, because of which delamination of layers should be avoided due to continuous change between properties of the constituents. By varying the percentage of volume fraction content of the two or more materials, FGM can be formed so that it achieves a desired gradient property in specific directions. Figure 1 shows schematic of continuously graded microstructure with metal-ceramic constituents [4].

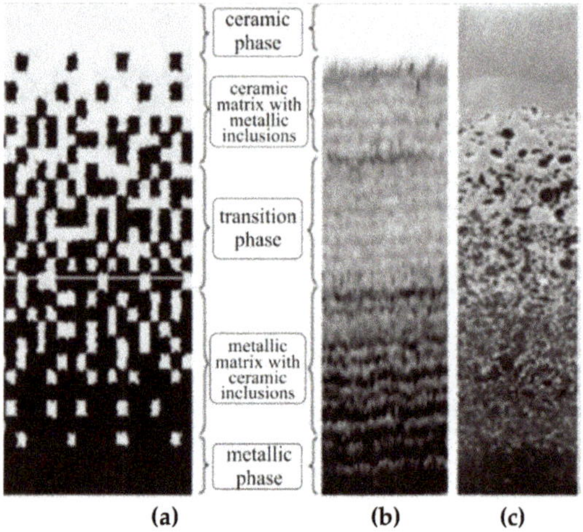

Figure 1. Schematic of continuously graded microstructure with metal-ceramic constituents: (**a**) smoothly graded microstructure; (**b**) enlarged view; (**c**) ceramic-metal functionally graded materials (FGM).

Depending on the nature of gradient, functionally graded materials may be grouped into fraction gradient type, shape gradient type, orientation gradient type and size gradient type (Figure 2) [5].

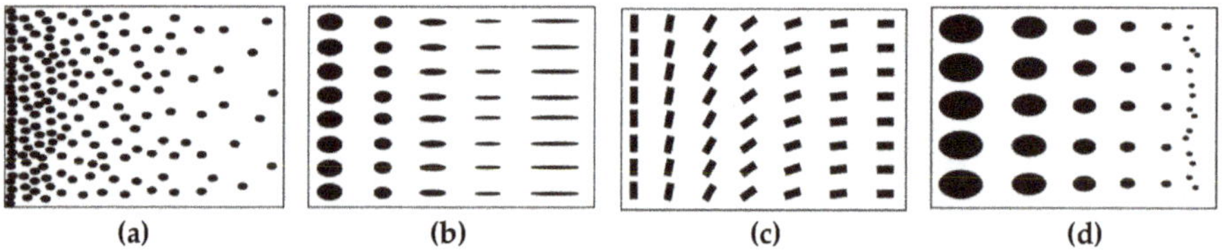

Figure 2. Different types of functionally graded materials based on nature of gradients: (**a**) fraction gradient type; (**b**) shape gradient type; (**c**) orientation gradient type; (**d**) size gradient type.

With the expansion of the FGM material application area, it was necessary to improve fabrication methods for mentioned materials. Various fabrication methods have been developed for the preparation of bulk FGMs and graded thin films. The processing methods are commonly classified into four groups like powder technology methods (dry powder processing, slip vesting, tape casting, infiltration process or electrochemical gradation, powder injection molding and self-propagating high temperature synthesis, etc.), deposition methods (chemical vapor deposition, physical vapor deposition, electrophoretic deposition, slurry deposition, pulsed laser deposition, plasma spraying, etc.), in-situ processing methods (laser cladding, spray forming, sedimentation and solidification, centrifugal casting, etc.), and rapid prototyping processes (multiphase jet solidification, 3D-printing, laser printing, laser sintering, etc.) [6]. The basic difference between the mentioned production methods can be made according to whether the obtained materials have a stepwise or continuous structure. The main disadvantage of the methods based on powder metallurgy is that it is very difficult to obtain FGM with a continuous change in properties. Continuous graded structures are produced by methods based on casting. Taking this fact into consideration, it was necessary to develop functions which would, with a smaller degree of approximation and in the best way possible, describe a gradient change of properties in a desired direction [7,8].

The majority of already existing software for the analysis of the composite materials are based on the classical plate theory (CPT) [9] and first-order shear deformation theories (FSDT), which were developed by Mindlin [10] and, in a similar way, by Reissner [11,12]. Although classical theory does not consider the effect of transverse shear stresses, it can provide acceptable predictions of the behavior and the results for thin FGM plates where the effects of shear and normal strains across the thickness of the plate are negligible.

Static problems of buckling and bending of FGM plates by using the CPT for different cases of boundary conditions were studied by the authors of the following papers [13–15]. Considering von Karman's type of geometric nonlinearity, FGM behavior was analyzed in [16,17]. The effect of a gradient distribution of materials in thin square and rectangular FGM plates was studied in terms of different cases of dynamic load. The papers [18,19] analyze free vibrations of the FGM plates using CPT for different boundary conditions in the area of geometric linearity. Von Karman's type of nonlinearity has been used in the papers [20,21].

Mindlin's and Reissner's theories take into consideration the effect of shear stresses across the thickness of the plate and require the use of correctional factors which generally depend on the shape and geometry. FSDT theory has been widely used in numerous papers mainly for solving nonlinear problems [22,23]. Static problems due to introducing geometric nonlinearity have been studied in [24], using Green's strain tensor, and in [25] using von Karman's strain tensor.

In order to avoid the use of shear correctional factors, high-order shear deformation theories (HSDT) have been introduced. HSDT theories can be developed by developing displacement components into power series at the coordinate of thickness. Generally, in the theories developed in this way, desired precision of the analysis can be achieved by introducing a sufficient number of terms in the power series. The most frequently used HSDT theory is the third-order shear deformation theory (TSDT) developed for composite laminates [26,27], which takes into consideration the effects of shear strains by satisfying the condition of keeping the upper and lower surface of the laminates free of stresses. Later, that theory was used in the analysis of FGM plates [28,29] for solving buckling problems [30,31], free vibrations and dynamic stability [32,33]. In addition to TSDT, there are HSDT theories based on the shape functions which represent a special group of HSDT theories introduced in order to eliminate the need for correctional factors [34,35]. Contrary to CPT and FSDT, the supposed displacement shapes in this theory do not foresee that the normal to the middle plane of the laminate plate remain a straight line, but that during deformation the normal will become curved. Generally, shape functions can be polynomial, hyperbolic, exponential, parabolic etc. Polynomial HSDTs usually diverge from other types of these theories and in accordance to the order of a polynomial at the thickness coordinate they are categorized into the group of second-order shear deformation theories (SSDT) or third-order shear deformation theories. Polynomial theories are those that are most common in the articles, which deal with FG plates' analysis using HSDT. According to [36,37] all polynomial HSDT of third order can be classified so that the supposed displacement fields contain eleven unknowns. The above-mentioned formulation has been expanded in [38,39] by supposing that the displacements are cubic functions of the thickness coordinate of the plate, that is, the supposed displacements contain twelve independent variables. In [40–42] the authors have proposed a shear deformation theory of n-series, which was obtained by modifying the displacement field of TSDT, in order to explain polynomial elements of n-series. Unlike HSDT based on polynomial shape functions, some authors have dealt with researching and introducing different hyperbolic, exponential, parabolic, and other shape functions [43–51]. Proposed functions were applied in the analysis of conventional laminate composites with the aim to describe the behavior of moderately thick and thick under different static and dynamic loads.

On the other hand, continuum-based 3D elasticity theory could be used for the analysis of these plates. However, 3D solution methods are mathematically complex which consequently results in prolonged calculation time and the need for high performance hardware. Taking the aforementioned into consideration, developing and using 2D shear deformation plate theories, which consider the effects of previously mentioned shear and normal strains and provide the precision in the same way as 3D models do, represents a trend in the process of analysis of FGM plates.

This paper presents, in detail, the methodology of the application of the HSDT theory based on the shape functions. A new shape function has been developed and the obtained results are compared to the results obtained with 13 different shape functions presented in the papers from the reference list. Also, the results have been verified through comparison with the results obtained with TSDT and 3D theories. In order to determine a procedure for the analysis and the prediction of behavior of FGM plates, the new program code in the software package MATLAB (MATrix LABoratory) has been developed based on theories studied in this paper.

Finally, the ultimate goal and the purpose of all the previously mentioned studies and analyses is the application of FGM in different areas of engineering and branches of industry. Although FGM were initially used as materials for thermal barrier in space shuttles, today they are becoming widely used in the field of medicine, dentistry, energy and nuclear sector, automotive industry, military, optoelectronics etc.

2. Description of the Problem

The subject of the analysis in this paper are FGM plate (Figure 3a) and FGM plate on elastic foundation (Figure 3b). The plate (length a, width b and height h) is made of functionally graded material consisting of the two constituents, namely, metal and ceramics.

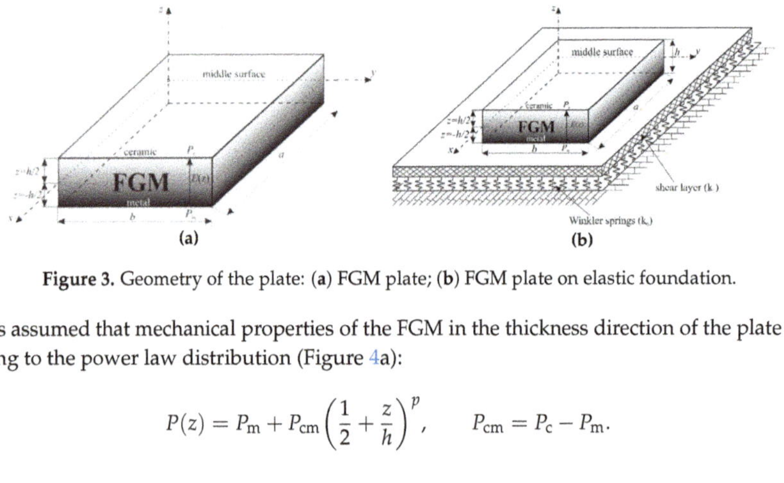

Figure 3. Geometry of the plate: (**a**) FGM plate; (**b**) FGM plate on elastic foundation.

It is assumed that mechanical properties of the FGM in the thickness direction of the plate change according to the power law distribution (Figure 4a):

$$P(z) = P_{\mathrm{m}} + P_{\mathrm{cm}}\left(\frac{1}{2} + \frac{z}{h}\right)^{p}, \qquad P_{\mathrm{cm}} = P_{\mathrm{c}} - P_{\mathrm{m}}. \tag{1}$$

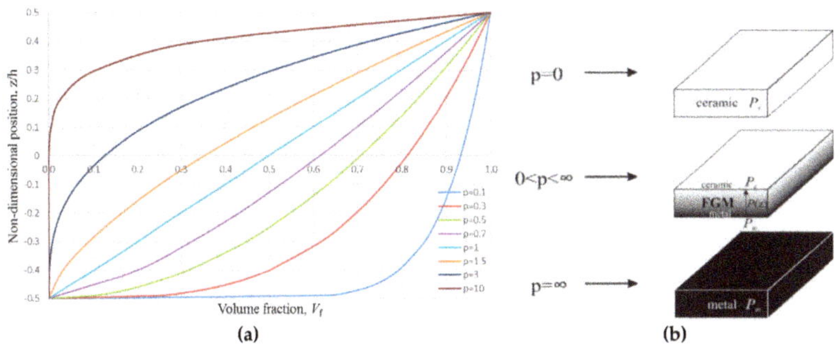

Figure 4. The comparison of homogenous plates (ceramic or metal) and FGM plates: (**a**) volume fraction of the material; (**b**) homogenous and FGM plates.

This law defines the change of the mechanical properties as the function of the volume fraction of the FGM constituents in the thickness direction of the plate.

In the Equation (1), h represents total thickness of the plate, and $P(z)$ represents a material property in an arbitrary cross-section z, $-h/2 < z < h/2$. P_c represents the material property at the top of the plate $z = h/2$ – ceramic, and P_m represents the material property at the bottom of the plate $z = -h/2$ – metal. Index p is the exponent of the equation which defines the volume fraction of the constituents in FGM. Practically, by varying the index p, homogenous plates as well as FGM plate with precisely determined gradient structure could be obtained, as it is presented in Figure 4b:

- when $p = 0$ the plate is homogenous, made of ceramics,
- when $0 < p < \infty$ the plate has a gradient structure,
- theoretically, when $p = \infty$ the plate becomes homogenous again, made of metal, although the plate can be considered homogenous even when $p > 20$.

3. Kinematic Displacement-Strain Relations and Constitutive Equation of Elasticity for FGM

According to HSDT based on the shape functions, displacements could be presented in the following way:

$$\begin{aligned} u(x,y,z,t) &= u_0(x,y,t) - z\frac{\partial w_0(x,y,t)}{\partial x} + f(z)\theta_x, \\ v(x,y,z,t) &= v_0(x,y,t) - z\frac{\partial w_0(x,y,t)}{\partial y} + f(z)\theta_y, \\ w(x,y,z,t) &= w_0(x,y,t), \end{aligned} \quad (2)$$

where: u_0, v_0, w_0 are displacement components in the middle plane of the plate, $\frac{\partial w_0}{\partial x}$, $\frac{\partial w_0}{\partial y}$ are rotation angles of transverse normal in relation to x and y axes, respectively, θ_x, θ_y are rotations of the transverse normal due to transverse shear and $f(z)$ is the shape function.

In the reference literature there are many shape functions which can be polynomial, trigonometric, exponential, hyperbolic. Some examples of the shape functions are given in Table 1.

Table 1. Shear deformation shape functions.

Number of Shape Function (SF)	Names of Authors	Shape Function $f(z)$
SF 1	Ambartsumain [52]	$(z/2)(h^2/4 - z^2/3)$
SF 2	Kaczkowski, Panc and Reissner [53]	$(5z/4)(1 - 4z^2/3h^2)$
SF 3	Levy, Stein, Touratier [54]	$(h/\pi)\sin(\pi z/h)$
SF 4	Mantari et al. [55]	$\sin(\pi z/h)e^{\cos(\pi z/h)/2} + (\pi z/2h)$
SF 5–6	Mantari et al. [45]	$\tan(mz) - zm\sec^2(mh/2),\ m = \{1/5h, \pi/2h\}$
SF 7	Karama et al. [56], Aydogdu [44]	$z\exp\left(-2(z/h)^2\right), z\exp\left(-2(z/h)^2/\ln\alpha\right),\ \forall\alpha > 0$
SF 8	Mantari et al. [46]	$z \cdot 2.85^{-2(z/h)^2} + 0.028z$
SF 9	El Meiche et al. [47]	$\xi[(h/\pi)\sin(\pi z/h) - z], \xi = \{1, 1/\cosh(\pi/2) - 1\}$
SF 10	Soldatos [43]	$h\sinh(z/h) - z\cosh(1/2)$
SF 11	Akavci and Tanrikulu [49]	$z\sec h(z^2/h^2) - z\sec h(\pi/4)[1 - (\pi/2)\tanh(\pi/4)]$
SF 12	Akavci and Tanrikulu [49]	$(3\pi/2)h\tanh(z/h) - (3\pi/2)z\sec h^2(1/2)$
SF 13	Mechab et al. [48]	$\frac{z\cos(1/2)}{-1+\cos(1/2)} - \frac{h\sin(z/h)}{-1+\cos(1/2)}$

This paper proposes a new shape function as follows:

$$f(z) = z\left(\cosh\left(\frac{z}{h}\right) - 1.388\right) \tag{3}$$

The introduced shape function is an odd function of the thickness coordinate z and satisfies zero stress conditions for out of plane shear stresses. Observing the shape functions in the Table 1, may see that the proposed function belongs to the group of simple mathematic functions. This fact makes the integration process easier and thus reduces considerably the calculation time. Having in mind that the function is analytically integrable, there is no need to switch to numeric integration, which additionally increases the precision of the obtained results. The verification of the above claims is shown in the comparative diagrams (Figure 5) of the newly introduced shape function and the shape functions given in the Table 1. These shape functions' diagrams can be categorized into two groups of functions. In both cases it can be seen in the diagram that, in the case of the ratio $z/h = 0.5$, all shape functions have extreme values, which are different (Figure 5a). The proposed new shape function (3) belongs to the second group (Figure 5b), together with the functions of Soldatos and Mechab which are also analytically integrable functions.

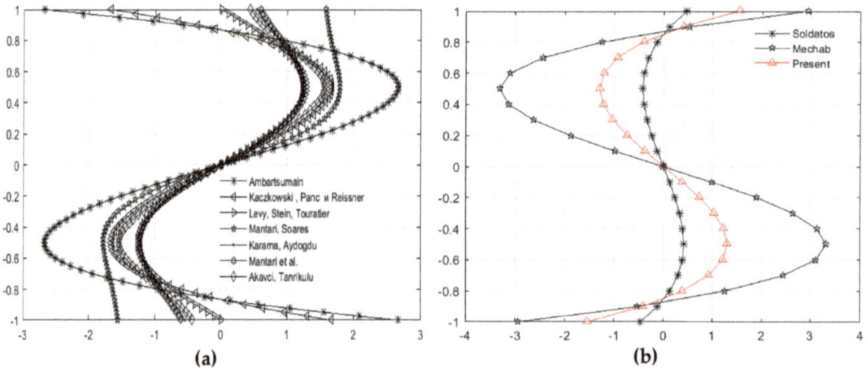

Figure 5. Shape function diagrams: (**a**) shape function from literature; (**b**) new proposed shape function.

For small displacements and moderate rotations of a transverse normal in relation to x axis and y axis, normal and shear strain components are obtained by well-known relations in linear elasticity between displacements and strains:

$$\varepsilon = \varepsilon^{(0)} + z\mathbf{k}^{(0)} + f(z)\mathbf{k}^{(1)}, \qquad \gamma = f'(z)\mathbf{k}^{(2)}, \tag{4}$$

where:

$$\varepsilon = \begin{Bmatrix} \varepsilon_{xx} \\ \varepsilon_{yy} \\ \gamma_{xy} \end{Bmatrix}, \quad \gamma = \begin{Bmatrix} \gamma_{xz} \\ \gamma_{yz} \end{Bmatrix}, \quad \varepsilon^{(0)} = \begin{Bmatrix} \varepsilon_{xx}^{(0)} \\ \varepsilon_{yy}^{(0)} \\ \gamma_{xy}^{(0)} \end{Bmatrix} = \begin{Bmatrix} u_{0,x} \\ v_{0,y} \\ u_{0,y} + v_{0,x} \end{Bmatrix},$$
$$\mathbf{k}^{(0)} = \begin{Bmatrix} k_{xx}^{(0)} \\ k_{yy}^{(0)} \\ k_{xy}^{(0)} \end{Bmatrix} = \begin{Bmatrix} -w_{0,xx} \\ -w_{0,yy} \\ -2w_{0,xy} \end{Bmatrix}, \quad \mathbf{k}^{(1)} = \begin{Bmatrix} k_{xx}^{(1)} \\ k_{yy}^{(1)} \\ k_{xy}^{(1)} \end{Bmatrix} = \begin{Bmatrix} \theta_{x,x} \\ \theta_{y,y} \\ \theta_{x,y} + \theta_{y,x} \end{Bmatrix}, \quad \mathbf{k}^{(2)} = \begin{Bmatrix} k_{xz}^{(2)} \\ k_{yz}^{(2)} \end{Bmatrix} = \begin{Bmatrix} \theta_x \\ \theta_y \end{Bmatrix}, \tag{5}$$

where $f'(z) = \frac{df(z)}{dz}$ is the first derivative of the shape function in the thickness direction of the plate.

The elastic constitutive relations for FGM are given as follows:

$$\begin{Bmatrix} \sigma_{xx} \\ \sigma_{yy} \\ \tau_{xz} \\ \tau_{yz} \\ \tau_{xy} \end{Bmatrix} = \begin{bmatrix} C_{11}(z) & C_{12}(z) & 0 & 0 & 0 \\ C_{12}(z) & C_{22}(z) & 0 & 0 & 0 \\ 0 & 0 & C_{44}(z) & 0 & 0 \\ 0 & 0 & 0 & C_{55}(z) & 0 \\ 0 & 0 & 0 & 0 & C_{66}(z) \end{bmatrix} \begin{Bmatrix} \varepsilon_{xx} \\ \varepsilon_{yy} \\ \gamma_{xz} \\ \gamma_{yz} \\ \gamma_{xy} \end{Bmatrix}, \tag{6}$$

where the coefficients of the constitutive elasticity tensor could be defined through engineering constants:

$$C_{11}(z) = C_{22}(z) = \frac{E(z)}{1-\nu^2}, \quad C_{44}(z) = C_{55}(z) = C_{66}(z) = \frac{E(z)}{2(1+\nu)}, \quad C_{12}(z) = \frac{\nu E(z)}{1-\nu^2}. \tag{7}$$

Due to the gradient change of the plate structure in the direction of the z coordinate, based on (1), the modulus of elasticity could be defined as:

$$E(z) = E_{\mathrm{m}} + E_{\mathrm{cm}}\left(\frac{1}{2} + \frac{z}{h}\right)^p, \qquad E_{\mathrm{cm}} = E_{\mathrm{c}} - E_{\mathrm{m}}, \tag{8}$$

while Poisson's ratio ν is considered constant due to a small value variation in the thickness direction of the plate, $\nu = const$.

As it could be seen, the coefficients of the constitutive tensor are functionally dependent on the z coordinate which practically means that for $p \neq 0$ there is a finite number of planes parallel to the middle plane, where each of these planes has different values of the constitutive tensor C_{ij}.

4. Bending of FGM Plates and FGM Plates on Elastic Foundation

It is assumed that the plate is loaded with an arbitrary transverse load $q(x,y)$. Work under external load is defined as:

$$V = -\frac{1}{2}\int_A qw dA, \tag{9}$$

where:

$$q(x,y) = q_0 \sin\left(\frac{\pi x}{a}\right)\sin\left(\frac{\pi y}{b}\right), \tag{10}$$

is the sinusoidal transverse load with an amplitude q_0.

Plate strain energy is defined as:

$$\begin{aligned} U &= \int_A (N_{xx}\varepsilon_{xx}^{(0)} + N_{yy}\varepsilon_{yy}^{(0)} + N_{xy}\gamma_{xy}^{(0)} + M_{xx}k_{xx}^{(0)} + M_{yy}k_{yy}^{(0)} + M_{xy}k_{xy}^{(0)} \\ &+ P_{xx}k_{xx}^{(1)} + P_{yy}k_{yy}^{(1)} + P_{xy}k_{xy}^{(1)} + R_x k_{xz}^{(2)} + R_y k_{yz}^{(2)})dA, \end{aligned} \tag{11}$$

where force, moments and higher order moments vectors are obtained in the following form:

$$\begin{aligned}
\mathbf{N} &= \int_{-h/2}^{h/2} \sigma dz = \int_{-h/2}^{h/2} \mathbf{C}_P \boldsymbol{\varepsilon} dz = \int_{-h/2}^{h/2} \mathbf{C}_P \boldsymbol{\varepsilon}^{(0)} dz + \int_{-h/2}^{h/2} \mathbf{C}_P \mathbf{k}^{(0)} z dz + \int_{-h/2}^{h/2} \mathbf{C}_P \mathbf{k}^{(1)} f(z) dz, \\
\mathbf{M} &= \int_{-h/2}^{h/2} \sigma z dz = \int_{-h/2}^{h/2} \mathbf{C}_P \boldsymbol{\varepsilon} z dz = \int_{-h/2}^{h/2} \mathbf{C}_P \boldsymbol{\varepsilon}^{(0)} z dz + \int_{-h/2}^{h/2} \mathbf{C}_P \mathbf{k}^{(0)} z^2 dz + \int_{-h/2}^{h/2} \mathbf{C}_P \mathbf{k}^{(1)} z f(z) dz, \\
\mathbf{P} &= \int_{-h/2}^{h/2} \sigma f(z) dz = \int_{-h/2}^{h/2} \mathbf{C}_P \boldsymbol{\varepsilon} f(z) dz = \int_{-h/2}^{h/2} \mathbf{C}_P \boldsymbol{\varepsilon}^{(0)} f(z) dz + \int_{-h/2}^{h/2} \mathbf{C}_P \mathbf{k}^{(0)} z f(z) dz + \int_{-h/2}^{h/2} \mathbf{C}_P \mathbf{k}^{(1)} (f(z))^2 dz, \\
\mathbf{R} &= \int_{-h/2}^{h/2} \tau f'(z) dz = \int_{-h/2}^{h/2} \mathbf{C}_S \mathbf{k}^{(2)} (f'(z))^2 dz,
\end{aligned} \tag{12}$$

Matrices in the developed form could be presented in the following way:

$$\begin{aligned}
&\mathbf{N} = \begin{Bmatrix} N_{xx} \\ N_{yy} \\ N_{xy} \end{Bmatrix}, \quad \mathbf{M} = \begin{Bmatrix} M_{xx} \\ M_{yy} \\ M_{xy} \end{Bmatrix}, \quad \mathbf{P} = \begin{Bmatrix} P_{xx} \\ P_{yy} \\ P_{xy} \end{Bmatrix}, \quad \mathbf{R} = \begin{Bmatrix} R_x \\ R_y \end{Bmatrix}, \\
&\mathbf{C}_P = \begin{bmatrix} C_{11} & C_{12} & 0 \\ C_{12} & C_{22} & 0 \\ 0 & 0 & C_{66} \end{bmatrix}, \quad \mathbf{C}_S = \begin{bmatrix} C_{44} & 0 \\ 0 & C_{55} \end{bmatrix}, \quad \sigma = \begin{Bmatrix} \sigma_{xx} \\ \sigma_{yy} \\ \tau_{xy} \end{Bmatrix}, \quad \tau = \begin{Bmatrix} \tau_{xz} \\ \tau_{yz} \end{Bmatrix}.
\end{aligned} \tag{13}$$

In the Equation (12) by grouping the terms with the elements of constitutive tensor, new matrices with the following components could be defined:

$$\begin{aligned}
A_{ij} &= \int_{-h/2}^{h/2} C_{ij} dz, & B_{ij} &= \int_{-h/2}^{h/2} C_{ij} z dz, & & \\
D_{ij} &= \int_{-h/2}^{h/2} C_{ij} f(z) dz, & E_{ij} &= \int_{-h/2}^{h/2} C_{ij} z^2 dz, & & (i,\ j) = (1,\ 2,\ 6), \\
F_{ij} &= \int_{-h/2}^{h/2} C_{ij} z f(z) dz, & G_{ij} &= \int_{-h/2}^{h/2} C_{ij} (f(z))^2 dz, & & \\
H_{lr} &= \int_{h^-}^{h^+} C_{lr} (f'(z))^2 dz, & & & & (l,\ r) = (4,\ 5),
\end{aligned} \tag{14}$$

Therefore, load vectors could now be defined in the following form:

$$\begin{Bmatrix} \mathbf{N} \\ \mathbf{M} \\ \mathbf{P} \end{Bmatrix} = \begin{bmatrix} A_{ij} & B_{ij} & D_{ij} \\ B_{ij} & E_{ij} & F_{ij} \\ D_{ij} & F_{ij} & G_{ij} \end{bmatrix} \begin{Bmatrix} \boldsymbol{\varepsilon}^{(0)} \\ \mathbf{k}^{(0)} \\ \mathbf{k}^{(1)} \end{Bmatrix}, \qquad \{\mathbf{R}\} = [H_{lr}] \left\{ \mathbf{k}^{(2)} \right\}, \tag{15}$$

By exchanging plate strain energy (11) and work under external load (9) into the equation which defines the minimum total potential energy principle:

$$\delta U + \delta V = \delta(U + V) \equiv \delta \Pi = 0, \tag{16}$$

The following form is obtained:

$$\begin{aligned}
\delta \Pi &= \int_A (N_{xx} \delta \varepsilon_{xx}^{(0)} + N_{yy} \delta \varepsilon_{yy}^{(0)} + N_{xy} \delta \gamma_{xy}^{(0)} + M_{xx} \delta k_{xx}^{(0)} + M_{yy} \delta k_{yy}^{(0)} + M_{xy} \delta k_{xy}^{(0)} \\
&+ P_{xx} \delta k_{xx}^{(1)} + P_{yy} \delta k_{yy}^{(1)} + P_{xy} \delta k_{xy}^{(1)} + R_x \delta k_{xz}^{(2)} + R_y \delta k_{yz}^{(2)}) dA - \int_A q \delta w dA = 0.
\end{aligned} \tag{17}$$

By exchanging the strain components (5) and by applying the calculus of variations, the following equilibrium equations are obtained:

$$\begin{aligned} \delta u_0: &\quad N_{xx,x} + N_{xy,y} = 0, \\ \delta v_0: &\quad N_{yy,y} + N_{xy,x} = 0, \\ \delta w_0: &\quad M_{xx,xx} + 2M_{xy,xy} + M_{yy,yy} + q = 0, \\ \delta \theta_x: &\quad P_{xx,x} + P_{xy,y} - R_x = 0, \\ \delta \theta_y: &\quad P_{xy,x} + P_{yy,y} - R_y = 0. \end{aligned} \tag{18}$$

which could be further solved through analytical and numerical methods.

In the case of a plate on elastic foundation, in the Equation (16) deformation energy of the elastic foundation should be taken into consideration, which is defined using Winkler–Pasternak model in the following way:

$$U_e = \frac{1}{2}\int_A \left\{ k_0 w^2 + k_1 \left[\left(\frac{\partial w}{\partial x} \right)^2 + \left(\frac{\partial w}{\partial x} \right)^2 \right] \right\} dA. \tag{19}$$

Using the previously mentioned the minimum total potential energy principle, the equilibrium equations of the plate on elastic foundation are the following:

$$\begin{aligned} \delta u_0: &\quad N_{xx,x} + N_{xy,y} = 0, \\ \delta v_0: &\quad N_{yy,y} + N_{xy,x} = 0, \\ \delta w_0: &\quad M_{xx,xx} + 2M_{xy,xy} + M_{yy,yy} + N_{xx} w_{0,xx} + 2N_{xy} w_{0,xy} + N_{yy} w_{0,yy} \\ &\quad + q - k_0 w_0 + k_1 \left(w_{0,xx} + w_{0,yy} \right) = 0, \\ \delta \theta_x: &\quad P_{xx,x} + P_{xy,y} - R_x = 0, \\ \delta \theta_y: &\quad P_{xy,x} + P_{yy,y} - R_y = 0. \end{aligned} \tag{20}$$

5. Analytical Solution of the Equilibrium Equations

Although analytical solution methods are limited to simple geometrical problems, boundary conditions and loads, they can provide a clear understanding of the physical aspect of the problem and its solutions are very precise. Since analytical solutions are extremely important for developing new theoretical models, primarily due to their understanding of the physical aspects of the problem, and considering that a new HSDT theory based on a new shape function has been developed in this paper, the analytical solution of the equilibrium equations for a rectangular plate will be presented in the following part of the paper. For complex engineering calculations, which include solving the system of a large number of equations, it is necessary to use numerical methods which provide approximate, but satisfactory results.

For a simply supported rectangular FGM plate, boundary conditions are defined based on [57] as:

$$\begin{aligned} &v_0 = w_0 = \theta_y = N_{xx} = M_{xx} = P_{xx} = 0, \text{ on the edges where } x = 0 \text{ or } x = a, \\ &u_0 = w_0 = \theta_x = N_{yy} = M_{yy} = P_{yy} = 0, \text{ on the edges where y} = 0 \text{ or } y = b. \end{aligned} \tag{21}$$

In order to satisfy these kinematic boundary conditions, assumed forms of Navier's solutions are introduced:

$$\begin{aligned} u_0(x,y,t) &= \sum_{m=1}^{\infty}\sum_{n=1}^{\infty} U_{mn} \cos\tfrac{m\pi x}{a} \sin\tfrac{n\pi y}{b}, & v_0(x,y,t) &= \sum_{m=1}^{\infty}\sum_{n=1}^{\infty} V_{mn} \sin\tfrac{m\pi x}{a} \cos\tfrac{n\pi y}{b}, \\ w_0(x,y,t) &= \sum_{m=1}^{\infty}\sum_{n=1}^{\infty} W_{mn} \sin\tfrac{m\pi x}{a} \sin\tfrac{n\pi y}{b}, \\ \theta_x(x,y,t) &= \sum_{m=1}^{\infty}\sum_{n=1}^{\infty} T_{xmn} \cos\tfrac{m\pi x}{a} \sin\tfrac{n\pi y}{b}, & \theta_y(x,y,t) &= \sum_{m=1}^{\infty}\sum_{n=1}^{\infty} T_{ymn} \sin\tfrac{m\pi x}{a} \cos\tfrac{n\pi y}{b}. \end{aligned} \tag{22}$$

The equilibrium equation is further developed into:

$$\underbrace{\begin{bmatrix} L_{11} & L_{12} & L_{13} & L_{14} & L_{15} \\ L_{12} & L_{22} & L_{23} & L_{24} & L_{25} \\ L_{13} & L_{23} & L_{33} & L_{34} & L_{35} \\ L_{14} & L_{24} & L_{34} & L_{44} & L_{45} \\ L_{15} & L_{25} & L_{35} & L_{45} & L_{55} \end{bmatrix}}_{\mathbf{L}} \underbrace{\begin{Bmatrix} U_{mn} \\ V_{mn} \\ W_{mn} \\ T_{xmn} \\ T_{ymn} \end{Bmatrix}}_{\overline{\mathbf{U}}} = \underbrace{\begin{Bmatrix} 0 \\ 0 \\ q_0 \\ 0 \\ 0 \end{Bmatrix}}_{\mathbf{P_P}} \tag{23}$$

or:

$$\mathbf{L}\overline{\mathbf{U}} = \mathbf{P_P} \tag{24}$$

Through the matrix multiplication of the Equation (24) with $\mathbf{L}^{-1}$, the following is obtained:

$$\underbrace{\mathbf{L}^{-1}\mathbf{L}}_{\mathbf{I}}\overline{\mathbf{U}} = \mathbf{L}^{-1}\mathbf{P_P} \quad \rightarrow \quad \overline{\mathbf{U}} = \mathbf{L}^{-1}\mathbf{P_P}. \tag{25}$$

The Equation (25) fully defines the amplitudes of the assumed displacement components. The displacement components are obtained if the displacement amplitude matrix is multiplied with the vector from trigonometric functions which depend on x and y.

6. Numerical Results

In order to apply the previously obtained theoretical results to a simulation of real problems, a new program code for static analysis of FGM plates has been developed within the software package MATLAB. Material properties of the used materials are shown in Table 2 [58].

Table 2. Material properties of FGM constituents.

Material	Material Properties	
	Elasticity Modulus, $E[GPa]$	Poisson's Ratio, ν
Aluminum (Al)	$E_m = 70$	$\nu = 0.3$
Alumina (Al_2O_3)	$E_c = 380$	$\nu = 0.3$

Normalized values of a vertical displacement $\overline{w}$ (deflection), normal stresses $\overline{\sigma}_{xx}$ and $\overline{\sigma}_{yy}$, shear stress $\overline{\tau}_{xy}$, and transverse shear stresses $\overline{\tau}_{xz}$ and $\overline{\tau}_{yz}$ are given by using HSDT theory based on the new shape function. Normalization of the aforementioned values has been conducted according to (26) as:

$$\begin{aligned} &\overline{w} = \tfrac{10E_c h^3}{q_0 a^4} w\left(\tfrac{a}{2}, \tfrac{b}{2}\right), && \overline{\sigma}_{xx}(z) = \tfrac{h}{q_0 a}\sigma_{xx}\left(\tfrac{a}{2}, \tfrac{b}{2}, z\right), && \overline{\sigma}_{yy}(z) = \tfrac{h}{q_0 a}\sigma_{yy}\left(\tfrac{a}{2}, \tfrac{b}{2}, z\right), \\ &\overline{\tau}_{xy}(z) = \tfrac{h}{q_0 a}\tau_{xy}(0, 0, z), && \overline{\tau}_{xz}(z) = \tfrac{h}{q_0 a}\tau_{xz}\left(0, \tfrac{b}{2}, z\right), && \overline{\tau}_{yz}(z) = \tfrac{h}{q_0 a}\tau_{yz}\left(\tfrac{a}{2}, 0, z\right). \end{aligned} \tag{26}$$

Table 3 shows comparative results of the normalized values of displacement and stresses of square plate for two different ratios of length and thickness of the plate (a/h = 5 and a/h = 10) and for different values of the index p. Verification of the results obtained in this paper has been conducted by comparing them to the results from the reference papers when a/h = 10. Based on that, the results when a/h = 5 are provided for different values of the index p, i.e., different volume fraction of the constituents in FGM. Using HSDT theory with the new shape function, the obtained results are compared to the results obtained using 13 different shape functions as well as to the results obtained using quasi 3D theory of elasticity [59] and TSDT theory [58]. The results based on the CPT theory are also presented [60] in order to find certain disadvantages of the theory. Based on the comparative results of displacement and stresses, which are provided in this paper and in previously mentioned theories, it could be seen that there is a match with both TSDT theory and quasi 3D theory of elasticity. On the other hand, it is clearly seen that there are some significant differences in the results obtained by CPT theory, especially related to the stress $\overline{\sigma}_{xx}$ which shows that CPT theory does not provide satisfying results in the analysis of thick and moderately thick FGM plates. A comparative review of these results with the results obtained using 13 different shape functions shows that the newly given shape function provides almost identical results. However, since these results are given for the plane on a certain height z (for example, stress $\overline{\sigma}_{xx}$ on the height of $h/3$ etc.), a real insight into the values obtained by varying the new function could be offered by presenting stress distribution across the thickness of the plate, which is done through appropriate diagrams.

Figure 6 shows the distribution of normal stresses $\overline{\sigma}_{xx}$ and $\overline{\sigma}_{yy}$ across the thickness of the plate for different values of the index p. By analyzing the diagrams, it could be noticed that the curves representing both stresses are identical. Also, the basic property of FGM could be noticed, namely, the shift of a neutral plane in relation to the plane $z/h = 0$. It can also be seen that for the planes at the height $z/h = 0.1$–0.15 (depending on the chosen value of the index p) normal stresses have a positive sign which clearly indicates extension, and then they change the sign. In case when $p = 0$, (homogenous material made of ceramics) stress distribution is a familiar linear function with the neutral plane when $z/h = 0$. Maximum values of normal stresses due to compression are on the lower edge of the plate while the maximum values of normal stresses due to extension are on the upper edge of the plate. It could be noticed that with the increase of the index p value, maximum values of stresses due to extension are significantly increased.

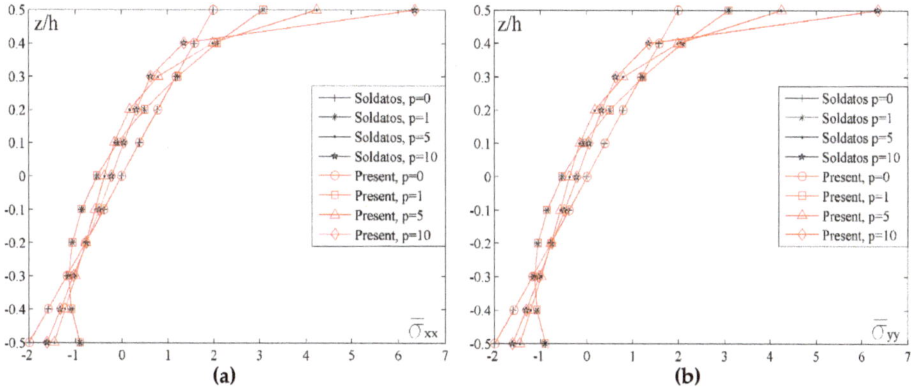

Figure 6. Distribution of the normalized values of the normal stresses $\overline{\sigma}_{xx}$ and $\overline{\sigma}_{yy}$ across the thickness of the plate for different values of the index p: (**a**) $a/h = 10$, $a/b = 1$; (**b**) $a/h = 10$, $a/b = 1$.

Table 3. Normalized values of displacement and stresses of square plate for different values of the index p and the ratio a/h ($a/b = 1$). CPT: classical plate theory; TSDT: third-order shear deformation theory.

p	Theory	$\bar{w}$		$\bar{\sigma}_{xx}(h/3)$		$\bar{\tau}_{xy}(-h/3)$		$\bar{\tau}_{xz}(h/6)$	
		$a/b = 1$							
		$a/h = 10$	$a/h = 5$	$a/h = 10$	$a/h = 5$	$a/h = 10$	$a/h = 5$	$a/h = 10$	$a/h = 5$
1	Present study	0.5889	0.6687	1.4899	0.7345	0.6111	0.3034	0.2604	0.2599
	CPT [60]	0.5623	—	2.0150	—	—	—	—	—
	Quasi 3D [59]	0.5876	—	1.5061	—	0.6112	—	0.2511	—
	TSDT [58]	0.5890	—	1.4898	—	0.6111	—	0.2599	—
	SF 1	0.5889	0.6687	1.4898	0.7344	0.6111	0.3034	0.2607	0.2602
	SF 2	0.5889	0.6687	1.4898	0.7344	0.6111	0.3034	0.2607	0.2602
	SF 3	0.5889	0.6685	1.4894	0.7336	0.6110	0.3033	0.2621	0.2615
	SF 4	0.5880	0.6648	1.4888	0.7323	0.6109	0.3030	0.2566	0.2554
	SF 5	0.5889	0.6687	1.4898	0.7344	0.6111	0.3034	0.2607	0.2601
	SF 6	0.5888	0.6683	1.4908	0.7363	0.6113	0.3038	0.2551	0.2547
	SF 7	0.5887	0.6679	1.4891	0.7330	0.6109	0.3031	0.2624	0.2616
	SF 8	0.5887	0.6679	1.4891	0.7330	0.6109	0.3031	0.2623	0.2615
	SF 9	0.5887	0.6679	1.4891	0.7330	0.6109	0.3031	0.2623	0.2615
	SF 10	0.5889	0.6687	1.4898	0.7344	0.6111	0.3034	0.2605	0.2600
	SF 11	0.5887	0.6679	1.4902	0.7352	0.6112	0.3036	0.2569	0.2566
	SF 12	0.5889	0.6686	1.4895	0.7338	0.6110	0.3033	0.2617	0.2611
	SF 13	0.5889	0.6687	1.4898	0.7343	0.6111	0.3034	0.2609	0.2603
2	Present study	0.7572	0.8670	1.3961	0.6838	0.5442	0.2696	0.2732	0.2726
	CPT [60]	—	—	—	—	—	—	—	—
	Quasi 3D [59]	0.7571	—	1.4133	—	0.5436	—	0.2495	—
	TSDT [58]	0.7573	—	1.3960	—	0.5442	—	0.2721	—
	SF 1	0.7573	0.8671	1.3960	0.6836	0.5442	0.2695	0.2736	0.2730
	SF 2	0.7573	0.8671	1.3960	0.6836	0.5442	0.2695	0.2736	0.2730
	SF 3	0.7573	0.8671	1.3954	0.6824	0.5440	0.2693	0.2763	0.2755
	SF 4	0.7563	0.8629	1.3940	0.6797	0.5437	0.2687	0.2741	0.2726
	SF 5	0.7572	0.8671	1.3961	0.6836	0.5442	0.2695	0.2735	0.2729
	SF 6	0.7568	0.8656	1.3975	0.6865	0.5444	0.2701	0.2653	0.2649
	SF 7	0.7572	0.8667	1.3949	0.6813	0.5439	0.2691	0.2777	0.2767
	SF 8	0.7572	0.8666	1.3948	0.6812	0.5439	0.2691	0.2777	0.2768
	SF 9	0.7572	0.8666	1.3948	0.6812	0.5439	0.2691	0.2777	0.2768
	SF 10	0.7572	0.8670	1.3961	0.6837	0.5442	0.2696	0.2733	0.2727
	SF 11	0.7567	0.8649	1.3969	0.6854	0.5444	0.2699	0.2667	0.2663
	SF 12	0.7573	0.8672	1.3956	0.6827	0.5441	0.2694	0.2755	0.2748
	SF 13	0.7573	0.8671	1.3960	0.6835	0.5442	0.2695	0.2739	0.2733

Table 3. *Cont.*

p	Theory	$\bar{w}$		$\bar{\sigma}_{xx}(h/3)$		$\bar{\tau}_{xy}(-h/3)$		$\bar{\tau}_{xz}(h/6)$	
		$a/b = 1$							
		$a/h = 10$	$a/h = 5$	$a/h = 10$	$a/h = 5$	$a/h = 10$	$a/h = 5$	$a/h = 10$	$a/h = 5$
4	Present study	0.8814	1.0406	1.1795	0.5707	0.5669	0.2799	0.2529	0.2523
	CPT [60]	0.8281	—	1.6049	—	—	—	—	—
	Quasi 3D [59]	0.8823	—	1.1841	—	0.5671	—	0.2362	—
	TSDT [58]	0.8815	—	1.1794	—	0.5669	—	0.2519	—
	SF 1	0.8814	1.0409	1.1794	0.5704	0.5669	0.2798	0.2537	0.2529
	SF 2	0.8814	1.0409	1.1794	0.5704	0.5669	0.2798	0.2537	0.2529
	SF 3	0.8818	1.0423	1.1783	0.5684	0.5667	0.2795	0.2580	0.2571
	SF 4	0.8815	1.0402	1.1756	0.5630	0.5662	0.2784	0.2623	0.2606
	SF 5	0.8814	1.0408	1.1794	0.5705	0.5669	0.2799	0.2535	0.2528
	SF 6	0.8802	1.0360	1.1816	0.5749	0.5673	0.2807	0.2421	0.2417
	SF 7	0.8820	1.0429	1.1774	0.5666	0.5665	0.2791	0.2612	0.2601
	SF 8	0.8820	1.0429	1.1773	0.5664	0.5665	0.2791	0.2614	0.2603
	SF 9	0.8820	1.0429	1.1773	0.5664	0.5665	0.2791	0.2614	0.2603
	SF 10	0.8814	1.0407	1.1795	0.5706	0.5669	0.2799	0.2532	0.2525
	SF 11	0.8798	1.0346	1.1811	0.5739	0.5672	0.2805	0.2427	0.2423
	SF 12	0.8817	1.0420	1.1786	0.5690	0.5668	0.2796	0.2568	0.2559
	SF 13	0.8815	1.0411	1.1793	0.5702	0.5669	0.2798	0.2541	0.2534
8	Present study	0.9745	1.1828	0.9478	0.4544	0.5858	0.2886	0.2082	0.2076
	CPT [60]	—	—	—	—	—	—	—	—
	Quasi 3D [59]	0.9739	—	0.9622	—	0.5883	—	0.2261	—
	TSDT [58]	0.9747	—	0.9747	—	0.5858	—	0.2087	—
	SF 1	0.9746	1.1832	0.9476	0.4541	0.5858	0.2886	0.2087	0.2081
	SF 2	0.9746	1.1832	0.9476	0.4541	0.5858	0.2886	0.2087	0.2081
	SF 3	0.9749	1.1845	0.9465	0.4520	0.5856	0.2881	0.2120	0.2113
	SF 4	0.9739	1.1794	0.9435	0.4461	0.5850	0.2871	0.2139	0.2125
	SF 5	0.9745	1.1831	0.9477	0.4542	0.5858	0.2886	0.2086	0.2080
	SF 6	0.9730	1.1774	0.9500	0.4589	0.5863	0.2895	0.1995	0.1991
	SF 7	0.9751	1.1848	0.9455	0.4500	0.5854	0.2877	0.2143	0.2134
	SF 8	0.9751	1.1848	0.9454	0.4498	0.5854	0.2877	0.2145	0.2135
	SF 9	0.9751	1.1848	0.9454	0.4498	0.5854	0.2877	0.2145	0.2135
	SF 10	0.9745	1.1830	0.9477	0.4543	0.5858	0.2886	0.2084	0.2078
	SF 11	0.9727	1.1763	0.9496	0.4581	0.5861	0.2893	0.2006	0.2003
	SF 12	0.9749	1.1842	0.9469	0.4526	0.5856	0.2883	0.2111	0.2104
	SF 13	0.9746	1.1833	0.9475	0.4539	0.5858	0.2885	0.2091	0.2084

Table 3. *Cont.*

p	Theory	$\bar{w}$		$\bar{\sigma}_{xx}(h/3)$		$\bar{\tau}_{xy}(-h/3)$		$\bar{\tau}_{xz}(h/6)$	
		a/b = 1							
		a/h = 10	*a/h* = 5	*a/h* = 10	*a/h* = 5	*a/h* = 10	*a/h* = 5	*a/h* = 10	*a/h* = 5
20	Present study	1.1377	1.3727	0.7710	0.3721	0.6079	0.2993	0.2011	0.2005
	CPT [60]	—	—	—	—	—	—	—	—
	Quasi 3D [59]	—	—	—	—	—	—	—	—
	TSDT [58]	—	—	—	—	—	—	—	—
	SF 1	1.1377	1.3727	0.7709	0.3720	0.6078	0.2993	0.2013	0.2008
	SF 2	1.1377	1.3727	0.7709	0.3720	0.6078	0.2993	0.2013	0.2008
	SF 3	1.1374	1.3712	0.7702	0.3707	0.6076	0.2989	0.2025	0.2019
	SF 4	1.1338	1.3561	0.7687	0.3677	0.6073	0.2982	0.1979	0.1966
	SF 5	1.1377	1.3727	0.7709	0.3720	0.6078	0.2993	0.2013	0.2007
	SF 6	1.1375	1.3723	0.7723	0.3748	0.6083	0.3002	0.1963	0.1960
	SF 7	1.1368	1.3686	0.7697	0.3696	0.6075	0.2986	0.2028	0.2019
	SF 8	1.1367	1.3683	0.7696	0.3695	0.6075	0.2986	0.2027	0.2019
	SF 9	1.1367	1.3683	0.7696	0.3695	0.6075	0.2986	0.2027	0.2019
	SF 10	1.1377	1.3727	0.7709	0.3721	0.6079	0.2993	0.2012	0.2006
	SF 11	1.1375	1.3722	0.7720	0.3741	0.6081	0.2998	0.1983	0.1979
	SF 12	1.1375	1.3718	0.7704	0.3711	0.6077	0.2990	0.2022	0.2016
	SF 13	1.1377	1.3726	0.7708	0.3718	0.6078	0.2993	0.2015	0.2009

Figure 7 shows the distribution of the shear stress $\overline{\tau}_{xy}$ across the thickness of the plate for different values of the index p (Figure 7a) and for different shape functions (Figure 7b), but for the unchanging values of $a/h = 10$ and $a/b = 1$. While analyzing the diagrams, it should be considered that when $p = 0$ the plate is homogenous made of ceramics, when $p = 20$ the plate is homogenous made of metal, and when $0 < p < 20$ the plate is made of FGM. By analyzing the diagram in the Figure 7a, it could be noticed that for all values of the index p, the stress $\overline{\tau}_{xy}$ achieves the maximum value on the upper edge of the plate. Ceramic plate has the lowest maximum value. Therefore, with an increase of the metal volume fraction when $p = 1$, maximum stress value also increases and the highest value is achieved when the plate is homogenous made of metal. Moreover, apart from affecting the maximum stress values, the variation of the index p value also affects the shape of the $\overline{\tau}_{xy}$ stress distribution curve across the thickness of the plate.

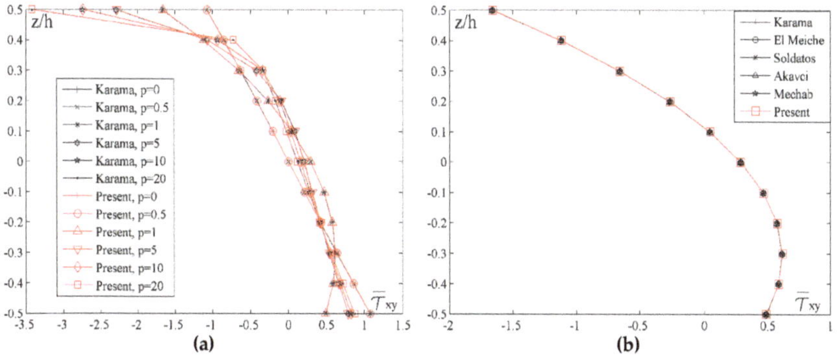

Figure 7. Distribution of the normalized values of the shear stress $\overline{\tau}_{xy}$ across the thickness of the plate for different values of the index p and different shape function: (**a**) $a/h = 10$, $a/b = 1$; (**b**) $a/h = 10$, $a/b = 1$, $p = 5$.

In order to conduct a comparative analysis of the results for different shape functions and to estimate the application of the new shape function to the given problems, Figure 7b shows the distribution of the shear stress $\overline{\tau}_{xy}$ by using newly developed shape function and the shape functions given in Table 1. It is clearly seen that all the previously mentioned shape functions give identical results to the results obtained with the new shape function.

Figure 8 shows the distribution of transverse shear stresses $\overline{\tau}_{xz}$ and $\overline{\tau}_{yz}$ across the thickness of the plate for different values of the index p and for different shape functions. By analyzing transverse shear stresses in Figure 8a,c, a basic distinction between homogenous and FGM plates can be noticed. When plates are made of ceramics ($p = 0$) or metal ($p = 20$), it can be noticed that both stresses achieve maximum values in the plane at the height $z/h = 0$, due to the homogeneity of the material. On the other hand, when FGM plates are considered, there is an asymmetry in relation to the plane $z/h = 0$, therefore, when $p = 1$ stresses achieve maximum values in the plane $z/h = 0.15$, and when $p = 5$ stresses achieve maximum values when $z/h = 0.3$. In contrast to the homogenous ceramic plate, where stress distribution curve is a parabola with the maximum value in the plane $z/h = 0$, plates with the larger volume fraction of metal ($p = 10$) also achieve the maximum value of stress when $z/h = 0$, but the distribution curve is not a parabola. With the further increase of the metal volume fraction ($p = 20$), and although the plate can be practically considered homogenous, the diagram still shows the curve which is not a parabola. Generally, due to insignificant but still present ceramic fractions in the upper part of the plate, there is a slight deformity of the curve.

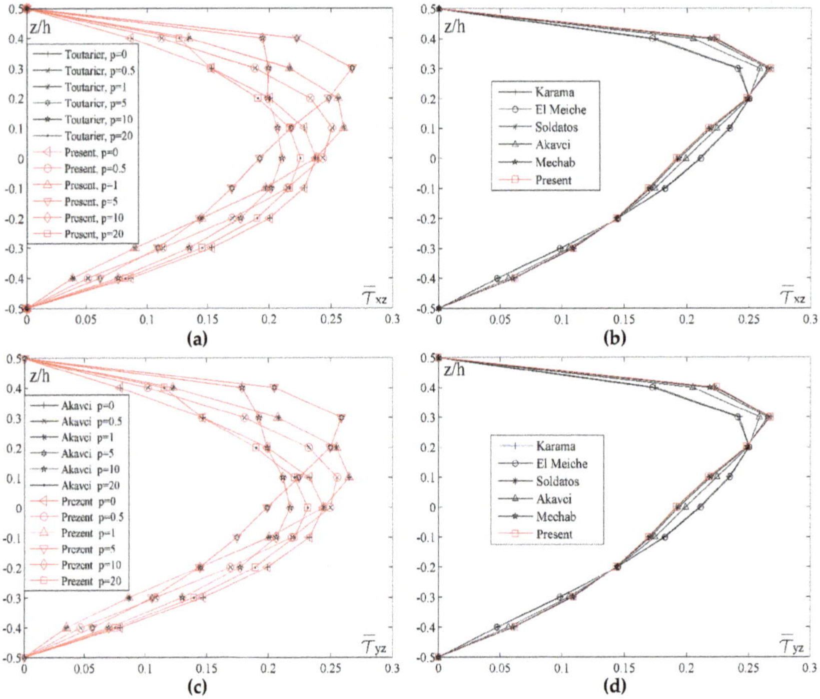

Figure 8. Distribution of the normalized values of the transverse shear stresses $\overline{\tau}_{xz}$ and $\overline{\tau}_{yz}$ across the thickness of the plate for different values of the index p and different shape functions: (**a**) $a/h = 10$, $a/b = 1$; (**b**) $a/h = 10$, $a/b = 1$, $p = 5$; (**c**) $a/h = 10$, $a/b = 1$; (**d**) $a/h = 10$, $a/b = 1$, $p = 5$.

By conducting comparative analysis of the stresses $\overline{\tau}_{xz}$ and $\overline{\tau}_{yz}$ for different shape functions, and with fixed values of $a/h = 10$, $a/b = 1$ and $p = 5$, it could be seen in Figure 8b,d that, unlike the stress $\overline{\tau}_{xy}$, the results do not match for all the shape functions. The most significant deviation could be noticed in the results for the El Meiche's and Karama's shape functions. The Akavci's function also shows a slight deviation and it achieves maximum stress value at the height $z/h = 0.25$, while the results for all the other shape functions are almost identical, achieving the maximum stress value in the plane $z/h = 0.25$.

In order to understand the effects of increasing the index p as well as the effect of the thickness and geometry, Figure 9 shows the diagram of the normalized values of the displacement $\overline{w}$ for different a/h and a/b ratios and values of the index p. By analyzing Figure 9a,b, it could be noticed that the displacement values $\overline{w}$ for the metal plate ($p = 20$) are the highest, for the ceramic plate they are the lowest, and for the FGM plate they are somewhere in between. Moreover, by varying the volume fraction of metal or ceramics, a desired bending rigidity of the plate could be achieved. In Figure 9a, it could be seen that the curves gradually become closer when $a/b > 4$. In contrast to that, Figure 9b shows that with an increase of the ratio a/h the curves do not become closer, namely, the difference of the displacement ratio remains constant regardless of the index p change. This conclusion comes from the fact that in thin plates it is less possible to vary the volume fraction of the FGM constituents in the thickness direction of the plate and, thus, the index p has no effects.

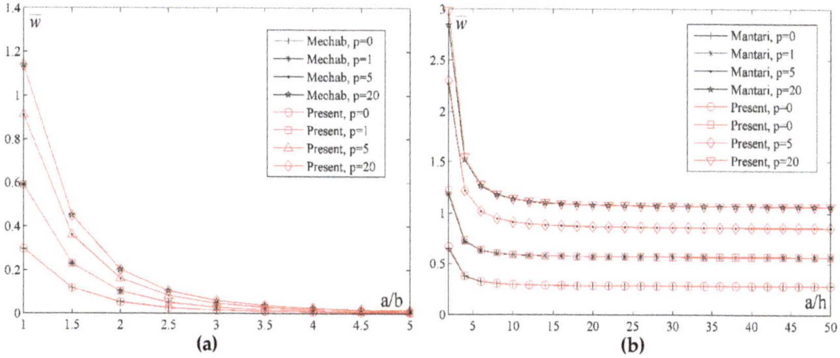

Figure 9. Normalized values of the displacement $\overline{w}$ for different a/h and a/b ratios and the values of the index p: (**a**) a/h = 10; (**b**) a/b = 1.

In order to determine the effect of the elastic foundation on the displacements and stresses of the FGM plate, the results of different combinations of the FGM constituents have been presented, as well as different combinations of the Winkler (k_0) and Pasternak (k_1) coefficient of the elastic foundation. Apart from the normalization given in (26), it is necessary to apply the normalization of the coefficients k_0 and k_1, in the following form:

$$\overline{k}_0 = k_0 \frac{a^4}{D} \quad i \quad \overline{k}_1 = k_1 \frac{a^2}{D}, \tag{27}$$

where the bending stiffness of the plate is $D = \frac{E_c h^3}{12(1-\nu^2)}$.

The Tables 4 and 5 show the results of the normalized values of displacements and stresses of the square plate on elastic foundation for p = 5, and p = 10, different values of k_0 and k_1 coefficients, as well as for two different ratios length/thickness of the plate (a/h = 10 and a/h = 5). In order to determine the effect of the elastic foundation on the displacements and stresses of the plate, the values of displacements and stresses for k_0 = 0 and k_1 = 0 are first shown, which practically matches the case of the plate without the elastic foundation. Afterwards, the values of the given coefficients are varied in order to conclude which of the two has greater influence. Based on the results, it is concluded that the introduction of the coefficient k_0 has less influence on the change of the displacements and stresses values then when only k_1 coefficient is introduced. By introducing k_0 and k_1 coefficients, bending stiffness of the plate increases, i.e., displacement and stresses values decrease and the influence of the Winkler coefficient is smaller than the influence of the Pasternak coefficient. This phenomenon is especially noticeable in the diagram dependency which is to be shown later.

Table 4. Normalized values of displacement and stresses of square plate on elastic foundation for $p = 5$, different values of the k_0 и k_1 and the ratio a/h ($a/b = 1$).

p	k_0	k_1	Theory	$\bar{w}$		$\bar{\sigma}_{xx}$		$\bar{\tau}_{xy}$		$\bar{\tau}_{xz}$	
				$a/b = 1$							
				$a/h = 10$	$a/h = 5$	$a/h = 10$	$a/h = 5$	$a/h = 10$	$a/h = 5$	$a/h = 10$	$a/h = 5$
5	0	0	Present study	0.9113	1.0882	4.2441	2.2107	0.5757	0.2840	0.1916	0.1911
			SF 1	0.9113	1.0885	4.2447	2.2118	0.5756	0.2839	0.1929	0.1924
			SF 2	0.9113	1.0885	4.2447	2.2118	0.5756	0.2839	0.1929	0.1924
			SF 3	0.9118	1.0902	4.2488	2.2199	0.5754	0.2835	0.2016	0.2009
			SF 4	0.9115	1.0885	4.2612	2.2443	0.5748	0.2824	0.2329	0.2313
			SF 5	0.9113	1.0884	4.2445	2.2116	0.5756	0.2839	0.1927	0.1921
			SF 6	0.9098	1.0826	4.2359	2.1945	0.5761	0.2848	0.1759	0.1756
			SF 7	0.9121	1.0911	4.2527	2.2276	0.5752	0.2831	0.2104	0.2095
			SF 8	0.9121	1.0911	4.2531	2.2284	0.5752	0.2831	0.2113	0.2104
			SF 9	0.9121	1.0911	4.2531	2.2284	0.5752	0.2831	0.2113	0.2104
			SF 10	0.9112	1.0883	4.2443	2.2110	0.5756	0.2839	0.1921	0.1916
			SF 11	0.9094	1.0810	4.2359	2.1945	0.5760	0.2846	0.1668	0.1665
			SF 12	0.9117	1.0898	4.2476	2.2175	0.5755	0.2836	0.1991	0.1985
			SF 13	0.9114	1.0887	4.2450	2.2126	0.5756	0.2839	0.1937	0.1932
	100	0	Present study	0.4967	0.5450	2.3135	1.1073	0.3138	0.1422	0.1045	0.0957
			SF 1	0.4967	0.5451	2.3137	1.1076	0.3137	0.1422	0.1051	0.0963
			SF 2	0.4967	0.5451	2.3137	1.1076	0.3137	0.1422	0.1051	0.0963
			SF 3	0.4969	0.5455	2.3154	1.1108	0.3136	0.1418	0.1098	0.1005
			SF 4	0.4968	0.5451	2.3225	1.1239	0.3133	0.1414	0.1269	0.1158
			SF 5	0.4967	0.5451	2.3136	1.1076	0.3137	0.1422	0.1050	0.0962
			SF 6	0.4963	0.5436	2.3107	1.1019	0.3142	0.1430	0.0960	0.0882
			SF 7	0.4969	0.5457	2.3172	1.1142	0.3134	0.1416	0.1146	0.1048
			SF 8	0.4969	0.5457	2.3174	1.1146	0.3134	0.1416	0.1151	0.1052
			SF 9	0.4969	0.5457	2.3174	1.1146	0.3134	0.1416	0.1151	0.1052
			SF 10	0.4967	0.5450	2.3135	1.1074	0.3138	0.1422	0.1047	0.0959
			SF 11	0.4961	0.5432	2.3112	1.1028	0.3143	0.1430	0.0910	0.0837
			SF 12	0.4968	0.5454	2.3148	1.1098	0.3136	0.1419	0.1085	0.0993
			SF 13	0.4967	0.5451	2.3138	1.6370	0.3137	0.1421	0.1056	0.0967
	0	10	Present study	0.3442	0.3668	1.6032	0.7451	0.2175	0.0957	0.0724	0.0644
			SF 1	0.3442	0.3667	1.6033	0.7453	0.2174	0.0956	0.0728	0.0648
			SF 2	0.3442	0.3667	1.6033	0.7453	0.2174	0.0956	0.0728	0.0648
			SF 3	0.3443	0.3669	1.6043	0.7472	0.2172	0.0954	0.0761	0.0676
			SF 4	0.3442	0.3667	1.6093	0.7562	0.2171	0.0951	0.0879	0.0779
			SF 5	0.3442	0.3667	1.6033	0.7452	0.2174	0.0956	0.0727	0.0647
			SF 6	0.3440	0.3661	1.6017	0.7421	0.2178	0.0963	0.0665	0.0594
			SF 7	0.3443	0.3670	1.6055	0.7494	0.2171	0.0952	0.0794	0.0704
			SF 8	0.3443	0.3671	1.6057	0.7496	0.2171	0.0952	0.0797	0.0707
			SF 9	0.3443	0.3671	1.6057	0.7496	0.2171	0.0952	0.0797	0.0707
			SF 10	0.3442	0.3667	1.6032	0.7451	0.2174	0.0957	0.0725	0.0645
			SF 11	0.3439	0.3659	1.6022	0.7428	0.2178	0.0963	0.0631	0.0563
			SF 12	0.3442	0.3669	1.6040	0.7466	0.2173	0.0955	0.0752	0.0668
			SF 13	0.3442	0.3668	1.6034	1.2283	0.2174	0.0956	0.0731	0.0650
	100	10	Present study	0.2617	0.2745	1.2190	0.5578	0.1654	0.0716	0.0550	0.0482
			SF 1	0.2617	0.2745	1.2190	0.5579	0.1653	0.0716	0.0554	0.0485
			SF 2	0.2617	0.2745	1.2190	0.5579	0.1653	0.0716	0.0554	0.0485
			SF 3	0.2617	0.2746	1.2197	0.5592	0.1652	0.0714	0.0578	0.0506
			SF 4	0.2617	0.2745	1.2236	0.5661	0.1650	0.0712	0.0668	0.0583
			SF 5	0.2617	0.2745	1.2190	0.5661	0.1653	0.0716	0.0553	0.0484
			SF 6	0.2616	0.2741	1.2180	0.5578	0.1656	0.0721	0.0506	0.0444
			SF 7	0.2617	0.2747	1.2206	0.5608	0.1651	0.0712	0.0604	0.0527
			SF 8	0.2617	0.2747	1.2207	0.5610	0.1651	0.0712	0.0606	0.0529
			SF 9	0.2617	0.2747	1.2207	0.5610	0.1651	0.0712	0.0606	0.0529
			SF 10	0.2617	0.2745	1.2190	0.5578	0.1653	0.0716	0.0551	0.0483
			SF 11	0.2615	0.2740	1.2184	0.5564	0.1656	0.0721	0.0479	0.0422
			SF 12	0.2617	0.2746	1.2195	0.5588	0.1652	0.0714	0.0571	0.0500
			SF 13	0.2617	0.2745	1.2191	0.9722	0.1653	0.0716	0.0556	0.0487

Table 5. Normalized values of displacement and stresses of square plate on elastic foundation for $p = 10$, different values of the k_0 и k_1 and the ratio a/h $(a/b = 1)$.

p	k_0	k_1	Theory	$\bar{w}$		$\bar{\sigma}_{xx}$		$\bar{\tau}_{xy}$		$\bar{\tau}_{xz}$	
				a/b =1							
				a/h = 10	*a/h* = 5	*a/h* = 10	*a/h* = 5	*a/h* = 10	*a/h* = 5	*a/h* = 10	*a/h* = 5
10			Present study	1.0086	1.2273	5.0843	2.6423	0.5896	0.2904	0.2101	0.2095
			SF 1	1.0087	1.2275	5.0848	2.6434	0.5895	0.2903	0.2113	0.2107
			SF 2	1.0087	1.2275	5.0848	2.6434	0.5895	0.2903	0.2113	0.2107
			SF 3	1.0089	1.2282	5.0890	2.6515	0.5893	0.2899	0.2198	0.2190
			SF 4	1.0071	1.2201	5.1006	2.6742	0.5888	0.2889	0.2488	0.2472
			SF 5	1.0086	1.2275	5.0847	2.6431	0.5895	0.2903	0.2111	0.2104
			SF 6	1.0074	1.2229	5.0758	2.6255	0.5900	0.2913	0.1944	0.1940
	0	0	SF 7	1.0088	1.2277	5.0928	2.6590	0.5891	0.2895	0.2281	0.2272
			SF 8	1.0088	1.2275	5.0931	2.6597	0.5891	0.2895	0.2290	0.2280
			SF 9	1.0088	1.2275	5.0931	2.6597	0.5891	0.2895	0.2290	0.2280
			SF 10	1.0086	1.2274	5.0845	2.6426	0.5896	0.2903	0.2105	0.2099
			SF 11	1.0072	1.2222	5.0762	2.6263	0.5899	0.2910	0.1852	0.1849
			SF 12	1.0088	1.2281	5.0877	2.6491	0.5894	0.2900	0.2174	0.2166
			SF 13	1.0087	1.2276	5.0852	2.6442	0.5895	0.2903	0.2121	0.2115
			Present study	0.5243	0.5779	2.6430	1.2440	0.3065	0.1367	0.1092	0.0986
			SF 1	0.5243	0.5779	2.6432	1.2444	0.3064	0.1366	0.1098	0.0992
			SF 2	0.5243	0.5779	2.6432	1.2444	0.3064	0.1366	0.1098	0.0992
			SF 3	0.5244	0.5780	2.6451	1.2479	0.3063	0.1364	0.1142	0.1030
			SF 4	0.5239	0.5762	2.6534	1.2630	0.3063	0.1364	0.1294	0.1167
			SF 5	0.5243	0.5779	2.6432	1.2443	0.3064	0.1367	0.1097	0.0990
			SF 6	0.5240	0.5768	2.6401	1.2385	0.3069	0.1374	0.1011	0.0915
	100	0	SF 7	0.5243	0.5779	2.6471	1.2517	0.3062	0.1363	0.1186	0.1069
			SF 8	0.5243	0.5779	2.6474	1.2521	0.3062	0.1363	0.1190	0.1073
			SF 9	0.5243	0.5779	2.6474	1.2521	0.3062	0.1363	0.1190	0.1073
			SF 10	0.5243	0.5778	2.6431	1.2441	0.3064	0.1367	0.1094	0.0988
			SF 11	0.5239	0.5767	2.6405	1.2392	0.3068	0.1373	0.0963	0.0872
			SF 12	0.5243	0.5780	2.6445	1.2468	0.3063	0.1365	0.1130	0.1019
			SF 13	0.5243	0.5779	2.6434	1.2447	0.3064	0.1366	0.1102	0.0995
			Present study	0.3573	0.3813	1.8008	0.8209	0.2088	0.0902	0.0744	0.0651
			SF 1	0.3572	0.3813	1.8010	0.8212	0.2088	0.0902	0.0748	0.0654
			SF 2	0.3572	0.3813	1.8010	0.8212	0.2088	0.0902	0.0748	0.0654
			SF 3	0.3572	0.3814	1.8022	0.8234	0.2087	0.0900	0.0778	0.0680
			SF 4	0.3570	0.3806	1.8084	0.8342	0.2087	0.0901	0.0882	0.0771
			SF 5	0.3572	0.3813	1.8009	0.8211	0.2088	0.0902	0.0747	0.0653
			SF 6	0.3571	0.3809	1.7992	0.8177	0.2091	0.0907	0.0689	0.0604
	0	10	SF 7	0.3572	0.3813	1.8036	0.8259	0.2086	0.0899	0.0808	0.0705
			SF 8	0.3572	0.3813	1.8038	0.8262	0.2086	0.0899	0.0811	0.0708
			SF 9	0.3572	0.3813	1.8038	0.8262	0.2086	0.0899	0.0811	0.0708
			SF 10	0.3572	0.3813	1.8009	0.8210	0.2088	0.0902	0.0745	0.0652
			SF 11	0.3570	0.3808	1.7995	0.8183	0.2091	0.0906	0.0656	0.0576
			SF 12	0.3572	0.3814	1.8018	0.8227	0.2087	0.0900	0.0770	0.0672
			SF 13	0.3572	0.3813	1.8011	0.9376	0.2088	0.0901	0.0751	0.0657
			Present study	0.2692	0.2826	1.3569	0.6084	0.1574	0.0669	0.0561	0.0482
			SF 1	0.2691	0.2826	1.3570	0.6086	0.1573	0.0668	0.0564	0.0485
			SF 2	0.2691	0.2826	1.3570	0.6086	0.1573	0.0668	0.0564	0.0485
			SF 3	0.2692	0.2826	1.3579	0.6102	0.1572	0.0667	0.0586	0.0504
			SF 4	0.2690	0.2822	1.3628	0.6186	0.1573	0.0668	0.0664	0.0571
			SF 5	0.2691	0.2826	1.3570	0.6086	0.1573	0.0668	0.0563	0.0484
			SF 6	0.2691	0.2823	1.3558	0.6062	0.1576	0.0672	0.0519	0.0448
	100	10	SF 7	0.2692	0.2826	1.3590	0.6121	0.1572	0.0666	0.0608	0.0523
			SF 8	0.2692	0.2826	1.3591	0.6123	0.1572	0.0666	0.0611	0.0525
			SF 9	0.2692	0.2826	1.3591	0.6123	0.1572	0.0666	0.0611	0.0525
			SF 10	0.2691	0.2826	1.3569	0.6085	0.1573	0.0668	0.0562	0.0483
			SF 11	0.2690	0.2823	1.3561	0.6067	0.1575	0.0672	0.0494	0.0427
			SF 12	0.2692	0.2826	1.3576	0.6097	0.1572	0.0667	0.0580	0.0498
			SF 13	0.2692	0.2826	1.3571	0.6087	0.1573	0.0668	0.0566	0.0486

Figure 10 shows the effect of the Winkler coefficient k_0 on the distribution of the normal stress $\overline{\sigma}_{xx}$, shear stress $\overline{\tau}_{xy}$ and transversal shear stresses $\overline{\tau}_{xz}$ and $\overline{\tau}_{yz}$ across the thickness of the plate on the elastic foundation. By analyzing the diagram, it can be seen that the value of the stresses $\overline{\sigma}_{xx}$ and $\overline{\tau}_{xy}$ equals zero for $z/h = 0.15$. On the other hand, the maximum values of $\overline{\tau}_{xz}$ and $\overline{\tau}_{yz}$ stresses are at $z/h = 0.2$ when the new proposed shape function is applied, while the maximum values of mentioned stresses is respectively at $z/h = 0.15$ i.e., $z/h = 0.25$ for Karama's shape function.

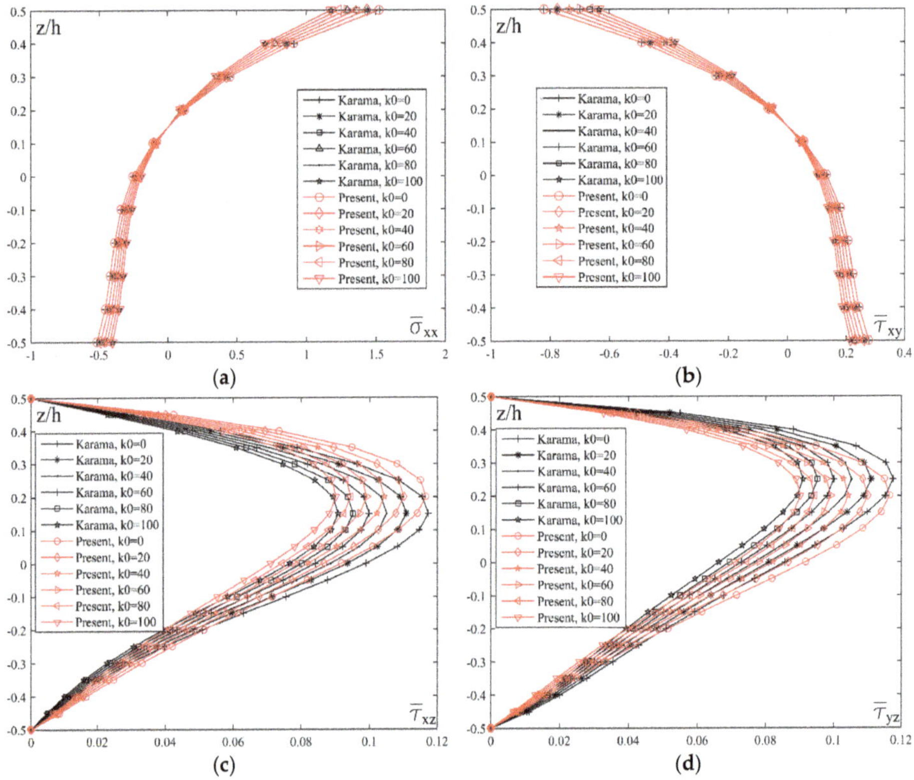

Figure 10. Distribution of the normalized values of the normal stresses $\overline{\sigma}_{xx}$, the shear stress $\overline{\tau}_{xy}$ and the transverse shear stresses $\overline{\tau}_{xz}$ and $\overline{\tau}_{yz}$ across the thickness of the plate on elastic foundation for different values of the coefficients k_0: (**a**) $a/h = 10$, $a/b = 1$, $p = 2$, $k_1 = 10$; (**b**) $a/h = 10$, $a/b = 1$, $p = 2$, $k_1 = 10$; (**c**) $a/h = 10$, $a/b = 1$, $p = 2$, $k_1 = 10$; (**d**) $a/h = 10$, $a/b = 1$, $p = 2$, $k_1 = 10$.

Figure 11 shows a comparative review of shear transversal stresses $\overline{\tau}_{xz}$ and $\overline{\tau}_{yz}$ distribution across the thickness of the plate on elastic foundation for different shape functions. As in the case of bending the plate without the elastic foundation, the shape functions do not give the same results. Therefore, it can be seen that for the Mantari's and Akavci's shape functions, stresses achieve their maximum values in the plane $z/h = 0.25$, and for El Meiche's function in the plane $z/h = 0.15$, while for all the other shape functions as well as new proposed function, maximum values of the stresses are in the plane $z/h = 0.2$.

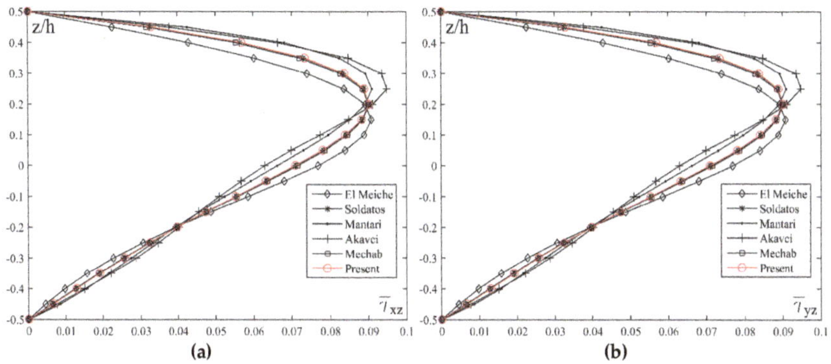

Figure 11. Distribution of the normalized values of the transverse shear stresses $\overline{\tau}_{xz}$ and $\overline{\tau}_{yz}$ across the thickness of the plate on elastic foundation for different shape functions: (**a**) $a/h = 10, a/b = 1, p = 2, k_0 = 100, k_1 = 10$; (**b**) $a/h = 10, a/b = 1, p = 2, k_0 = 100, k_1 = 10$.

In order to get a clear insight on the effect of Winkler and Pasternak coefficients of the elastic foundation, Figure 12 shows the diagram of the normalized values of the displacement $\overline{w}$ plate on the elastic foundation for different values of the index p and coefficients k_0 and k_1. By comparing the two diagrams, it can be seen that the change of the displacement value $\overline{w}$ is higher with the increase of the coefficient k_1 value than with the increase of the coefficient k_0. For example, for the FGM plate when $p = 5$, and the increase of the coefficient from $k_0 = 0$ to $k_0 = 100$, the value of deflection changes twice its value. In the other case, with the change of the coefficient from $k_1 = 0$ to $k_1 = 100$, the value of deflection changes 8 times its value.

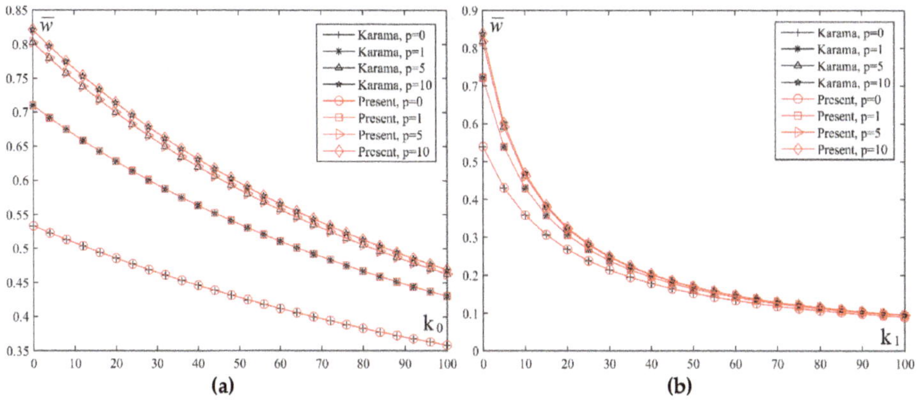

Figure 12. Normalized values of the displacement $\overline{w}$ of plate on elastic foundation for different values index p and coefficients k_0 and k_1: (**a**) $a/h = 10, a/b = 0.2, k_1 = 10$; (**b**) $a/h = 10, a/b = 0.2, k_0 = 10$.

7. Conclusions

The results obtained in the previously published papers have been a starting point for developing and applying the new shape function. They have emphasized the importance and topicality of the research on the application of the functionally graded materials. A thorough and comprehensive systematization and investigation of the literature on the matter have been conducted according to the problem type which authors tried to solve during FGM plate analysis. Special attention and focus have been given to different deformation theories which authors had used in their analyses. The new shape function has been presented along with the comparative review of it with 13 different shape functions which were primarily developed by different authors for the analysis of composite laminates but, in this paper, they have been adjusted and implemented in appropriate relations for the analysis of FGM plates. Based on the obtained results of the static analysis of moderately thick and thick plates, it can be concluded that the newly developed shape function could be applied in the analysis of FGM plates.

By analyzing the obtained results, the following could be concluded:

- the values of the vertical displacement $\overline{w}$ (deflection) and the corresponding stresses, which were obtained in this paper by using HSDT theory based on the new shape function, match the results of the same values obtained in the reference papers by using TSDT theory [58], quasi 3D theory of elasticity [59] and HSDT theories based on 13 different shape functions. However, in contrast to that, there are significant deviations of the results obtained for the values of the vertical displacement, especially for stresses $\overline{\sigma}_{xx}$, from the results obtained by CPT theory from the reference papers [60].
- the diagram of the distribution of transverse shear stresses $\overline{\tau}_{xz}$ and $\overline{\tau}_{yz}$ across the thickness of the plate shows the difference in behavior between a homogenous, ceramic or metal, plate and FGM plate. A basic property of FGM can be clearly seen, and that is the asymmetry of the stress distribution in relation to the middle plane of the plate ($z = 0$). The maximum values of stresses, depending on the volume fraction of certain constituents, are shifted in relation to the plane $z = 0$, which represents a neutral plane in homogenous plates.
- the highest values of the displacement $\overline{w}$ are obtained in a metal plate, the lowest in a ceramic plate and in an FGM plate, the values are somewhere in between and they depend on the volume fraction of the constituents. Based on that, it can be concluded that by varying the volume fraction of metal and ceramic, a desired bending rigidity of the plate can be achieved.
- a comparative analysis of the change of transverse shear stresses $\overline{\tau}_{xz}$ and $\overline{\tau}_{yz}$ across the thickness of the plate shows that, unlike the stress $\overline{\tau}_{xy}$, their values do not match for all the shape functions.
- by introducing FG plate on Winkler–Pasternak model of elastic foundation is shown that the influence of the Winkler coefficient (k_0) is smaller than the influence of the Pasternak coefficient (k_1).

Author Contributions: D.Č. and G.B. proposed studied problem and applied solving method; A.R. and M.M. conducted the theoretical derivation; H.R. and D.D. conceived and performed the numerical examples; D.Č. and A.R. analyzed the results and wrote the manuscript.

Funding: This research received no external funding.

Acknowledgments: Research presented in this paper was supported by Ministry of Education, Science and Technological Development of Republic of Serbia, TR32036, TR33015 and multidisciplinary project III 44007.

Conflicts of Interest: The authors declare no conflict of interest.

References

1. Golak, S.; Dolata, A.J. Fabrication of functionally graded composites using a homogenised low-frequency electromagnetic field. *J. Compos. Mater.* **2016**, *50*, 1751–1760. [CrossRef]
2. Watanabe, Y.; Inaguma, Y.; Sato, H.; Miura-Fujiwara, E. A Novel Fabrication Method for Functionally Graded Materials under Centrifugal Force: The Centrifugal Mixed-Powder Method. *Materials* **2009**, *2*, 2510–2525. [CrossRef]
3. El-Hadad, S.; Sato, H.; Miura-Fujiwara, E.; Watanabe, Y. Fabrication of Al-Al_3Ti/Ti_3Al Functionally Graded Materials under a Centrifugal Force. *Materials* **2010**, *3*, 4639–4656. [CrossRef] [PubMed]
4. Jha, D.K.; Kant, T.; Singh, R.K. A critical review of recent research on functionally graded plates. *Compos. Struct.* **2013**, *96*, 833–849. [CrossRef]
5. Bharti, I.; Gupta, N.; Gupta, K.M. Novel Applications of Functionally Graded Nano, Optoelectronic and Thermoelectric Materials. *IJMMM Int. J. Mater. Mech. Manuf.* **2013**, *1*, 221–224. [CrossRef]

6. Saiyathibrahim, A.; Nazirudeen, M.S.S.; Dhanapal, P. Processing Techniques of Functionally Graded Materials—A Review. In Proceedings of the International Conference on Systems, Science, Control, Communication, Engineering and Technology 2015, Coimbatore, India, 10–11 August 2015.
7. Birman, V.; Byrd, L.W. Modeling and analysis of functionally graded materials and structures. *Appl. Mech. Rev.* **2007**, *60*, 195–216. [CrossRef]
8. Akbarzadeh, A.H.; Abedini, A.; Chen, Z.T. Effect of micromechanical models on structural responses of functionally graded plates. *Compos. Struct.* **2015**, *119*, 598–609. [CrossRef]
9. Kirchhoff, G. Uber das gleichgewicht und die bewegung einer elastischen scheibe. *J. Die Reine Angew. Math.* **1850**, *1850*, 51–88. [CrossRef]
10. Mindlin, R.D. Influence of rotatory inertia and shear on flexural motions of isotropic, elastic plates. *J. Appl. Mech.* **1951**, *18*, 31–38.
11. Reissner, E. On bending of elastic plates. *Q. Appl. Math.* **1947**, *5*, 55–68. [CrossRef]
12. Reissner, E. The effect of transverse shear deformation on the bending of elastic plates. *J. Appl. Mech.* **1945**, *12*, 69–72.
13. Zenkour, A.M. An exact solution for the bending of thin rectangular plates with uniform, linear, and quadratic thickness variations. *Int. J. Mech. Sci.* **2003**, *45*, 295–315. [CrossRef]
14. Mohammadi, M.; Saidi, A.R.; Jomehzadeh, E. Levy solution for buckling analysis of functionally graded rectangular plates. *Appl. Compos. Mater.* **2010**, *17*, 81–93. [CrossRef]
15. Bhandari, M.; Purohit, K. Static Response of Functionally Graded Material Plate under Transverse Load for Varying Aspect Ratio. *Int. J. Met.* **2014**, *2014*, 980563. [CrossRef]
16. Yang, J.; Shen, H.S. Non-linear analysis of FGM plates under transverse and in plane loads. *Int. J. Nonlinear Mech.* **2003**, *38*, 467–482. [CrossRef]
17. Alinia, M.M.; Ghannadpour, S.A.M. Nonlinear analysis of pressure loaded FGM plates. *Compos. Struct.* **2009**, *88*, 354–359.
18. Liu, D.Y.; Wang, C.Y.; Chen, W.Q. Free vibration of FGM plates with in-plane material inhomogeneity. *Compos. Struct.* **2010**, *92*, 1047–1051. [CrossRef]
19. Ruan, M.; Wang, Z.M. Transverse vibrations of moving skew plates made of functionally graded material. *J. Vib. Control* **2016**, *22*, 3504–3517. [CrossRef]
20. Woo, J.; Meguid, S.A.; Ong, L.S. Nonlinear free vibration behavior of functionally graded plates. *J. Sound Vib.* **2006**, *289*, 595–611. [CrossRef]
21. Hu, Y.; Zhang, X. Parametric vibrations and stability of a functionally graded plate. *Mech. Based Des. Struct.* **2011**, *39*, 367–377.
22. Taczała, M.; Buczkowski, R.; Kleiber, M. Nonlinear free vibration of pre- and post-buckled FGM plates on two-parameter foundation in the thermal environment. *Compos. Struct.* **2016**, *137*, 85–92. [CrossRef]
23. Xing, C.; Wang, Y.; Waisman, H. Fracture analysis of cracked thin-walled structures using a high-order XFEM and Irwin's integral. *Comput. Struct.* **2019**, *212*, 1–19. [CrossRef]
24. Kim, K.D.; Lomboy, G.R.; Han, S.C. Geometrically non-linear analysis of functionally graded material (FGM) plates and shells using a four-node quasi-conforming shell element. *J. Compos. Mater.* **2008**, *42*, 485–511. [CrossRef]
25. Wu, T.L.; Shukla, K.K.; Huang, J.H. Nonlinear static and dynamic analysis of functionally graded plates. *IJAME Int. J. Appl. Mech. Eng.* **2006**, *11*, 679–698.
26. Reddy, J.N. A simple higher-order theory for laminated composite plates. *J. Appl. Mech.* **1984**, *51*, 745–752. [CrossRef]
27. Phan, N.D.; Reddy, J.N. Analysis of laminated composite plates using a higher-order shear deformation theory. *Int. Numer. Methods Eng.* **1985**, *21*, 2201–2219. [CrossRef]
28. Reddy, J.N. Analysis of functionally graded plates. *Int. Numer. Methods Eng.* **2000**, *47*, 663–684. [CrossRef]
29. Kim, J.; Reddy, J.N. A general third-order theory of functionally graded plates with modified couple stress effect and the von Kármán nonlinearity: Theory and finite element analysis. *Acta Mech.* **2015**, *226*, 2973–2998. [CrossRef]
30. Dung, D.V.; Nga, N.T. Buckling and postbuckling nonlinear analysis of imperfect FGM plates reinforced by FGM stiffeners with temperature-dependent properties based on TSDT. *Acta Mech.* **2016**, *227*, 2377–2401. [CrossRef]
31. Bodaghi, M.; Saidi, A.R. Thermoelastic buckling behavior of thick functionally graded rectangular plates. *Arch. Appl. Mech.* **2011**, *81*, 1555–1572. [CrossRef]

32. Yang, J.; Liew, K.M.; Kitipornchai, S. Dynamic stability of laminated FGM plates based on higher-order shear deformation theory. *Comput. Mech.* **2004**, *33*, 305–315. [CrossRef]
33. Akbarzadeh, A.H.; Zad, S.H.; Eslami, M.R.; Sadighi, M. Mechanical behaviour of functionally graded plates under static and dynamic loading. *Proc. Inst. Mech. Eng. Part C J. Mech. Eng. Sci.* **2011**, *225*, 326–333. [CrossRef]
34. Thai, H.T.; Choi, D.H. Efficient higher-order shear deformation theories for bending and free vibration analyses of functionally graded plates. *Arch. Appl. Mech.* **2013**, *83*, 1755–1771. [CrossRef]
35. Thai, H.T.; Park, T.; Choi, D.H. An efficient shear deformation theory for vibration of functionally graded plates. *Arch. Appl. Mech.* **2013**, *83*, 137–149.
36. Lo, K.H.; Christensen, R.M.; Wu, E.M. A high-order theory of plate deformation part 1: Homogeneous plates. *J. Appl. Mech.* **1977**, *44*, 663–668. [CrossRef]
37. Lo, K.H.; Christensen, R.M.; Wu, E.M. A high-order theory of plate deformation part 2: Laminated plates. *J. Appl. Mech.* **1977**, *44*, 669–676. [CrossRef]
38. Kant, T.; Swaminathan, K. Analytical solutions for free vibration of laminated composite and sandwich plates based on a higher-order refined theory. *Compos. Struct.* **2001**, *53*, 73–85. [CrossRef]
39. Kant, T.; Swaminathan, K. Analytical solutions for the static analysis of laminated composite and sandwich plates based on a higher order refined theory. *Compos. Struct.* **2002**, *56*, 329–344. [CrossRef]
40. Xiang, S.; Kang, G.; Xing, B. A nth-order shear deformation theory for the free vibration analysis on the isotropic plates. *Meccanica* **2012**, *47*, 1913–1921. [CrossRef]
41. Song, X.; Kang, G. A nth-order shear deformation theory for the bending analysis on the functionally graded plates. *Eur. J. Mech. A-Solid* **2013**, *37*, 336–343.
42. Song, X.; Kang, G.; Liu, Y. A nth-order shear deformation theory for natural frequency of the functionally graded plates on elastic foundations. *Compos. Struct.* **2014**, *11*, 224–231.
43. Soldatos, K. A transverse shear deformation theory for homogeneous monoclinic plates. *Acta Mech.* **1992**, *94*, 195–220. [CrossRef]
44. Aydogdu, M. A new shear deformation theory for laminated composite plates. *Compos. Struct.* **2009**, *89*, 94–101. [CrossRef]
45. Mantari, J.L.; Oktem, A.S.; Soares, S.G. A new trigonometric shear deformation theory for isotropic, laminated composite and sandwich plates. *Int. J. Solids Struct.* **2012**, *49*, 43–53. [CrossRef]
46. Mantari, J.L.; Bonilla, E.M.; Soares, C.G. A new tangential-exponential higher order shear deformation theory for advanced composite plates. *Compos. Part B-Eng.* **2014**, *60*, 319–328. [CrossRef]
47. El Meichea, N.; Tounsia, A.; Zianea, N.; Mechaba, I.; El Abbes, A.B. A new hyperbolic shear deformation theory for buckling and vibration of functionally graded sandwich plate. *Int. J. Mech. Sci.* **2011**, *53*, 237–247. [CrossRef]
48. Mechab, B.; Mechab, I.; Benaissa, S. Analysis of thick orthotropic laminated composite plates based on higher order shear deformation theory by the new function under thermo-mechanical loading. *Compos. Part B-Eng.* **2014**, *43*, 1453–1458. [CrossRef]
49. Akavci, S.S. Two new hyperbolic shear displacement models for orthotropic laminated composite plates. *Mech. Compos. Mater.* **2010**, *46*, 215–226. [CrossRef]
50. Viola, E.; Tornabene, F.; Fantuzzi, N. General higher-order shear deformation theories for the free vibration analysis of completely doubly-curved laminated shells and panels. *Compos. Struct.* **2013**, *95*, 639–666. [CrossRef]
51. Grover, N.; Maiti, D.K.; Singh, B.N. Flexural behavior of general laminated composite and sandwich plates using a secant function based shear deformation theory. *Lat. Am. J. Solids Strut.* **2014**, *11*, 1275–1297. [CrossRef]
52. Ambartsumyan, A.S. On the Theory of Anisotropic Shells and Plates. In Proceedings of the Non-Homogeneity in Elasticity and Plasticity: Symposium, Warsaw, Poland, 2–9 September 1958; Olszak, W., Ed.; Pergamon Press: London, UK, 1958.
53. Reissner, E.; Stavsky, Y. Bending and Stretching of Certain Types of Heterogeneous Aeolotropic Elastic Plates. *J. Appl. Mech.* **1961**, *28*, 402–408. [CrossRef]
54. Stein, M. Nonlinear theory for plates and shells including the effects of transverse shearing. *AIAA J.* **1986**, *24*, 1537–1544. [CrossRef]

55. Mantari, J.L.; Oktem, A.S.; Soares, G.C. Bending and free vibration analysis of isotropic and multilayered plates and shells by using a new accurate higherorder shear deformation theory. *Compos. Part B-Eng.* **2012**, *43*, 3348–3360. [CrossRef]
56. Karama, M.; Afaq, K.S.; Mistou, S. Mechanical behaviour of laminated composite beam by the new multi-layered laminated composite structures model with transverse shear stress continuity. *Int. J. Solids Struct.* **2003**, *40*, 1525–1546. [CrossRef]
57. Reddy, J.N. *Mechanics of Laminated Composite Plates and Shells: Theory and Analysis*; CRC Press LLC: New York, NY, USA, 2004.
58. Wu, C.P.; Li, H.Y. An RMVT-based third-order shear deformation theory of multilayered functionally graded material plates. *Compos. Struct.* **2010**, *92*, 2591–2605.
59. Wu, C.P.; Chiu, K.H.; Wang, Y.M. RMVT-based meshless collocation and element-free Galerkin methods for the quasi-3D analysis of multilayered composite and FGM plates. *Compos. Struct.* **2011**, *93*, 923–943. [CrossRef]
60. Carrera, E.; Brischetto, S.; Robaldo, A. Variable kinematic model for the analysis of functionally graded material plates. *AIAA J.* **2008**, *46*, 194–203. [CrossRef]

materials

MDPI

Article

Application of First-Order Shear Deformation Theory on Vibration Analysis of Stepped Functionally Graded Paraboloidal Shell with General Edge Constraints

Fuzhen Pang, Haichao Li *, Fengmei Jing * and Yuan Du

College of Shipbuilding Engineering, Harbin Engineering University, Harbin 150001, China; pangfuzhen@hrbeu.edu.cn (F.P.); duyuan@hrbeu.edu.cn (Y.D.)

* Correspondence: lihaichao@hrbeu.edu.cn (H.L.); jingfengmei@hrbeu.edu.cn (F.J.); Tel.: +86-451-82589161 (H.L.)

Received: 25 November 2018; Accepted: 21 December 2018; Published: 25 December 2018

Abstract: The paper introduces a semi-analytical approach to analyze free vibration characteristics of stepped functionally graded (FG) paraboloidal shell with general edge conditions. The analytical model is established based on multi-segment partitioning strategy and first-order shear deformation theory. The displacement components along axial direction are represented by Jacobi polynomials, and the Fourier series are utilized to express displacement components in circumferential direction. Based on penalty method about spring stiffness technique, the general edge conditions of doubly curved paraboloidal shell can be easily simulated. The solutions about doubly curved paraboloidal shell were solved by approach of Rayleigh–Ritz. Convergence study about boundary parameters, Jacobi parameters et al. are carried out, respectively. The comparison with published literatures, FEM and experiment results show that the present method has good convergence ability and excellent accuracy.

Keywords: stepped FG paraboloidal shell; general edge conditions; spring stiffness technique; free vibration characteristics

1. Introduction

The stepped FG paraboloidal shells are very useful in the engineering. The vibration problems of the structures have always been the concern of the research: Fantuzzi et al. [1] investigated free vibration behavior of FG cylindrical and spherical shells. On the base of FSDT, Tornabene and Reddy [2] used the GDQ approach to investigate the vibration behavior of FGM shells and panels. Based on higher-order finite element method, Pradyumna and Bandyopadhyay [3] studied the vibration behavior of FG structures. Jouneghani et al. [4] also investigated the characteristics of FG doubly curved shells. Chen et al. [5] obtained the vibration characteristics of FG sandwich structure based on shear deformation theory. Wang et al. [6–9] investigated the approach of Improved Fourier to study vibration phenomenon of various structures. Tornabene et al. [10–12] used the GDQ method to research four parameter FG composite structures. Fazzolari and Carrera [13] solved the vibration issues of FG structures based on Ritz minimum energy approach. Kar and Panda [14] studied vibration characteristics of FG spherical shell by FEM. Tornabene [15] focused on the dynamic behavior of FG structures. Zghal [16] investigated the vibration characteristics of FG shells. Kulikov et al. [17] dealt with a recently developed approach to analyze free vibration behavior of FG plates by the formulations of sampling surfaces. Kapuria et al. [18] developed a four-node quadrilateral element method to analyze dynamic vibration of FGM shallow shells.

In field of FG stepped shells, Hosseini-Hashemi et al. [19] proposed an accurate solution to study vibration characteristics of stepped FG plates. Bambill et al. [20] solved vibrations behavior of axially FG beams with stepped changes in geometry. Vinyas and Kattimani [21,22] carried out the static analysis of stepped FG beam and plates with various loads. Su et al. [23] presented an effective method to study free vibration of stepped FG beams.

From literatures reviewed, we can find that many scholars applied Rayleigh Ritz method, GDQ method, Improved Fourier series method, FEM and Haar Wavelet Discretization method etc. to study vibration characteristics of FG doubly curved structures. There are no literatures put attentions on free vibration problems of stepped FG paraboloidal shell. So, it is very important to propose a unified formulation to study free vibration behaviors of stepped FG paraboloidal shell subject to general edge conditions.

2. Fundamental Theory

2.1. The Description of the Model

The model of stepped FG paraboloidal shell is described in Figure 1. h_i represents the thickness of the structure. The stepped structure is obtained by the curve c_1c_2. The model is established on the basis of orthogonal coordinate system (φ, θ, z), which represent axial, circumferential and normal directions, respectively. The displacements are represented by u, v and w, respectively.

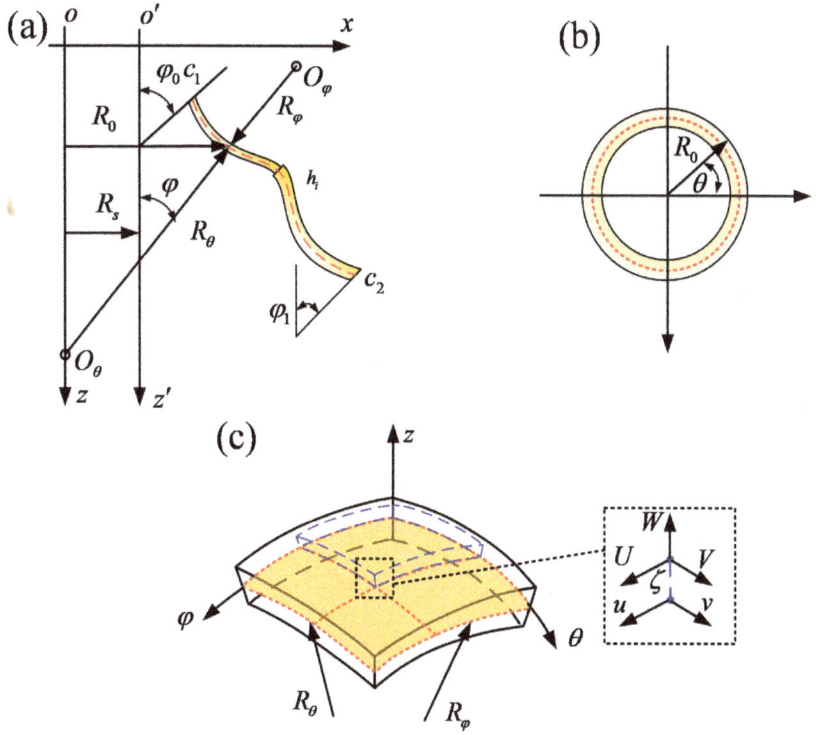

Figure 1. Geometry notations and coordinate system of stepped FG paraboloidal shell. (**a**) Geometric relationship; (**b**) cross-section; (**c**) coordinate system.

The doubly-curved paraboloidal shell is shown in Figure 2. The displacement components of stepped FG paraboloidal shell are represented by U, V and W. In addition, the doubly curved paraboloidal shell is divided into H shell segments along axial direction [24,25].

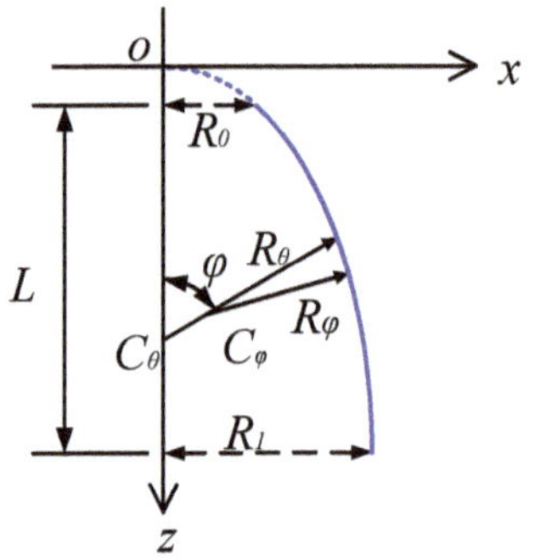

Figure 2. The geometric of doubly curved paraboloidal shell.

The Young's modulus E, Poisson's ratios ν and mass density ρ of two typical FG models are shown as follow [26–32]:

$$E(z) = (E_c - E_m)V_c + E_m \tag{1a}$$

$$\rho(z) = (\rho_c - \rho_m)V_c + \rho_m \tag{1b}$$

$$\nu(z) = (\nu_c - \nu_m)V_c + \nu_m \tag{1c}$$

where c and m denote the ceramic and metallic constituents, respectively. The volume fractions V_c are shown as follow [33]:

$$\text{FGM}_{\text{I}}(a/b/c/p) : V_c = \left[1 - a\left(\frac{1}{2} + \frac{z}{h}\right) + b\left(\frac{1}{2} + \frac{z}{h}\right)^c\right]^p \tag{2a}$$

$$\text{FGM}_{\text{II}}(a/b/c/p) : V_c = \left[1 - a\left(\frac{1}{2} - \frac{z}{h}\right) + b\left(\frac{1}{2} - \frac{z}{h}\right)^c\right]^p \tag{2b}$$

where z and p represent the thickness and power law exponent of the structure, respectively. We should note that the value of parameter p takes only positive values. The symbols a, b and c are the key parameters which affect the property of FG material largely. As the volume fraction, the total value of which should be the one. From Equations (1) and (2), we can easily get that the functionally graded material will be the isotropic material when the power law exponent equal to infinity or zero. The variations V_c about various values of a, b, c and p are showed in Figure 3. In addition, we should note that the distributions of volume fraction (2a) and (2b) are mirror reflections. Thus, the Variations V_c of FGM_{II} are ignored in Figure 3. The detailed descriptions of FG material are reported in Refs. [34–36].

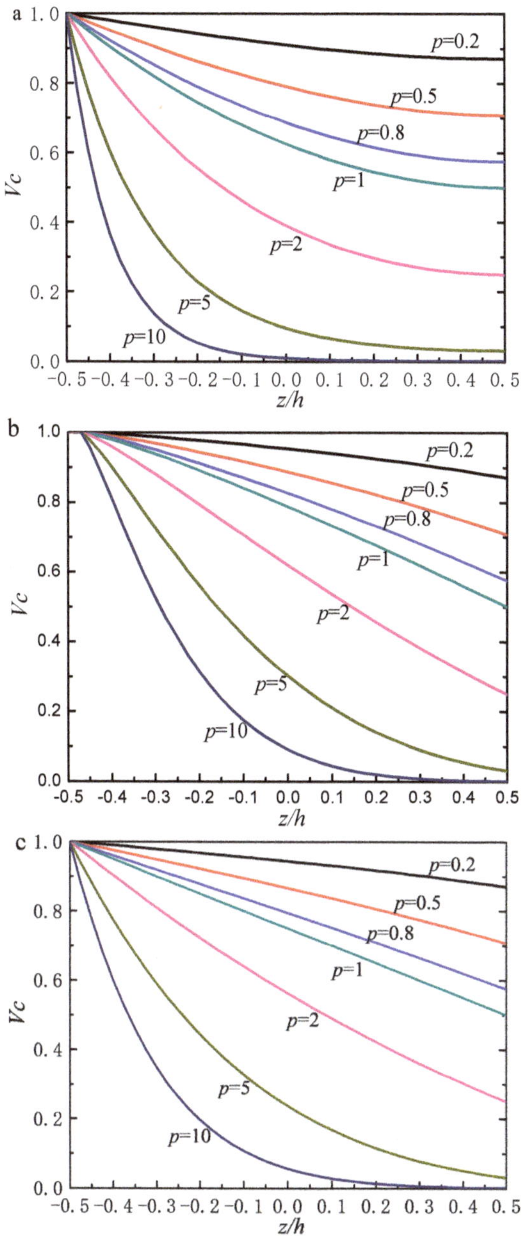

Figure 3. The variations V_c about various values of a, b, c and p: (**a**) FGM_I ($a = 0$; $b = 0.5$; $c = 2$; p); (**b**) FGM_I ($a = 1$; $b = 0.5$; $c = 0.8$; p); (**c**) FGM_I ($a = 0$; $b = -0.5$; $c = 1$; p).

2.2. Energy Equations of Stepped FG Paraboloidal Shell

The displacements of ith segment in stepped FG paraboloidal shell are shown as below:

$$\overline{U}^i(\varphi,\theta,z,t) = u^i(\varphi,\theta,t) + z\psi^i_\varphi(\varphi,\theta,t) \tag{3a}$$

$$\overline{V}^i(\varphi,\theta,z,t) = v^i(\varphi,\theta,t) + z\psi_\theta^i(\varphi,\theta,t) \tag{3b}$$

$$\overline{W}^i(\varphi,\theta,z,t) = w^i(\varphi,\theta,t) \tag{3c}$$

The strains of stepped FG paraboloidal shell are shown as follow

$$\varepsilon_\varphi^i = \varepsilon_\varphi^{0,i} + z\kappa_\varphi^{0,i}\varepsilon_\theta^i = \varepsilon_\theta^{0,i} + z\kappa_\theta^{0,i} \tag{4a}$$

$$\gamma_{\varphi\theta}^i = \gamma_{\varphi\theta}^{0,i} + z\kappa_{\varphi\theta}^{0,i}\gamma_{\varphi z}^i = \gamma_{\varphi z}^{0,i}\gamma_{\theta z}^i = \gamma_{\theta z}^{0,i} \tag{4b}$$

where ε_φ^i, ε_θ^i, $\varepsilon_{\varphi\theta}^i$, $\gamma_{\varphi z}^{0,i}$, $\gamma_{\theta z}^{0,i}$, k_φ^i, k_θ^i and $k_{\varphi\theta}^i$ are given as:

$$\varepsilon_\varphi^{0,i} = \frac{1}{A}\frac{\partial u^i}{\partial\varphi} + \frac{v^i}{AB}\frac{\partial A}{\partial\theta} + \frac{w^i}{R_\varphi} \tag{5a}$$

$$\varepsilon_\theta^{0,i} = \frac{1}{B}\frac{\partial v^i}{\partial\theta} + \frac{u^i}{AB}\frac{\partial B}{\partial\varphi} + \frac{w^i}{R_\theta} \tag{5b}$$

$$\gamma_{\varphi\theta}^{0,i} = \frac{A}{B}\frac{\partial}{\partial\theta}\left(\frac{u^i}{A}\right) + \frac{B}{A}\frac{\partial}{\partial\varphi}\left(\frac{v^i}{B}\right) \tag{5c}$$

$$\kappa_\varphi^i = \frac{1}{A}\frac{\partial\psi_\varphi^i}{\partial\varphi} + \frac{\psi_\theta^i}{AB}\frac{\partial A}{\partial\theta} \tag{5d}$$

$$\kappa_\theta^i = \frac{1}{B}\frac{\partial\psi_\theta^i}{\partial\theta} + \frac{\psi_\varphi^i}{AB}\frac{\partial B}{\partial\varphi} \tag{5e}$$

$$\kappa_{\varphi\theta}^i = \frac{A}{B}\frac{\partial}{\partial\theta}\left(\frac{\psi_\varphi^i}{A}\right) + \frac{B}{A}\frac{\partial}{\partial\varphi}\left(\frac{\psi_\theta^i}{B}\right) \tag{5f}$$

$$\gamma_{\varphi z}^{0,i} = \frac{1}{A}\frac{\partial w^i}{\partial\varphi} - \frac{u^i}{R_\varphi} + \psi_\varphi^i \tag{5g}$$

$$\gamma_{\theta z}^{0,i} = \frac{1}{B}\frac{\partial w^i}{\partial\theta} - \frac{v^i}{R_\theta} + \psi_\theta^i \tag{5h}$$

For doubly curved paraboloidal shell, the symbols A and B are shown as below [37,38]:

$$A = R_\varphi, B = R_\theta \sin\varphi \tag{6}$$

Based on Hooke's law, the stresses corresponding to strains can be expressed as:

$$\begin{Bmatrix} \sigma_\varphi^i \\ \sigma_\theta^i \\ \tau_{\varphi\theta}^i \\ \tau_{\varphi z}^i \\ \tau_{\theta z}^i \end{Bmatrix} = \begin{bmatrix} Q_{11}(z) & Q_{12}(z) & 0 & 0 & 0 \\ Q_{12}(z) & Q_{11}(z) & 0 & 0 & 0 \\ 0 & 0 & Q_{66}(z) & 0 & 0 \\ 0 & 0 & 0 & Q_{66}(z) & 0 \\ 0 & 0 & 0 & 0 & Q_{66}(z) \end{bmatrix} \begin{Bmatrix} \varepsilon_\varphi^i \\ \varepsilon_\theta^i \\ \gamma_{\varphi\theta}^i \\ \gamma_{\varphi z}^i \\ \gamma_{\theta z}^i \end{Bmatrix} \tag{7}$$

where σ_φ^i and σ_θ^i are normal stresses; $\tau_{\varphi\theta}^i$, $\tau_{\varphi z}^i$ and $\tau_{\theta z}^i$ are shear stresses. The $Q_{ij}(z)$ are defined as follows:

$$Q_{11}(z) = \frac{E(z)}{1-\nu^2(z)}, \; Q_{12}(z) = \frac{\nu(z)E(z)}{1-\nu^2(z)}, \; Q_{66}(z) = \frac{E(z)}{2[1+\nu(z)]} \tag{8}$$

The force and moment resultants can be obtained as follow:

$$\left\{\begin{array}{c} N_{\varphi}^{i} \\ N_{\theta}^{i} \\ N_{\varphi\theta}^{i} \end{array}\right\}=\left[\begin{array}{ccc} A_{11} & A_{12} & 0 \\ A_{12} & A_{22} & 0 \\ 0 & 0 & A_{66} \end{array}\right]\left\{\begin{array}{c} \varepsilon_{\varphi}^{0,i} \\ \varepsilon_{\theta}^{0,i} \\ \gamma_{\varphi\theta}^{0,i} \end{array}\right\}+\left[\begin{array}{ccc} B_{11} & B_{12} & 0 \\ B_{12} & B_{22} & 0 \\ 0 & 0 & B_{66} \end{array}\right]\left\{\begin{array}{c} \varepsilon_{\varphi}^{0,i} \\ \varepsilon_{\theta}^{0,i} \\ \gamma_{\varphi\theta}^{0,i} \end{array}\right\} \tag{9a}$$

$$\left\{\begin{array}{c} M_{\varphi}^{i} \\ M_{\theta}^{i} \\ M_{\varphi\theta}^{i} \end{array}\right\}=\left[\begin{array}{ccc} B_{11} & B_{12} & 0 \\ B_{12} & B_{22} & 0 \\ 0 & 0 & B_{66} \end{array}\right]\left\{\begin{array}{c} \varepsilon_{\varphi}^{0,i} \\ \varepsilon_{\theta}^{0,i} \\ \gamma_{\varphi\theta}^{0,i} \end{array}\right\}+\left[\begin{array}{ccc} D_{11} & D_{12} & 0 \\ D_{12} & D_{22} & 0 \\ 0 & 0 & D_{66} \end{array}\right]\left\{\begin{array}{c} \kappa_{\varphi}^{i} \\ \kappa_{\theta}^{i} \\ \kappa_{\varphi\theta}^{i} \end{array}\right\} \tag{9b}$$

$$\left\{\begin{array}{c} Q_{\varphi}^{i} \\ Q_{\theta}^{i} \end{array}\right\}=\overline{\kappa}\left[\begin{array}{cc} A_{66} & 0 \\ 0 & A_{66} \end{array}\right]\left[\begin{array}{c} \gamma_{\varphi z}^{0,i} \\ \gamma_{\theta z}^{0,i} \end{array}\right] \tag{9c}$$

where $\overline{\kappa}$ is shear correction factor. A_{ij}, B_{ij} and D_{ij} are obtained by following integral:

$$(A_{ij}, B_{ij}, D_{ij}) = \int_{-h/2}^{h/2} Q_{ij}(z)(1, z, z^2)dz \tag{10}$$

The strain energy of the select segment can be expressed from Equation (11) as shown:

$$U^i = \frac{1}{2}\iiint_V \left(\begin{array}{c} N_{\varphi}^{i}\varepsilon_{\varphi}^{0,i} + N_{\theta}^{i}\varepsilon_{\theta}^{0,i} + N_{\varphi\theta}^{i}\gamma_{\varphi\theta}^{0,i} + M_{\varphi}^{i}k_{\varphi}^{i} + \\ M_{\theta}^{i}k_{\theta}^{i} + M_{\varphi\theta}^{i}k_{\varphi\theta}^{i} + Q_{\varphi}^{i}\gamma_{\varphi z}^{0,i} + Q_{\theta}^{i}\gamma_{\theta z}^{0,i} \end{array} \right) ABd\varphi d\theta dz \tag{11}$$

To save the space of this paper, the Equation (11) can be expressed as $U^i = U_S^i + U_B^i + U_{BC}^i$. The detailed description of U_S^i, U_B^i and U_{BC}^i are shown in Appendix A.

The maximum kinetic energy of the select segment can be obtained from Equation (12) as shown:

$$\begin{array}{l} T^i = \frac{1}{2}\iiint_V \rho(z)\left[\left(\dot{\overline{U}}^i\right)^2 + \left(\dot{\overline{V}}^i\right)^2 + \left(\dot{\overline{W}}^i\right)^2\right]\left(1+\frac{z}{R_\varphi}\right)\left(1+\frac{z}{R_\theta}\right)ABd\varphi d\theta dz = [\ ] \\ = \frac{1}{2}\int_{\varphi_0}^{\varphi_1}\int_0^{2\pi}\left\{I_0\left[\left(u^i\right)^2 + \left(v^i\right)^2 + \left(w^i\right)^2\right] + 2I_1\left(u^i\psi_\varphi^i + v^i\psi_\theta^i\right) + I_2\left[\left(\psi_\varphi^i\right)^2 + \left(\psi_\theta^i\right)^2\right]\right\}ABd\varphi d\theta \end{array} \tag{12}$$

where the dot denotes the differentiation about time, whereas three integrals are defined as follows:

$$(I_0, I_1, I_2) = \int_{-h/2}^{h/2} \rho(z)\left(1+\frac{z}{R_\varphi}\right)\left(1+\frac{z}{R_\theta}\right)\left(1,\ z,\ z^2\right)dz \tag{13}$$

The energy in two sides of boundary springs can be expressed as:

$$\begin{array}{l} U_b = \frac{1}{2}\int_0^{2\pi}\int_{-h/2}^{h/2}\left\{k_{u,0}u^2 + k_{v,0}v^2 + k_{w,0}w^2 + k_{\varphi,0}\psi_\varphi^2 + k_{\theta,0}\psi_\theta^2\right\}_{\varphi=\varphi_{r,0}} Bd\theta dz \\ +\frac{1}{2}\int_0^{2\pi}\int_{-h/2}^{h/2}\left\{k_{u,1}u^2 + k_{v,1}v^2 + k_{w,1}w^2 + k_{\varphi,1}\psi_\varphi^2 + k_{\theta,1}\psi_\theta^2\right\}_{\varphi=\varphi_{r,1}} Bd\theta dz \end{array} \tag{14}$$

where $k_{t,0}$ $(t = u, v, w, \varphi, \theta)$ and $k_{t,1}$ denote the value of springs at two sides.

The energy in connective springs of two neighbor segments is expressed as:

$$U_s^i = \frac{1}{2}\int_0^{2\pi}\int_{-h/2}^{h/2}\left\{\begin{array}{c} k_u\left(u^i - u^{i+1}\right)^2 + k_v\left(v^i - v^{i+1}\right)^2 + k_w\left(w^i - w^{i+1}\right)^2 \\ +k_\varphi\left(\psi_\varphi^i - \psi_\varphi^{i+1}\right)^2 + k_\theta\left(\psi_\theta^i - \psi_\theta^{i+1}\right)^2 \end{array}\right\}_{i,i+1} Bd\theta dz \tag{15}$$

The total energy of the constraint conditions can be expressed as:

$$U_{BC} = U_b + \sum_{i=1}^{H-1} U_s^i \tag{16}$$

2.3. Displacement Functions and Solution

Proper selection of the admissible displacement function is a critical factor for the accuracy of final solution [39–43]. As displayed in literatures [44,45], classical Jacobi polynomials are valued in range of $\phi \in [-1,1]$. Typical Jacobi polynomials $P_i^{(\alpha,\beta)}(\phi)$ of degree i are shown as below in present method.

$$P_0^{(\alpha,\beta)}(\phi) = 1 \tag{17a}$$

$$P_1^{(\alpha,\beta)}(\phi) = \frac{\alpha+\beta+2}{2}\phi - \frac{\alpha-\beta}{2} \tag{17b}$$

$$\begin{aligned} P_i^{(\alpha,\beta)}(\phi) = {} & \frac{(\alpha+\beta+2i-1)\{\alpha^2-\beta^2+\phi(\alpha+\beta+2i)(\alpha+\beta+2i-2)\}}{2i(\alpha+\beta+i)(\alpha+\beta+2i-2)} P_{i-1}^{(\alpha,\beta)}(\phi) \\ & - \frac{(\alpha+i-1)(\beta+i-1)(\alpha+\beta+2i)}{i(\alpha+\beta+i)(\alpha+\beta+2i-2)} P_{i-2}^{(\alpha,\beta)}(\phi) \end{aligned} \tag{17c}$$

where $\alpha, \beta > -1$ and $i = 2, 3, \ldots$

Thus, the displacement functions of shell segments can be written in form of Equation (18) as shown:

$$u = \sum_{m=0}^{M} U_m P_m^{(\alpha,\beta)}(\phi)\cos(n\theta)e^{i\omega t} \tag{18a}$$

$$v = \sum_{m=0}^{M} V_m P_m^{(\alpha,\beta)}(\phi)\sin(n\theta)e^{i\omega t} \tag{18b}$$

$$w = \sum_{m=0}^{M} W_m P_m^{(\alpha,\beta)}(\phi)\cos(n\theta)e^{i\omega t} \tag{18c}$$

$$\psi_\varphi = \sum_{m=0}^{M} \psi_{\varphi m} P_m^{(\alpha,\beta)}(\phi)\cos(n\theta)e^{i\omega t} \tag{18d}$$

$$\psi_\theta = \sum_{m=0}^{M} \psi_{\theta m} P_m^{(\alpha,\beta)}(\phi)\cos(n\theta)e^{i\omega t} \tag{18e}$$

where U_m, V_m, W_m, $\psi_{\varphi m}$ and $\psi_{\theta m}$ are unknown coefficients. n and m denote the semi wave number in axial and circumferential direction, respectively. M is highest degrees of semi wave number m. The total Lagrangian energy functions L can be obtained as it is shown in Equation (19):

$$L = \sum_{i=1}^{H}\left(T^i - U^i\right) - U_{BC} \tag{19}$$

The total Lagrangian energy function L is shown in Equation (20):

$$\frac{\partial L}{\partial \vartheta} = 0 \quad \vartheta = U_m, V_m, W_m, \psi_{\varphi m}, \psi_{\theta m} \tag{20}$$

Substituting Equations (11), (12), (16), (18), (19) into Equation (20), then Equation (21) can be obtained as:

$$\left(\mathbf{K} - \omega^2\mathbf{M}\right)\mathbf{Q} = 0 \tag{21}$$

where **K** and **M** denote stiffness and mass matrixes, respectively. **Q** is unknown coefficient matrix.

3. Analysis of Examples

The general boundary conditions are denoted by the abbreviations. Thus the abbreviations F, C, SD, SS and Ei respectively represent free, clamped, shear diaphragm, shear support and elastic boundary conditions. The material properties are chosen as: $E_m = 70$ GPa, $E_c = 168$ GPa, $\rho_c = 5700$ kg/m^3, $\rho_m = 2707$ kg/m^3, $v_m = v_c = 0.3$, $M = 8$, $\alpha = 0$, $\beta = -0.5$, $H = 5$. The geometrical dimensions are chosen as follows: R_0 = 0.2 m, R_1 = 1 m, L_p = 1 m, h_1:h_2:h_3:h_4:h_5 = 0.04:0.045:0.05:0.055:0.06 . The results of this paper are handle by: $\Omega = \omega R_1 \sqrt{\rho_c / E_c}$.

3.1. Convergence Analysis

Figure 4 shows the frequency parameter of stepped FGM$_I$ (a = 1; b = −0.5; c = 2; p = 2) doubly curved paraboloidal shell with different boundary parameters. We can get that the spring stiffness values in range of 10–10^{10} E_c can converge to stable, regardless of the kinds of spring. In other words, for clamped boundary condition, the spring stiffness can be assigned within the range of 10–10^{10} E_c. Based on the boundary parameters analysis, the general edge constraints are be provided as shown Table 1.

Table 1. Spring stiffness values.

BC	$k_{u,0}$, $k_{u,1}$	$k_{v,0}$, $k_{v,1}$	$k_{w,0}$, $k_{w,1}$	$k_{\varphi,0}$, $k_{\varphi,1}$	$k_{\theta,0}$, $k_{\theta,1}$
F	0	0	0	0	0
SD	0	10^3 E_c	10^3 E_c	0	0
SS	10^3 E_c	10^3 E_c	10^3 E_c	0	10^3 E_c
C	10^3 E_c	10^3 E_c	10^3 E_c	10^3 E_c	10^3 E_c
E1	10^{-3} E_c	10^3 E_c	10^3 E_c	10^3 E_c	10^3 E_c
E2	10^3 E_c	10^{-3} E_c	10^3 E_c	10^3 E_c	10^3 E_c
E3	10^{-3} E_c	10^{-3} E_c	10^3 E_c	10^3 E_c	10^3 E_c

The relative percentage errors of stepped FGM$_I$ (a = 1; b = −0.5; c = 2; p = 2) paraboloidal shell with various Jacobi parameters are presented in Figure 5. The results of $\alpha = \beta = 0$ are selected as the reference values. We can easily conclude from Figure 5 that different Jacobi parameters will lead to almost the same results when n is a fixed value. The maximum relative error is less than 8×10^{-8}. Thus, we can conclude that displacement functions consisting with Jacobi polynomial and Fourier series are perfectly appropriate. The most advantages of proposed method are the unified Jacobi polynomials, which make the displacement functions easier to select in contrast with other approaches. Figure 6 exhibits the results of stepped FG paraboloidal shell about truncation. We can get that the convergent results can be guaranteed when M is higher than 5. M is defined as the value of eight in this paper.

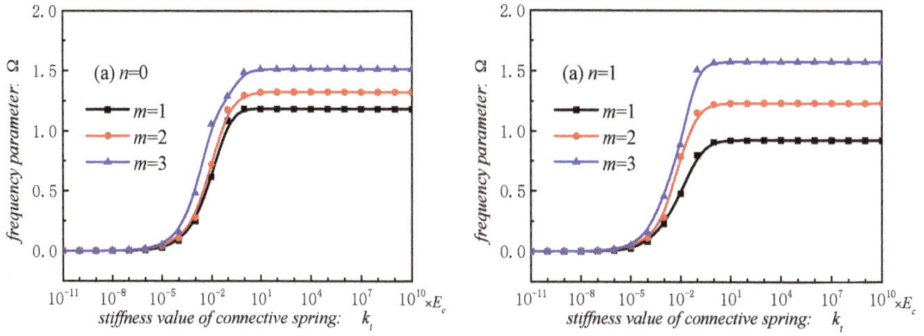

Figure 4. *Cont.*

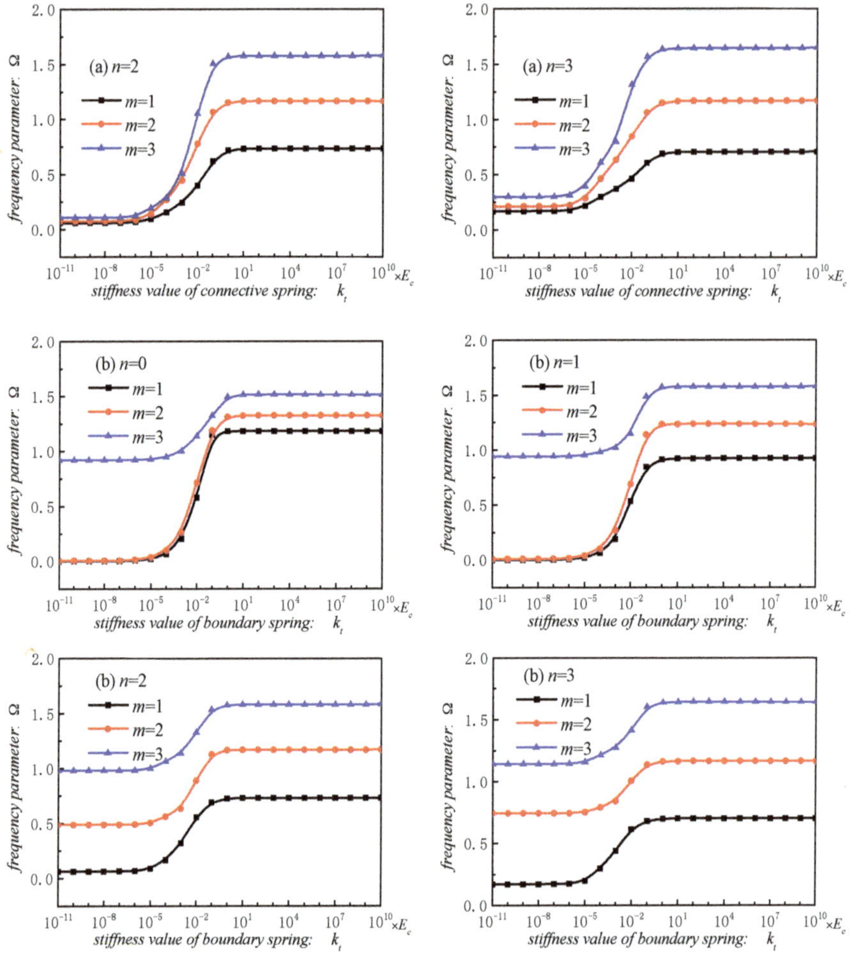

Figure 4. Frequency parameters Ω of stepped FG paraboloidal shell with various boundary parameters.

Table 2 exhibits the frequency parameter Ω of FGM_I ($a = 1$; $b = 0$; c; p) about the value of H, and the verification model is a spherical shell. The results are compared with those in literature [46]. From Table 2, we can conclude that the results will converge quickly as the value of H increase. We can also conclude that very high value of M is unnecessary. In addition, it can be obtained from Table 2 that the present method is strongly agreed with reference data.

Table 2. Frequency parameter Ω of the FGM_I ($a = 1$; $b = 0$; c; p) spherical shell structure (BC; C–C, $m = 1$).

Power-Law Exponent	Number of the Segment (H_e)							Ref [46]
	n	2	3	4	5	6	7	8	
$p = 0.6$	1	1.0569	1.0569	1.0568	1.0568	1.0568	1.0568	1.0568	1.0538
	2	1.0379	1.0376	1.0374	1.0372	1.0371	1.0371	1.0370	1.0354
	3	1.0319	1.0317	1.0314	1.0312	1.0312	1.0310	1.0310	1.0294
	4	1.0760	1.0757	1.0755	1.0752	1.0751	1.0750	1.0749	1.0733
	5	1.1588	1.1586	1.1584	1.1581	1.1581	1.1580	1.1580	1.1559

Table 2. *Cont.*

Power-Law Exponent	Number of the Segment (H_e)								Ref [46]
	n	2	3	4	5	6	7	8	
$p = 5$	1	1.0446	1.0446	1.0446	1.0445	1.0445	1.0445	1.0445	1.0411
	2	1.0116	1.0115	1.0113	1.0111	1.0110	1.0109	1.0108	1.0085
	3	1.0085	1.0083	1.0082	1.0080	1.0079	1.0079	1.0078	1.0053
	4	1.0572	1.0571	1.0569	1.0568	1.0566	1.0565	1.0563	1.0539
	5	1.1470	1.1468	1.1467	1.1465	1.1464	1.1464	1.1463	1.1433
$p = 20$	1	1.0282	1.0282	1.0281	1.0281	1.0281	1.0281	1.0281	1.0266
	2	0.9958	0.9957	0.9956	0.9954	0.9953	0.9953	0.9952	0.9945
	3	0.9927	0.9926	0.9924	0.9923	0.9922	0.9921	0.9920	0.9913
	4	1.0407	1.0405	1.0404	1.0403	1.0403	1.0402	1.0399	1.0392
	5	1.1290	1.1289	1.1287	1.1286	1.1285	1.1284	1.1284	1.1273

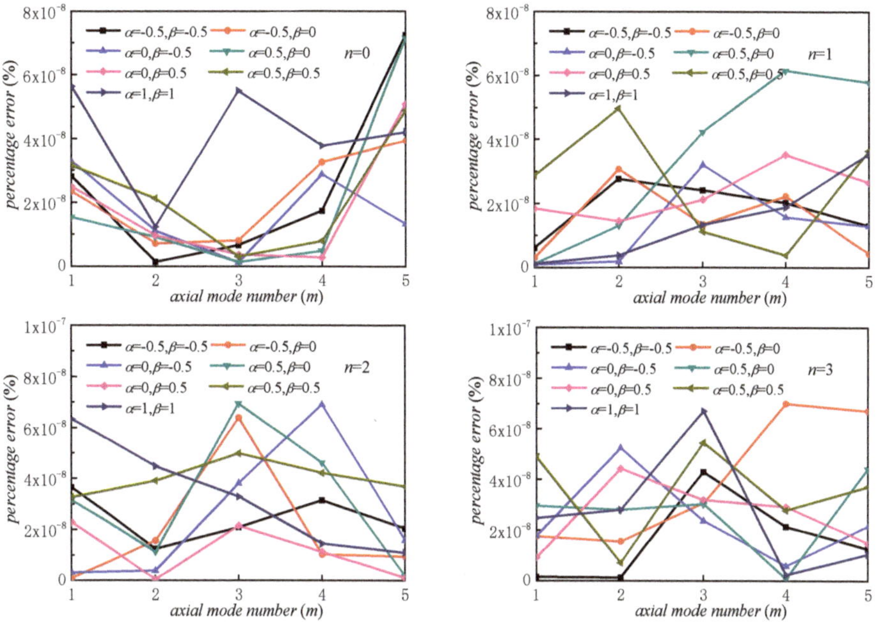

Figure 5. Relative error of frequency parameters Ω in stepped FG paraboloidal shell (BC: C–C).

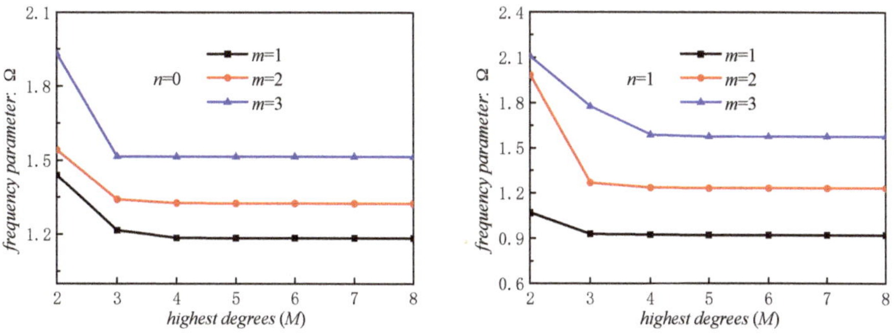

Figure 6. *Cont.*

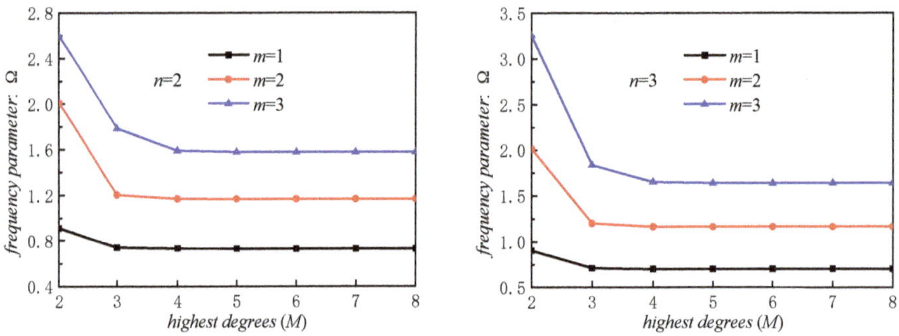

Figure 6. Frequency parameters Ω for various truncation in stepped FG paraboloidal shell.

3.2. Free vibration Behavior of Stepped FG Paraboloidal Shell

Table 3 shows the precision of the approach in solving free vibration behavior of stepped FG paraboloidal shell with clamed boundary condition, and all the FEM commercial program ABAQUS (S4R model) results have converged to stable when the element size is chosen as 0.03 m. In addition, it should be note that the homogeneous elements not graded elements [47] were used in this paper. From the comparison study, we can conclude that the present method is capable to analyze the vibration behaviors of stepped doubly curved paraboloidal shell with general boundary conditions.

Table 3. Comparison of frequency parameter Ω for stepped doubly curved paraboloidal shell (FGM_I $(a, b, c, p = 0)$).

n	m	Proposed Method	FEM
0	1	1.2139	1.2144
	2	1.3579	1.3586
	3	1.5621	1.5645
	4	1.6154	1.6183
1	1	0.9499	0.9504
	2	1.2605	1.2615
	3	1.6030	1.6070
	4	1.9770	1.9725
2	1	0.7521	0.7524
	2	1.1907	1.1924
	3	1.6002	1.6056
	4	2.1071	2.1083
3	1	0.7171	0.7176
	2	1.1811	1.1835
	3	1.6590	1.6566
	4	2.2217	2.2251

To further prove the effectiveness of this method, the experiment test focused on free vibration of cylindrical shell was carried out. It should be note that the cylindrical shell is isotropic. The material properties and geometrical parameters are chosen as: $E = 210$ GPa, $\rho = 7850$ kg/m^3, $\nu = 0.3$, $R = 0.06$ m, $L = 0.3$ m, $h = 0.005$ m. The boundary condition is free for isotropic cylindrical shell due to the of the restraints test environment. Figure 7 shows the test instrument and model. In experiment, the hammer was used to strike different positions of cylindrical shells in turn, and acceleration sensors with sensitivity of 100 mv/g were used to collect the vibration response at the same point. Then the time domain signals obtained by test were transformed into frequency domain signals by Fourier transform. The final results of frequencies are shown in Table 4. For natural frequencies obtained by FEM commercial program ABAQUS (S4R model), it is obvious that the structure and material parameters are the same as the experiment, and it should be note that the results have converge to stable when the mesh size is 0.03 m. From Table 4, it is easy to find that the present results closely agreed with experiment and FEM. For selected five modes, the maximum error of present method and experiment is 2.35%, and the maximum error of present method and FEM is 0.38%. The reason for the large error of present method with the test results are mainly the influence of elastic hoisting boundary and random error. The mode shapes obtained by three different methods are presented in Figure 8.

(**a**) (**b**)

Figure 7. Testing instruments and model. (**a**) The test system; (**b**) the test model.

Table 4. Comparison study of the frequencies for cylindrical shell.

n, m	Present	Experimental	Error (%)	FEM	Error (%)
0, 1	545.89	551.97	1.11	547.49	0.29
2, 2	582.13	588.39	1.08	581.98	0.03
0, 3	1561.93	1572.53	0.68	1567.90	0.38
2, 3	1618.37	1656.42	2.35	1613.70	0.29
3, 3	2143.98	2169.05	1.17	2150.70	0.31

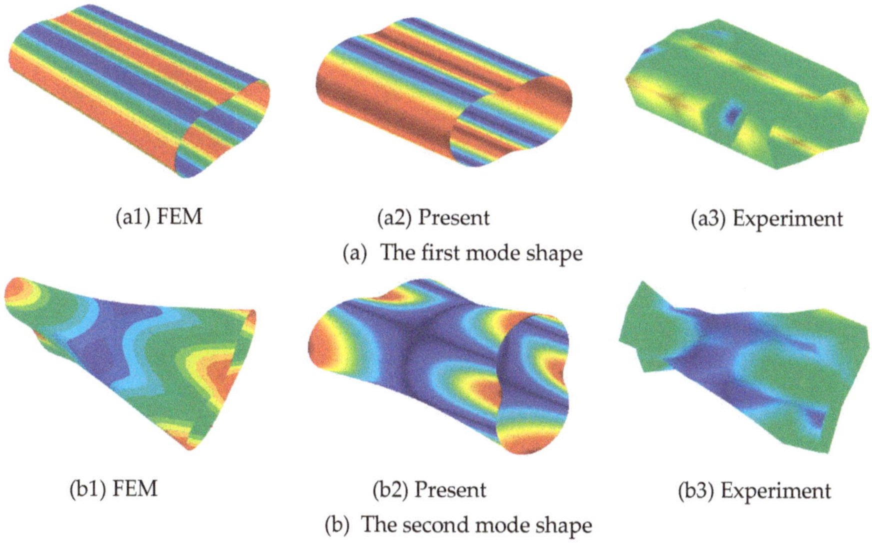

Figure 8. The selected mode shapes of three kinds of method.

Table 5 exhibits the results of free vibration behaviors for stepped FG paraboloidal shell with various boundary conditions. From Table 5, it is easy to find that the free vibration characteristics are not only influence by boundary conditions, but material parameters. To better reveal the vibration characteristics of the shell, some mode shapes are given in Figure 9.

Table 5. Frequency parameters Ω of stepped paraboloidal shell.

Type	n	m	Boundary Restraints									
			F–C	C–C	SD–SD	SS–SS	E1–E1	E2–E2	E3–E3	F–E1	F–E2	F–SS
FGM$_I$ (a = 1; b = −0.5; c = 2; p = 2)	1	1	0.7470	0.9238	0.6151	0.8886	0.6736	0.5307	0.2076	0.2171	0.4519	0.7301
		2	1.1767	1.2318	0.8874	1.1522	0.9171	1.2104	0.5468	0.8334	1.1066	1.0967
		3	1.4199	1.5746	1.1503	1.4582	1.2343	1.4709	1.1710	1.1866	1.4113	1.3661
		4	1.6262	1.9378	1.4683	1.8208	1.5746	1.7016	1.4638	1.4535	1.4271	1.6000
		5	1.7717	2.0975	1.8376	2.0156	2.0259	1.9062	1.5645	1.7696	1.7209	1.6947
	2	1	0.5453	0.7334	0.6855	0.6973	0.7196	0.6154	0.5819	0.4742	0.4983	0.5329
		2	0.9432	1.1666	1.0460	1.0800	1.1426	1.1078	0.9250	0.9212	0.8446	0.8894
		3	1.3195	1.5767	1.3190	1.4539	1.3368	1.5389	1.3106	1.3023	1.3013	1.2305
		4	1.7651	2.0848	1.4434	1.9063	1.5690	2.0412	1.4790	1.3588	1.7237	1.6529
		5	2.2154	2.5207	1.8933	2.4718	2.0823	2.2287	1.9788	1.7737	2.1426	2.1436
	3	1	0.6918	0.7037	0.6469	0.6597	0.6939	0.6630	0.6588	0.6816	0.6549	0.6516
		2	1.1043	1.1629	1.0674	1.0763	1.1594	1.1255	1.0998	1.1029	1.0725	1.0362
		3	1.4964	1.6405	1.5063	1.5173	1.6362	1.6151	1.5855	1.4963	1.4798	1.4098
		4	1.9638	2.2032	1.9827	2.0370	2.0021	2.1782	1.8904	1.9536	1.9463	1.8532
		5	2.5376	2.8897	2.0261	2.6731	2.1986	2.8331	2.1627	2.0105	2.5267	2.4127
FGM$_{II}$ (a = 1; b = −0.5; c = 2; p = 2)	1	1	0.7418	0.9171	0.6086	0.8875	0.6689	0.5274	0.2064	0.2161	0.4491	0.7153
		2	1.1663	1.2205	0.8821	1.1388	0.9110	1.1991	0.5427	0.8263	1.0974	1.1007
		3	1.4059	1.5589	1.1408	1.4569	1.2232	1.4590	1.1604	1.1759	1.4010	1.3328
		4	1.6117	1.9209	1.4556	1.7561	1.5588	1.6896	1.4532	1.4381	1.4141	1.6112
		5	1.7523	2.0820	1.8189	2.0554	2.0045	1.8919	1.5520	1.7494	1.7022	1.6794
	2	1	0.5378	0.7271	0.6821	0.6895	0.7136	0.6103	0.5768	0.4670	0.4920	0.5171
		2	0.9331	1.1553	1.0399	1.0802	1.1315	1.0968	0.9159	0.9117	0.8351	0.8909
		3	1.3062	1.5593	1.3101	1.4468	1.3285	1.5218	1.3017	1.2916	1.2878	1.2172
		4	1.7468	2.0615	1.4329	1.8881	1.5518	2.0196	1.4630	1.3468	1.7066	1.6445
		5	2.1931	2.5059	1.8788	2.4820	2.0593	2.2170	1.9600	1.7554	2.1281	2.0916
	3	1	0.6846	0.6968	0.6436	0.6477	0.6872	0.6566	0.6524	0.6746	0.6482	0.6389
		2	1.0906	1.1506	1.0609	1.0731	1.1471	1.1136	1.0877	1.0892	1.0592	1.0300
		3	1.4762	1.6214	1.4944	1.5073	1.6171	1.5961	1.5665	1.4761	1.4597	1.3933
		4	1.9404	2.1772	1.9727	2.0216	1.9913	2.1526	1.8796	1.9316	1.9231	1.8339
		5	2.5057	2.8601	2.0097	2.6553	2.1728	2.8113	2.1386	1.9983	2.4952	2.3801

Table 6 shows the results of stepped FG paraboloidal shell with different power-law exponents, in which four values are included. From Table 6, we can get that the boundary conditions and power-law exponents all will have important impact on the results of the structure.

Table 7 shows the results of stepped FG paraboloidal shell with different thickness distributions. Four kinds of thickness distributions, i.e., h_1:h_2:h_3:h_4:h_5 = 0.04:0.045:0.05:0.055:0.06 are included. It is obvious that the thickness distributions affect the vibration behavior of stepped FG paraboloidal shell largely.

Figures 10–12 exhibit the frequency parameters Ω of stepped FG paraboloidal shell with various parameters a, b, c and p. From selected data, it could be found that a, b and c have a great deal of impact on the results of Ω. In addition, for parameters a and c, the smaller value will obtain the larger results. Figure 13 exhibits the results of stepped FG paraboloidal shell with various stiffness ratios and parameter p. It can be seen that no matter what value of parameter p, the vibration characteristics will decrease with E_c/E_m increasing.

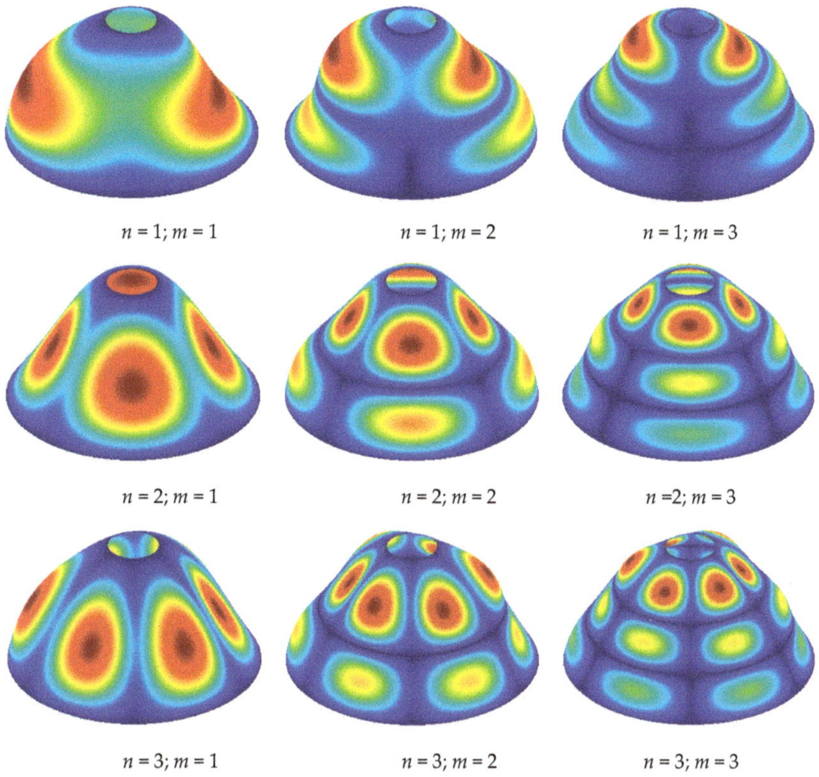

Figure 9. Mode shapes of stepped FG paraboloidal shell (BC: SS–SS).

Table 6. Frequency parameters Ω for stepped FGM_I (a = 1; b = 0.5; c = 2; p) shell with different power-law exponents.

Power-Law Exponents	n	m	C–C	SD–SD	F–SS
p = 0.2	1	1	0.9480	0.6315	0.7461
		2	1.2583	0.9133	1.1301
		3	1.6007	1.1769	1.3859
	2	1	0.7507	0.7047	0.5374
		2	1.1888	1.0703	0.9119
		3	1.5983	1.3557	1.2499
	3	1	0.7161	0.6612	0.6606
		2	1.1796	1.0868	1.0534
		3	1.6574	1.5264	1.4231
p = 0.5	1	1	0.9451	0.6297	0.7444
		2	1.2550	0.9104	1.1259
		3	1.5971	1.1737	1.3836
	2	1	0.7486	0.7025	0.5369
		2	1.1859	1.0673	0.9090
		3	1.5951	1.3514	1.2472
	3	1	0.7144	0.6593	0.6595
		2	1.1772	1.0841	1.0508
		3	1.6545	1.5232	1.4207

Table 6. *Cont.*

Power-Law Exponents	n	m	C–C	SD–SD	F–SS
$p = 2$	1	1	0.9321	0.6210	0.7359
		2	1.2395	0.8969	1.1078
		3	1.5796	1.1585	1.3711
	2	1	0.7389	0.6922	0.5332
		2	1.1721	1.0532	0.8959
		3	1.5789	1.3322	1.2339
	3	1	0.7064	0.6508	0.6534
		2	1.1650	1.0712	1.0389
		3	1.6394	1.5074	1.4081
$p = 5$	1	1	0.9164	0.6103	0.7236
		2	1.2221	0.8807	1.0892
		3	1.5625	1.1413	1.3542
	2	1	0.7276	0.6803	0.5280
		2	1.1575	1.0383	0.8832
		3	1.5646	1.3084	1.2210
	3	1	0.6983	0.6422	0.6460
		2	1.1540	1.0597	1.0285
		3	1.6282	1.4954	1.3991

Table 7. Frequency parameters Ω for stepped FGM_I ($a = 1$; $b = 0.5$; $c = 2$; $p = 2$) shell with different thickness distributions.

h_1:h_2:h_3:h_4:h_5	n	m	C–C	SD–SD	F–SS
0.04:0.05:0.06:0.07:0.08	1	1	0.9579	0.5884	0.7655
		2	1.3085	0.9008	1.1470
		3	1.6903	1.2140	1.4461
	2	1	0.7667	0.6952	0.5476
		2	1.2454	1.1009	0.9267
		3	1.7064	1.2969	1.3145
	3	1	0.7590	0.6841	0.6925
		2	1.2600	1.1513	1.1067
		3	1.7917	1.6410	1.5140
0.08:0.07:0.06:0.05:0.04	1	1	0.8600	0.6915	0.6176
		2	1.1979	0.9477	1.0841
		3	1.6283	1.1045	1.2400
	2	1	0.7026	0.6680	0.5916
		2	1.1782	1.0529	0.9674
		3	1.6697	1.4977	1.3075
	3	1	0.6992	0.6584	0.6578
		2	1.2297	1.1176	1.1192
		3	1.8009	1.6410	1.6103
0.04:0.06:0.08:0.07:0.05	1	1	0.8483	0.5982	0.6988
		2	1.2766	0.8162	1.1448
		3	1.6965	1.1861	1.3817
	2	1	0.6747	0.6343	0.5059
		2	1.2266	1.0603	0.9036
		3	1.7104	1.4039	1.2943
	3	1	0.6993	0.6493	0.6457
		2	1.2530	1.1403	1.1109
		3	1.8046	1.6523	1.5196

Table 7. *Cont.*

h_1:h_2:h_3:h_4:h_5	n	m	C–C	SD–SD	F–SS
0.07:0.05:0.04:0.06:0.08	1	1	1.0086	0.6664	0.7052
		2	1.2356	0.9912	1.0789
		3	1.6421	1.1382	1.4098
	2	1	0.8278	0.7647	0.6457
		2	1.1948	1.0710	0.9595
		3	1.6748	1.2893	1.3483
	3	1	0.8000	0.7185	0.7324
		2	1.2351	1.1212	1.1102
		3	1.7878	1.6336	1.5970

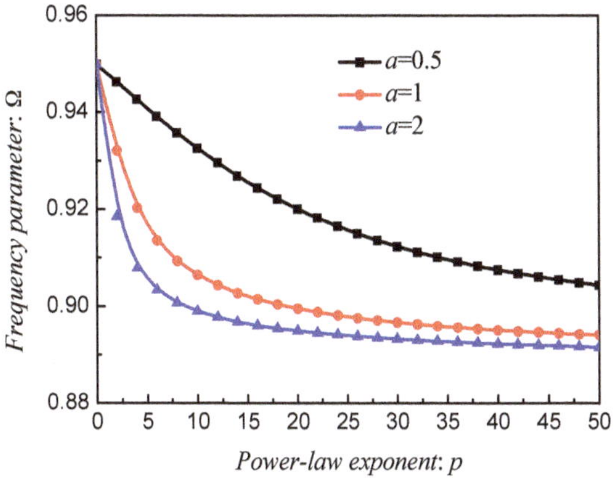

Figure 10. Results about different p and a of stepped FGM_I (a, b = 0.5; c = 2; p) paraboloidal shell.

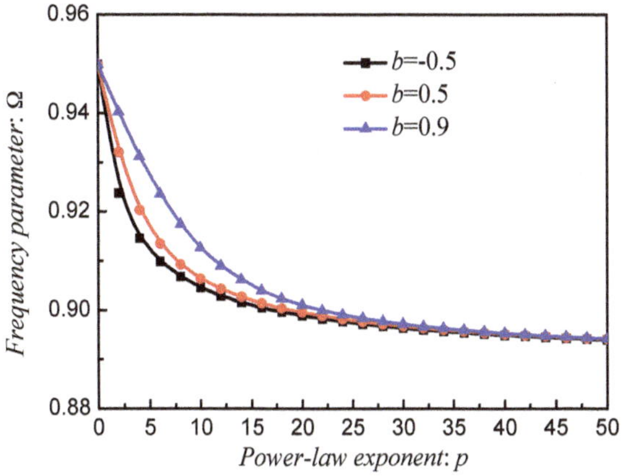

Figure 11. Results about different p and b of stepped FGM_I (a = 1; b, c = 2; p) paraboloidal shell.

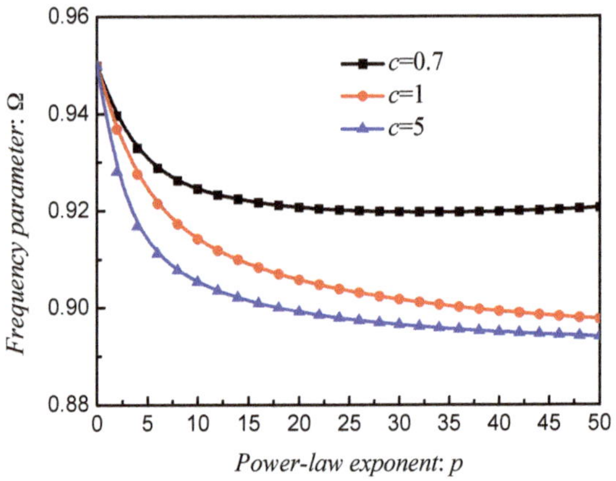

Figure 12. Results about different p and c of stepped FGM_I ($a = 1; b = 0.5; c; p$) paraboloidal shell.

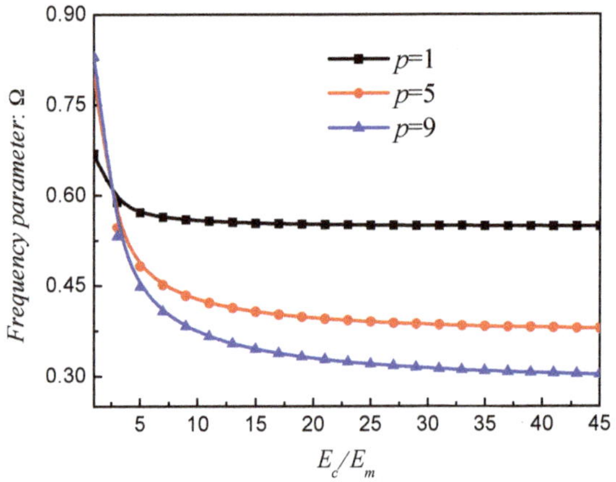

Figure 13. Results about different E_c/E_m and p of stepped FGM_I ($a = 1; b = 0.5; c = 2; p$) paraboloidal shell.

4. Conclusions

The paper proposed a solving formulation to investigate the free vibration behaviors of stepped FG paraboloidal shell with general boundary conditions. The paper is based on multi-segment strategy and FSDT. The displacement functions are simulated by Jacobi polynomials and Fourier series. To obtain the general boundary conditions of stepped FG paraboloidal shell, the penalty method was adopted. The final modes solutions about FG paraboloidal shell were obtained by Rayleigh–Ritz method. The most discoveries of proposed method are unified Jacobi polynomials, which make the displacement functions easier to select. For convergence analysis, the influence of boundary parameters, numbers of shell segments etc. are examined. The accuracy of this method was verified by the comparison study with those obtained by published literature, FEM, and the experiment. The results of this paper can provide the reference data for future research.

Author Contributions: Conceptualization, F.P. and H.L.; Data curation, H.L. and F.X.; Formal analysis, H.L.; Funding acquisition, F.P.; Investigation, H.L.; Project administration, H.L. and Y.D.

Funding: This study was funded by Naval pre-research project, Ph.D. Student Research and Innovation Fund of the Fundamental Research Funds for the Central Universities (HEUGIP201801), Major innovation projects Of High Technology Ship Funds of Ministry of Industry and Information of China, High Technology Ship Funds of Ministry of Industry and Information of China, National key Research and Development program (2016YFC0303406), Assembly Advanced Research Fund Of China (6140210020105).

Conflicts of Interest: The authors declare that there is no conflict of interest regarding the publication of this paper.

Appendix A

$$\mathbf{M} = diag\left[\mathbf{M}^1, \mathbf{M}^2, \cdots, \mathbf{M}^H\right] \tag{A1}$$

$$\mathbf{M}^i = \int_{\varphi_i}^{\varphi_{i+1}} \int_0^{2\pi} \begin{bmatrix} M_{uu} & 0 & 0 & M_{u\varphi} & 0 \\ 0 & M_{vv} & 0 & 0 & M_{v\theta} \\ 0 & 0 & M_{ww} & 0 & 0 \\ M_{u\varphi} & 0 & 0 & M_{\varphi\varphi} & 0 \\ 0 & M_{v\theta} & 0 & 0 & M_{\theta\theta} \end{bmatrix} ABd\varphi d\theta \tag{A2}$$

$$\mathbf{M}_{uu} = I_0\mathbf{U}^T\mathbf{U}, \mathbf{M}_{vv} = I_0\mathbf{V}^T\mathbf{V}, \mathbf{M}_{ww} = I_0\mathbf{W}^T\mathbf{W}, \mathbf{M}_{\varphi\varphi} = I_2\mathbf{\Phi}^T\mathbf{\Phi} \tag{A3}$$

$$\mathbf{M}_{\theta\theta} = I_2\mathbf{\Theta}^T\mathbf{\Theta}, \mathbf{M}_{u\varphi} = I_1\mathbf{U}^T\mathbf{\Phi}, \mathbf{M}_{v\theta} = I_1\mathbf{V}^T\mathbf{\Theta} \tag{A4}$$

$$\mathbf{U} = \mathbf{P}_m \otimes \mathbf{C}_n,\ \mathbf{V} = \mathbf{P}_m \otimes \mathbf{S}_n,\ \mathbf{W} = \mathbf{P}_m \otimes \mathbf{C}_n,\ \mathbf{\Phi} = \mathbf{P}_m \otimes \mathbf{C}_n,\ \mathbf{\Theta} = \mathbf{P}_m \otimes \mathbf{S}_n \tag{A5}$$

$$\mathbf{P}_m = [P_0^{(\alpha,\beta)}(\phi), P_1^{(\alpha,\beta)}(\phi), \cdots, P_m^{(\alpha,\beta)}(\phi), \cdots, P_M^{(\alpha,\beta)}(\phi)] \tag{A6}$$

$$\mathbf{C}_n = [\cos(0\theta), \cos(1\theta), \cdots \cos(n\theta), \cdots, \cos(N\theta)] \tag{A7}$$

$$\mathbf{S}_n = [\sin(0\theta), \sin(1\theta), \cdots \sin(n\theta), \cdots, \sin(N\theta)] \tag{A8}$$

$$\mathbf{K} = \mathbf{K}_\xi + \mathbf{K}_b + \mathbf{K}_s \tag{A9}$$

$$\mathbf{K}_\xi = diag\left[\mathbf{K}_\xi^1, \mathbf{K}_\xi^2, \cdots, \mathbf{K}_\xi^H\right] \tag{A10}$$

$$\mathbf{K}_\xi^i = \int_{\varphi_{\xi,i}}^{\varphi_{\xi,i+1}} \int_0^{2\pi} \begin{bmatrix} K_{\xi,uu} & K_{\xi,uv} & K_{\xi,uw} & K_{\xi,u\varphi} & K_{\xi,u\theta} \\ K_{\xi,uv}^T & K_{\xi,vv} & K_{\xi,vw} & K_{\xi,v\varphi} & K_{\xi,v\theta} \\ K_{\xi,uw}^T & K_{\xi,vw}^T & K_{\xi,ww} & K_{\xi,w\varphi} & K_{\xi,w\theta} \\ K_{\xi,u\varphi}^T & K_{\xi,v\varphi}^T & K_{\xi,w\varphi}^T & K_{\xi,\varphi\varphi} & K_{\xi,\varphi\theta} \\ K_{\xi,u\theta}^T & K_{\xi,v\theta}^T & K_{\xi,w\theta}^T & K_{\xi,\varphi\theta}^T & K_{\xi,\theta\theta} \end{bmatrix} ABd\varphi d\theta \tag{A11}$$

$$\mathbf{K}_b = diag[\mathbf{K}_{bl}, 0, \cdots, \mathbf{K}_{br}] \tag{A12}$$

$$\mathbf{K}_{bl} = \int_0^{2\pi} diag\left[\mathbf{K}_{bl,uu}, K_{bl,vv}, K_{bl,ww}, K_{bl,\varphi\varphi}, K_{bl,\theta\theta}\right]\Big|_{\varphi=\varphi_0} Bd\theta \tag{A13}$$

$$\mathbf{K}_{br} = \int_0^{2\pi} diag\left[K_{br,uu}, K_{br,vv}, K_{br,ww}, K_{br,\varphi\varphi}, K_{br,\theta\theta}\right]\Big|_{\varphi=\varphi_1} Bd\theta \tag{A14}$$

$$\mathbf{K}_s = diag\left[\mathbf{K}_s^1, \mathbf{K}_s^2, \cdots, \mathbf{K}_s^H\right] \tag{A15}$$

$$\mathbf{K}_s^i = \int_0^{2\pi} \begin{bmatrix} K_{s0} & K_{s1} \\ K_{s1}^T & K_{s2} \end{bmatrix} Bd\theta \tag{A16}$$

$$\mathbf{K}_{s0} = diag\left[\mathbf{K}_{u_iu_i}, \mathbf{K}_{v_iv_i}, \mathbf{K}_{w_iw_i}, \mathbf{K}_{\varphi_i\varphi_i}, \mathbf{K}_{\varphi_i\varphi_i}\right] \tag{A17}$$

$$\mathbf{K}_{s1} = diag\left[\mathbf{K}_{u_iu_{i+1}}, \mathbf{K}_{v_iv_{i+1}}, \mathbf{K}_{w_iw_{i+1}}, \mathbf{K}_{\varphi_i\varphi_{i+1}}, \mathbf{K}_{\varphi_i\varphi_{i+1}}\right] \tag{A18}$$

$$\mathbf{K}_{s2} = diag\left[\mathbf{K}_{u_{i+1}u_{i+1}}, \mathbf{K}_{v_{i+1}v_{i+1}}, \mathbf{K}_{w_{i+1}w_{i+1}}, \mathbf{K}_{\varphi_{i+1}\varphi_{i+1}}, \mathbf{K}_{\varphi_{i+1}\varphi_{i+1}}\right] \tag{A19}$$

$$U_S^i = \frac{1}{2}\iiint \left\{ \begin{array}{l} A_{11}\left(\frac{1}{A}\frac{\partial u^i}{\partial\varphi} + \frac{v^i}{AB}\frac{\partial A}{\partial\theta} + \frac{w^i}{R_\varphi}\right)^2 + A_{22}\left(\frac{1}{B}\frac{\partial v^i}{\partial\theta} + \frac{u^i}{AB}\frac{\partial B}{\partial\varphi} + \frac{w^i}{R_\theta}\right)^2 \\ +A_{66}\left(\frac{A}{B}\frac{\partial}{\partial\theta}\left(\frac{u^i}{A}\right) + \frac{B}{A}\frac{\partial}{\partial\varphi}\left(\frac{v^i}{B}\right)\right)^2 + \\ 2A_{12}\left(\frac{1}{A}\frac{\partial u^i}{\partial\varphi} + \frac{v^i}{AB}\frac{\partial A}{\partial\theta} + \frac{w^i}{R_\varphi}\right)\left(\frac{1}{B}\frac{\partial v^i}{\partial\theta} + \frac{u^i}{AB}\frac{\partial B}{\partial\varphi} + \frac{w^i}{R_\theta}\right) + \\ +\overline{\kappa}A_{66}\left(\frac{1}{A}\frac{\partial w^i}{\partial\varphi} - \frac{u^i}{R_\varphi} + \psi_\varphi^i\right)^2 + \overline{\kappa}A_{66}\left(\frac{1}{B}\frac{\partial w^i}{\partial\theta} - \frac{v^i}{R_\theta} + \psi_\theta^i\right)^2 \end{array} \right\} ABd\varphi d\theta dz \quad \text{(A20)}$$

$$U_B^i = \frac{1}{2}\iiint \left\{ \begin{array}{l} D_{11}\left(\frac{1}{A}\frac{\partial\psi_\varphi^i}{\partial\varphi} + \frac{\psi_\theta^i}{AB}\frac{\partial A}{\partial\theta}\right)^2 + D_{22}\left(\frac{1}{B}\frac{\partial\psi_\theta^i}{\partial\theta} + \frac{\psi_\varphi^i}{AB}\frac{\partial B}{\partial\varphi}\right)^2 \\ +D_{66}\left(\frac{A}{B}\frac{\partial}{\partial\theta}\left(\frac{\psi_\varphi^i}{A}\right) + \frac{B}{A}\frac{\partial}{\partial\varphi}\left(\frac{\psi_\theta^i}{B}\right)\right)^2 \\ +2D_{12}\left(\frac{1}{A}\frac{\partial\psi_\varphi^i}{\partial\varphi} + \frac{\psi_\theta^i}{AB}\frac{\partial A}{\partial\theta}\right)\left(\frac{1}{B}\frac{\partial\psi_\theta^i}{\partial\theta} + \frac{\psi_\varphi^i}{AB}\frac{\partial B}{\partial\varphi}\right) \end{array} \right\} ABd\varphi d\theta dz \quad \text{(A21)}$$

$$U_{BS}^i = \iiint \left\{ \begin{array}{l} B_{11}\left(\frac{1}{A}\frac{\partial u^i}{\partial\varphi} + \frac{v^i}{AB}\frac{\partial A}{\partial\theta} + \frac{w^i}{R_\varphi}\right)\left(\frac{1}{A}\frac{\partial\psi_\varphi^i}{\partial\varphi} + \frac{\psi_\theta^i}{AB}\frac{\partial A}{\partial\theta}\right) \\ +B_{12}\left(\frac{1}{A}\frac{\partial u^i}{\partial\varphi} + \frac{v^i}{AB}\frac{\partial A}{\partial\theta} + \frac{w^i}{R_\varphi}\right)\left(\frac{1}{B}\frac{\partial\psi_\theta^i}{\partial\theta} + \frac{\psi_\varphi^i}{AB}\frac{\partial B}{\partial\varphi}\right) \\ +B_{12}\left(\frac{1}{B}\frac{\partial v^i}{\partial\theta} + \frac{u^i}{AB}\frac{\partial B}{\partial\varphi} + \frac{w^i}{R_\theta}\right)\left(\frac{1}{A}\frac{\partial\psi_\varphi^i}{\partial\varphi} + \frac{\psi_\theta^i}{AB}\frac{\partial A}{\partial\theta}\right) + \\ B_{66}\left(\frac{1}{A}\frac{\partial u^i}{\partial\varphi} + \frac{v^i}{AB}\frac{\partial A}{\partial\theta} + \frac{w^i}{R_\varphi}\right)\left(\frac{A}{B}\frac{\partial}{\partial\theta}\left(\frac{\psi_\varphi^i}{A}\right) + \frac{B}{A}\frac{\partial}{\partial\varphi}\left(\frac{\psi_\theta^i}{B}\right)\right) \\ +B_{22}\left(\frac{1}{B}\frac{\partial v^i}{\partial\theta} + \frac{u^i}{AB}\frac{\partial B}{\partial\varphi} + \frac{w^i}{R_\theta}\right)\left(\frac{1}{B}\frac{\partial\psi_\theta^i}{\partial\theta} + \frac{\psi_\varphi^i}{AB}\frac{\partial B}{\partial\varphi}\right) \end{array} \right\} ABd\varphi d\theta dz \quad \text{(A22)}$$

References

1. Fantuzzi, N.; Brischetto, S.; Tornabene, F.; Viola, E. 2D and 3D shell models for the free vibration investigation of functionally graded cylindrical and spherical panels. *Compos. Struct.* **2016**, *154*, 573–590. [CrossRef]
2. Tornabene, F.; Reddy, J.N. FGM and Laminated Doubly-Curved and Degenerate Shells Resting on Nonlinear Elastic Foundations: A GDQ Solution for Static Analysis with a Posteriori Stress and Strain Recovery. *J. Indian Inst. Sci.* **2013**, *93*, 635–688.
3. Pradyumna, S.; Bandyopadhyay, J.N. Free vibration analysis of functionally graded curved panels using a higher-order finite element formulation. *J. Sound Vib.* **2008**, *318*, 176–192. [CrossRef]
4. Jouneghani, F.Z.; Dimitri, R.; Bacciocchi, M.; Tornabene, F. Free Vibration Analysis of Functionally Graded Porous Doubly-Curved Shells Based on the First-Order Shear Deformation Theory. *Appl. Sci.* **2017**, *7*, 1252. [CrossRef]
5. Chen, H.Y.; Wang, A.W.; Hao, Y.X.; Zhang, W. Free vibration of FGM sandwich doubly-curved shallow shell based on a new shear deformation theory with stretching effects. *Compos. Struct.* **2017**, *179*, 50–60. [CrossRef]
6. Wang, Q.S.; Cui, X.H.; Qin, B.; Liang, Q.; Tang, J.Y. A semi-analytical method for vibration analysis of functionally graded (FG) sandwich doubly-curved panels and shells of revolution. *Int. J. Mech. Sci.* **2017**, *134*, 479–499. [CrossRef]
7. Wang, Q.S.; Qin, B.; Shi, D.Y.; Liang, Q. A semi-analytical method for vibration analysis of functionally graded carbon nanotube reinforced composite doubly-curved panels and shells of revolution. *Compos. Struct.* **2017**, *174*, 87–109. [CrossRef]
8. Wang, Q.S.; Cui, X.H.; Qin, B.; Liang, Q. Vibration analysis of the functionally graded carbon nanotube reinforced composite shallow shells with arbitrary boundary conditions. *Compos. Struct.* **2017**, *182*, 364–379. [CrossRef]
9. Wang, Q.S.; Shi, D.Y.; Liang, Q.; Pang, F.Z. Free vibration of moderately thick functionally graded parabolic and circular panels and shells of revolution with general boundary conditions. *Eng. Comput.* **2017**, *34*, 1598–1641. [CrossRef]
10. Tornabene, F.; Viola, E. Free vibrations of four-parameter functionally graded parabolic panels and shells of revolution. *Eur. J. Mech. A Solids* **2009**, *28*, 991–1013. [CrossRef]
11. Viola, E.; Tornabene, F. Free vibrations of three parameter functionally graded parabolic panels of revolution. *Mech. Res. Commun.* **2009**, *36*, 587–594. [CrossRef]

12. Tornabene, F.; Viola, E. Static analysis of functionally graded doubly-curved shells and panels of revolution. *Meccanica* **2013**, *48*, 901–930. [CrossRef]
13. Fazzolari, F.A.; Carrera, E. Refined hierarchical kinematics quasi-3D Ritz models for free vibration analysis of doubly curved FGM shells and sandwich shells with FGM core. *J. Sound Vib.* **2014**, *333*, 1485–1508. [CrossRef]
14. Kar, V.R.; Panda, S.K. Free vibration responses of functionally graded spherical shell panels using finite element method. In Proceedings of the ASME 2013 Gas Turbine India Conference, Bangalore, India, 5–6 December 2013; p. V001T005A014.
15. Tornabene, F. Free vibration analysis of functionally graded conical, cylindrical shell and annular plate structures with a four-parameter power-law distribution. *Comput. Methods Appl. Mech. Eng.* **2009**, *198*, 2911–2935. [CrossRef]
16. Zghal, S.; Frikha, A.; Dammak, F. Free vibration analysis of carbon nanotube-reinforced functionally graded composite shell structures. *Appl. Math. Model.* **2018**, *53*, 132–155. [CrossRef]
17. Kulikov, G.M.; Plotnikova, S.V.; Kulikov, M.G.; Monastyrev, P.V. Three-dimensional vibration analysis of layered and functionally graded plates through sampling surfaces formulation. *Compos. Struct.* **2016**, *152*, 349–361. [CrossRef]
18. Kapuria, S.; Patni, M.; Yasin, M.Y. A quadrilateral shallow shell element based on the third-order theory for functionally graded plates and shells and the inaccuracy of rule of mixtures. *Eur. J. Mech. A Solids* **2015**, *49*, 268–282. [CrossRef]
19. Hosseini-Hashemi, S.; Derakhshani, M.; Fadaee, M. An accurate mathematical study on the free vibration of stepped thickness circular/annular Mindlin functionally graded plates. *Appl. Math. Model.* **2013**, *37*, 4147–4164. [CrossRef]
20. Bambill, D.V.; Rossit, C.A.; Felix, D.H. Free vibrations of stepped axially functionally graded Timoshenko beams. *Meccanica* **2015**, *50*, 1073–1087. [CrossRef]
21. Vinyas, M.; Kattimani, S.C. Static analysis of stepped functionally graded magneto-electro-elastic plates in thermal environment: A finite element study. *Compos. Struct.* **2017**, *178*, 63–86. [CrossRef]
22. Vinyas, M.; Kattimani, S.C. Static studies of stepped functionally graded magneto-electro-elastic beam subjected to different thermal loads. *Compos. Struct.* **2017**, *163*, 216–237. [CrossRef]
23. Su, Z.; Jin, G.Y.; Ye, T.G. Vibration analysis of multiple-stepped functionally graded beams with general boundary conditions. *Compos. Struct.* **2018**, *186*, 315–323. [CrossRef]
24. Li, H.; Pang, F.; Chen, H.; Du, Y. Vibration analysis of functionally graded porous cylindrical shell with arbitrary boundary restraints by using a semi analytical method. *Compos. Part B Eng.* **2019**, *164*, 249–264. [CrossRef]
25. Li, H.; Pang, F.; Wang, X.; Du, Y.; Chen, H. Free vibration analysis for composite laminated doubly-curved shells of revolution by a semi analytical method. *Compos. Struct.* **2018**, *201*, 86–111. [CrossRef]
26. Tornabene, F.; Fantuzzi, N.; Bacciocchi, M.; Reddy, J.N. An Equivalent Layer-Wise Approach for the Free Vibration Analysis of Thick and Thin Laminated and Sandwich Shells. *Appl. Sci.* **2017**, *7*, 17. [CrossRef]
27. Tornabene, F.; Fantuzzi, N.; Bacciocchi, M. A new doubly-curved shell element for the free vibrations of arbitrarily shaped laminated structures based on Weak Formulation IsoGeometric Analysis. *Compos. Struct.* **2017**, *171*, 429–461. [CrossRef]
28. Tornabene, F.; Fantuzzi, N.; Bacciocchi, M.; Viola, E. Effect of agglomeration on the natural frequencies of functionally graded carbon nanotube-reinforced laminated composite doubly-curved shells. *Compos. Part B Eng.* **2016**, *89*, 187–218. [CrossRef]
29. Tornabene, F.; Fantuzzi, N.; Viola, E.; Batra, R.C. Stress and strain recovery for functionally graded free-form and doubly-curved sandwich shells using higher-order equivalent single layer theory. *Compos. Struct.* **2015**, *119*, 67–89. [CrossRef]
30. Li, H.; Pang, F.; Wang, X.; Li, S. Benchmark Solution for Free Vibration of Moderately Thick Functionally Graded Sandwich Sector Plates on Two-Parameter Elastic Foundation with General Boundary Conditions. *Shock Vib.* **2017**, *2017*, 4018629. [CrossRef]
31. Li, H.; Liu, N.; Pang, F.; Du, Y.; Li, S. An Accurate Solution Method for the Static and Vibration Analysis of Functionally Graded Reissner-Mindlin Rectangular Plate with General Boundary Conditions. *Shock Vib.* **2018**, *2018*, 4535871. [CrossRef]

32. Zhong, R.; Wang, Q.; Tang, J.; Shuai, C.; Qin, B. Vibration analysis of functionally graded carbon nanotube reinforced composites (FG-CNTRC) circular, annular and sector plates. *Compos. Struct.* **2018**, *194*, 49–67. [CrossRef]
33. Fantuzzi, N.; Tornabene, F.; Viola, E. Four-parameter functionally graded cracked plates of arbitrary shape: A GDQFEM solution for free vibrations. *Mech. Adv. Mater. Struct.* **2016**, *23*, 89–107. [CrossRef]
34. Choe, K.; Tang, J.; Shui, C.; Wang, A.; Wang, Q. Free vibration analysis of coupled functionally graded (FG) doubly-curved revolution shell structures with general boundary conditions. *Compos. Struct.* **2018**, *194*, 413–432. [CrossRef]
35. Zhao, J.; Zhang, Y.; Choe, K.; Qu, X.; Wang, A.; Wang, Q. Three-dimensional exact solution for the free vibration of thick functionally graded annular sector plates with arbitrary boundary conditions. *Compos. Part B Eng.* **2019**, *159*, 418–436. [CrossRef]
36. Zhao, J.; Choe, K.; Xie, F.; Wang, A.; Shuai, C.; Wang, Q. Three-dimensional exact solution for vibration analysis of thick functionally graded porous (FGP) rectangular plates with arbitrary boundary conditions. *Compos. Part B Eng.* **2018**, *155*, 369–381. [CrossRef]
37. Guo, J.; Shi, D.; Wang, Q.; Tang, J.; Shuai, C. Dynamic analysis of laminated doubly-curved shells with general boundary conditions by means of a domain decomposition method. *Int. J. Mech. Sci.* **2018**, *138–139*, 159–186. [CrossRef]
38. Pang, F.; Li, H.; Du, Y.; Li, S.; Chen, H.; Liu, N. A Series Solution for the Vibration of Mindlin Rectangular Plates with Elastic Point Supports around the Edges. *Shock Vib.* **2018**, *2018*, 8562079. [CrossRef]
39. Li, H.; Pang, F.; Chen, H. A semi-analytical approach to analyze vibration characteristics of uniform and stepped annular-spherical shells with general boundary conditions. *Eur. J. Mech. A Solids* **2019**, *74*, 48–65. [CrossRef]
40. Li, H.; Pang, F.; Miao, X.; Du, Y.; Tian, H. A semi-analytical method for vibration analysis of stepped doubly-curved shells of revolution with arbitrary boundary conditions. *Thin-Walled Struct.* **2018**, *129*, 125–144. [CrossRef]
41. Li, H.; Pang, F.; Miao, X.; Li, Y. Jacobi–Ritz method for free vibration analysis of uniform and stepped circular cylindrical shells with arbitrary boundary conditions: A unified formulation. *Comput. Math. Appl.* **2018**. [CrossRef]
42. Li, H.; Pang, F.; Wang, X.; Du, Y.; Chen, H. Free vibration analysis of uniform and stepped combined paraboloidal, cylindrical and spherical shells with arbitrary boundary conditions. *Int. J. Mech. Sci.* **2018**, *145*, 64–82. [CrossRef]
43. Pang, F.; Li, H.; Wang, X.; Miao, X.; Li, S. A semi analytical method for the free vibration of doubly-curved shells of revolution. *Comput. Math. Appl.* **2018**, *75*, 3249–3268. [CrossRef]
44. Bhrawy, A.H.; Taha, T.M.; Machado, J.A.T. A review of operational matrices and spectral techniques for fractional calculus. *Nonlinear Dyn.* **2015**, *81*, 1023–1052. [CrossRef]
45. Pang, F.; Li, H.; Choe, K.; Shi, D.; Kim, K. Free and Forced Vibration Analysis of Airtight Cylindrical Vessels with Doubly Curved Shells of Revolution by Using Jacobi-Ritz Method. *Shock Vib.* **2017**, *2017*, 4538540. [CrossRef]
46. Qu, Y.; Long, X.; Yuan, G.; Meng, G. A unified formulation for vibration analysis of functionally graded shells of revolution with arbitrary boundary conditions. *Compos. Part B Eng.* **2013**, *50*, 381–402. [CrossRef]
47. Martínez-Pañeda, E.; Gallego, R. Numerical analysis of quasi-static fracture in functionally graded materials. *Int. J. Mech. Mater. Des.* **2015**, *11*, 405–424. [CrossRef]

materials

MDPI

Article

A Theory of Elastic/Plastic Plane Strain Pure Bending of FGM Sheets at Large Strain

Sergey Alexandrov [1,2], Yun-Che Wang [3,*] and Lihui Lang [1]

1 School of Mechanical Engineering and Automation, Beihang University, No. 37 Xueyuan Road, Beijing 100191, China; sergei_alexandrov@spartak.ru (S.A.); lang@buaa.edu.cn (L.L.)
2 Ishlinsky Institute for Problems in Mechanics, 101-1 Prospect Vernadskogo, 119526 Moscow, Russia
3 Department of Civil Engineering, National Cheng Kung University, 1 University Road, Tainan 70101, Taiwan
* Correspondence: yunche@mail.ncku.edu.tw

Received: 31 December 2018; Accepted: 28 January 2019; Published: 1 February 2019

Abstract: An efficient analytical/numerical method has been developed and programmed to predict the distribution of residual stresses and springback in plane strain pure bending of functionally graded sheets at large strain, followed by unloading. The solution is facilitated by using a Lagrangian coordinate system. The study is concentrated on a power law through thickness distribution of material properties. However, the general method can be used in conjunction with any other through thickness distributions assuming that plastic yielding initiates at one of the surfaces of the sheet. Effects of material properties on the distribution of residual stresses are investigated.

Keywords: functionally graded materials; elastoplastic analysis; pure bending; residual stress; large strain

---

1. Introduction

Structures made of functionally graded materials (FGM) are advantageous for many applications. A difficulty with theoretical analysis and design is that structures made of FGM are classified by a much greater number of parameters than similar structures made of homogeneous materials. For this reason, it is desirable to perform parametric studies by analytic or semi-analytic methods as much as possible. A review of results related to the analysis of FGM and published before 2007 is presented in [1]. This review focuses on structures with through-thickness variation of material properties. Analytic solutions derived in [1–5] belong to this class of FGM as well. In [2–4], elastic and elastic/plastic spherical vessels subjected to various loading conditions are considered. Thermo-elastic simply supported and clamped circular plates are studied in [5]. Many analytic and semi-analytic solutions are available for FGM discs and cylinders assuming that material properties vary in the radial direction but are independent of the circumferential and axial directions. Purely elastic solutions for a hollow disc or cylinder subjected to internal or/and external pressure are derived in [6–8]. An axisymmetric thermo-elastic solution for a hollow cylinder subjected quite a general system of thermo-mechanical loading is presented in [9]. It is assumed that the temperature varies along the radial coordinate. A plane strain analytic elastic/plastic solution for pressurized tubes is found in [10]. The solution is based on the Tresca yield criterion. Many solutions are proposed for functionally graded solid and hollow rotating discs. Purely elastic solutions for solid discs of constant thickness are given in [11,12], a purely elastic solution for a hollow disc of variable thickness in [13], a purely elastic solution for hollow polar orthotropic discs in [14], and a solution for hollow cylinders using the theory of electrothermoelasticity in [15]. An elastic perfectly plastic stress solution for hollow discs is derived in [16] using the von Mises yield criterion.

All of the aforementioned solutions deal with infinitesimal strain. A distinguished feature of the solution provided in the present paper is that strains are large. The process considered is pure bending of a FGM sheet under plane strain conditions. A review on bending of functionally graded sheets and beams at infinitesimal strains is given in [17]. The present solution is based on the approach proposed in [18]. It is shown in this paper that the use of Lagrangian coordinates facilitates the solution. Moreover, the equations describing kinematics can be solved independently of stress equations in the case of isotropic incompressible material. This is an advantage as compared to the classic approach developed in [19] where the stress equations are solved first. The classic approach is restricted to perfectly plastic materials, whereas the mapping in Equation (1) is valid for a large class of constitutive equations. The approach proposed in [18] has already been successfully extended to more general constitutive equations in [20–23]. It is shown in the present paper that the approach is also efficient for FGM sheets. It is worth noting that a rigid plastic solution for pure bending of laminated sheets (such sheets can also be referred to as functionally graded sheets) at large strain is given in [24].

2. Basic Equations

The process of plane strain pure bending is illustrated in Figure 1. The approach proposed in [18] for solving the corresponding boundary value problem is based on the following transformation equations:

$$\frac{x}{H} = \sqrt{\frac{\zeta}{a} + \frac{s}{a^2}} \cos(2a\eta) - \frac{\sqrt{s}}{a}, \quad \frac{y}{H} = \sqrt{\frac{\zeta}{a} + \frac{s}{a^2}} \sin(2a\eta). \tag{1}$$

where (x, y) is an Eulerian–Cartesian coordinate system and (ζ, η) is a Lagrangian coordinate system. Without loss of generality, it is possible to assume that the origin of the Cartesian coordinate system is located at the intersection of the axis of symmetry of the process and the outer surface AB and that the x-axis coincides with the axis of symmetry. The Lagrangian coordinate system is chosen such that

$$\zeta = x/H \quad \text{and} \quad \eta = y/H \tag{2}$$

at the initial instant where H is the initial thickness of the sheet. It is evident from these relations and the geometry in Figure 1 that $\zeta = 0$ on AB, $\zeta = -1$ on CD, $\eta = L/H$ on CB and $\eta = -L/H$ on AD throughout the process of deformation. Here, L is the initial width of the sheet. In Equation (1), a is a time-like variable. In particular, $a = 0$ at the initial instant. In Equation (1), s is a function of a. This function should be found from the stress solution and therefore depends on constitutive equations. The condition in Equation (2) is satisfied if

$$s = \frac{1}{4} \tag{3}$$

at $a = 0$. It is possible to verify by inspection that the mapping in Equation (1) satisfies the equation of incompressibility. Moreover, this mapping transforms initially straight lines A_1B_1 and C_1D_1 into circular arcs AB and CD and initially straight lines C_1B_1 and A_1D_1 into circular arcs CB and AD after any amount of deformation (Figure 1). Furthermore, coordinate curves of the Lagrangian coordinate system coincide with trajectories of the principal strain rates and, for coaxial models, with trajectories of the principal stresses. Thus, the shear stress vanishes in the Lagrangian coordinates. In particular, the contour ABCD is free of shear stresses. Let σ_ζ and σ_η be the physical stress components referred to the Lagrangian coordinates. The stress solution should satisfy the boundary conditions

$$\sigma_\zeta = 0 \tag{4}$$

for $\zeta = -1$ and $\zeta = 0$. The only non-trivial equilibrium equation in the Lagrangian coordinates has been derived in [18] as

$$\frac{\partial \sigma_\zeta}{\partial \zeta} + \frac{a(\sigma_\zeta - \sigma_\eta)}{2(\zeta a + s)} = 0. \tag{5}$$

The initial plane strain yield criterion of the functionally graded sheet is supposed to be

$$|\sigma_\zeta - \sigma_\eta| = \frac{2}{\sqrt{3}}\sigma_0\beta\left(\frac{x}{H}\right). \tag{6}$$

where σ_0 is a material constant and $\beta(x/H)$ is an arbitrary function of its argument. It is assumed that material properties are not affected by plastic deformation. Therefore, Equation (6) can be rewritten in the form

$$|\sigma_\zeta - \sigma_\eta| = \frac{2}{\sqrt{3}}\sigma_0\beta(\zeta). \tag{7}$$

In this case, the yield locus is invariant along the motion. The importance of this property of material models has been emphasized in [25]. Let τ_ζ and τ_η be the deviatoric portions of σ_ζ and σ_η, respectively. Since the material is incompressible, $\tau_\zeta + \tau_\eta = 0$ under plane strain conditions. Then, the yield criterion in Equation (7) is equivalent to

$$|\tau_\zeta| = |\tau_\eta| = \frac{\sigma_0\beta(\zeta)}{\sqrt{3}}. \tag{8}$$

Hooke's law generalized on functionally graded materials reads

$$\tau_\zeta = 2G_0 g(\zeta)\varepsilon_\zeta^e, \quad \tau_\eta = 2G_0 g(\zeta)\varepsilon_\eta^e. \tag{9}$$

It has been taken into account here that Poisson's ratio is equal to 1/2 for incompressible materials. In addition, ε_ζ^e and ε_η^e are the total strain components in elastic regions and the elastic portions of the total strain components in plastic regions referred to the Lagrangian coordinate system, G_0 is a material constant and $g(\zeta)$ is an arbitrary function of its argument.

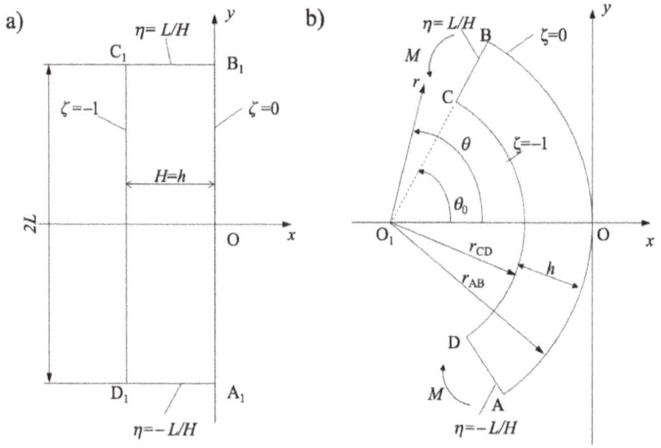

Figure 1. Geometric configuration of the bending problem: (**a**) before deformation; and (**b**) after deformation.

Geometric parameters shown in Figure 1 depend on a and are expressed as [18]

$$\frac{r_{AB}}{H} = \frac{\sqrt{s}}{a}, \quad \frac{r_{CD}}{H} = \frac{\sqrt{s-a}}{a}, \quad \theta_0 = \frac{2aL}{H}, \quad \frac{h}{H} = \frac{\sqrt{s}-\sqrt{s-a}}{a}. \tag{10}$$

Once s has been found as a function of a, these parameters are immediate from Equation (10).

3. Stress Solution at Loading

It is assumed that the functions $\beta(\zeta)$ and $g(\zeta)$ involved in Equations (7) and (9) are such that plastic yielding can only initiate at $\zeta = 0$ or $\zeta = -1$. This assumption can be verified using the purely elastic solution with no difficulty. At the very beginning of the process, the entire sheet is elastic. As deformation proceeds, one of the following three cases arises: (i) plastic yielding initiates at the surface $\zeta = -1$; (ii) plastic yielding initiates at the surface $\zeta = 0$; and (iii) plastic yielding initiates simultaneously at the surfaces $\zeta = -1$ and $\zeta = 0$. These cases should be treated separately. In the following, ζ_1 is the elastic/plastic boundary between the plastic region that propagates from the surface $\zeta = 0$ and the elastic region and ζ_2 is the elastic/plastic boundary between the plastic region that propagates from the surface $\zeta = -1$ and the elastic region. It is evident that both ζ_1 and ζ_2 depend on a. The general structure of the solution with two plastic regions is illustrated in Figure 2. Let M be the bending moment. Then, its dimensionless representation is in terms of the Lagrangian coordinates given by [18]

$$m = \frac{2\sqrt{3}M}{\sigma_0 H^2} = \frac{\sqrt{3}}{a}\int_{-1}^{0}\frac{\sigma_\eta}{\sigma_0}d\zeta. \tag{11}$$

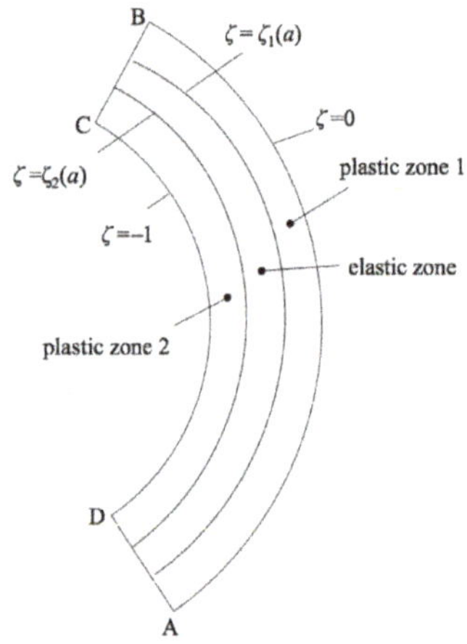

Figure 2. Schematics of elastic and plastic zones.

In the elastic region, the whole strain is elastic. Therefore, it follows from Equation (1) that the principal logarithmic strains are

$$2\varepsilon_\zeta^e = -2\varepsilon_\eta^e = -\ln\left[4\left(\zeta a + s\right)\right]. \tag{12}$$

Since $\sigma_\zeta - \sigma_\eta = \tau_\zeta - \tau_\eta$, Equations (5) and (9) combine to give

$$\frac{\partial \sigma_\zeta}{\partial \zeta} + \frac{G_0 a g\left(\zeta\right)}{\left(\zeta a + s\right)}\left(\varepsilon_\zeta^e - \varepsilon_\eta^e\right) = 0. \tag{13}$$

Eliminating the strain components in Equation (13) by means of Equation (12) results in

$$\frac{\partial \sigma_\zeta}{\partial \zeta} - \frac{G_0 a g\left(\zeta\right)}{\left(\zeta a + s\right)}\ln\left[4\left(\zeta a + s\right)\right] = 0. \tag{14}$$

Integrating this equation with respect to ζ and using the boundary condition in Equation (4) at $\zeta = 0$ leads to

$$\frac{\sigma_\zeta}{\sigma_0} = \frac{a}{3k} \int_0^\zeta \frac{g(\chi) \ln[4(\chi a + s)]}{(\chi a + s)} d\chi, \quad \frac{\sigma_\eta}{\sigma_0} = \frac{\sigma_\zeta}{\sigma_0} + \frac{2}{3k} g(\zeta) \ln[4(\zeta a + s)], \tag{15}$$

where $k = \sigma_0/(3G_0)$ and χ is a dummy variable of integration. The expression for σ_η in Equation (15) has been derived using the identity $\sigma_\eta = \sigma_\zeta - \tau_\zeta + \tau_\eta$, and Equations (9) and (12). In the case of the purely elastic solution, Equation (15) must satisfy the boundary condition in Equation (4) at $\zeta = -1$. Then, the equation for the function s(a) is

$$\int_{-1}^0 \frac{g(\chi) \ln[4(\chi a + s)]}{(\chi a + s)} d\chi = 0. \tag{16}$$

Using Equation (15), in which s should be eliminated by means of the solution of Equation (16), and the yield criterion in Equation (8), it is possible to determine which of the three cases mentioned above occurs for given material properties. Simultaneously, the value of a at which plastic yielding initiates is determined. This value of a is denoted as a_e. In the following, it is assumed that $a \geq a_e$. It is now necessary to consider Cases (i), (ii) and (iii) separately.

Case (i). There are two regions. A plastic region occupies the domain $-1 \leq \zeta \leq \zeta_2$ and an elastic region the domain $\zeta_2 \leq \zeta \leq 0$. Equation (15) is valid in the elastic region. However, the function $s(a)$ is not determined from Equation (16). It is reasonable to assume that $\sigma_\eta < \sigma_\zeta$ in the plastic region. Therefore, the yield criterion in Equation (7) becomes

$$\sigma_\zeta - \sigma_\eta = \frac{2}{\sqrt{3}} \sigma_0 \beta(\zeta). \tag{17}$$

Substituting Equation (17) into Equation (5) and integrating yields the dependence of the stress σ_ζ on ζ. Using Equation (17) again provides the dependence of the stress σ_η on ζ. As a result,

$$\frac{\sigma_\zeta}{\sigma_0} = -\frac{a}{\sqrt{3}} \int_{-1}^\zeta \frac{\beta(\chi)}{(\chi a + s)} d\chi, \quad \frac{\sigma_\eta}{\sigma_0} = \frac{\sigma_\zeta}{\sigma_0} - \frac{2}{\sqrt{3}} \beta(\zeta). \tag{18}$$

It is evident that this solution satisfies the boundary condition in Equation (4) at $\zeta = -1$. Both σ_ζ and σ_η should be continuous across $\zeta = \zeta_2$. Consequently, τ_ζ is continuous across $\zeta = \zeta_2$. The stress τ_ζ on the elastic side of the elastic/plastic boundary is determined from Equation (15) and on the plastic side from Equation (8). Then, the condition of continuity of τ_ζ across the surface $\zeta = \zeta_2$ is represented as

$$g(\zeta_2) \ln[4(\zeta_2 a + s)] = -\sqrt{3} k \beta(\zeta_2). \tag{19}$$

Solving this equation for s yields

$$s = \frac{1}{4} \exp\left[-\frac{\sqrt{3} k \beta(\zeta_2)}{g(\zeta_2)}\right] - \zeta_2 a. \tag{20}$$

Using Equations (15) and (18), the condition of continuity of σ_ζ across the surface $\zeta = \zeta_2$ is represented as

$$\int_0^{\zeta_2} \frac{g(\chi)\ln[4(\chi a+s)]}{(\chi a+s)} d\chi = -\sqrt{3}k \int_{-1}^{\zeta_2} \frac{\beta(\chi)}{(\chi a+s)} d\chi. \tag{21}$$

In this equation, s can be eliminated by means of Equation (20). The resulting equation should be solved numerically to find ζ_2 as a function of a. Then, s as a function of a is readily found from Equation (20). The yield criterion should be checked in the elastic region using the solution in Equation (15). The calculation should be stopped when the yield condition is satisfied at one point of the elastic region. Denote the corresponding value of a as a_2.

In Case (i), Equation (11) becomes

$$m = \frac{\sqrt{3}}{a} \int_{-1}^{\zeta_2} \left(\frac{\sigma_\eta}{\sigma_0}\right) d\zeta + \frac{\sqrt{3}}{a} \int_{\zeta_2}^{0} \left(\frac{\sigma_\eta}{\sigma_0}\right) d\zeta. \tag{22}$$

In the first integrand, σ_η/σ_0 should be eliminated by means of Equation (18) and in the second by means of Equation (15).

Case (ii). There are two regions. A plastic region occupies the domain $\zeta_1 \le \zeta \le 0$ and an elastic region the domain $-1 \le \zeta \le \zeta_1$. The elastic solution in Equation (15) satisfies the boundary condition in Equation (4) at $\zeta = 0$. Therefore, it is convenient to rewrite this solution as

$$\frac{\sigma_\zeta}{\sigma_0} = \frac{a}{3k} \int_{-1}^{\zeta} \frac{g(\chi)\ln[4(\chi a+s)]}{(\chi a+s)} d\chi, \quad \frac{\sigma_\eta}{\sigma_0} = \frac{\sigma_\zeta}{\sigma_0} + \frac{2}{3k} g(\zeta)\ln[4(\zeta a+s)]. \tag{23}$$

The elastic solution in this form satisfies the boundary condition in Equation (4) at $\zeta = -1$. It is reasonable to assume that $\sigma_\eta > \sigma_\zeta$ in the plastic region. Therefore, the yield criterion in Equation (7) becomes

$$\sigma_\zeta - \sigma_\eta = -\frac{2}{\sqrt{3}}\sigma_0 \beta(\zeta). \tag{24}$$

Substituting Equation (24) into Equation (5) and integrating yields the dependence of the stress σ_ζ on ζ. Using Equation (24) again provides the dependence of the stress σ_η on ζ. As a result,

$$\frac{\sigma_\zeta}{\sigma_0} = \frac{a}{\sqrt{3}} \int_0^{\zeta} \frac{\beta(\chi)}{(\chi a+s)} d\chi, \quad \frac{\sigma_\eta}{\sigma_0} = \frac{\sigma_\zeta}{\sigma_0} + \frac{2}{\sqrt{3}}\beta(\zeta). \tag{25}$$

It is evident that this solution satisfies the boundary condition in Equation (4) at $\zeta = 0$. Both σ_ζ and σ_η should be continuous across $\zeta = \zeta_1$. Consequently, τ_ζ is continuous across $\zeta = \zeta_1$. The stress τ_ζ on the elastic side of the elastic/plastic boundary is determined from Equation (23) and on the plastic side from Equation (8). Then, the condition of continuity of τ_ζ across the surface $\zeta = \zeta_1$ is represented as

$$g(\zeta_1)\ln[4(\zeta_1 a+s)] = \sqrt{3}k\beta(\zeta_1). \tag{26}$$

Solving this equation for s yields

$$s = \frac{1}{4}\exp\left[-\frac{\sqrt{3}k\beta(\zeta_1)}{g(\zeta_1)}\right] - \zeta_1 a. \tag{27}$$

Using Equations (23) and (25), the condition of continuity of σ_ζ across the surface $\zeta = \zeta_1$ is represented as

$$\int_{-1}^{\zeta_1} \frac{g(\chi)\ln[4(\chi a + s)]}{(\chi a + s)} d\chi = \sqrt{3}k \int_{0}^{\zeta_1} \frac{\beta(\chi)}{(\chi a + s)} d\chi. \tag{28}$$

In this equation, s can be eliminated by means of Equation (27). The resulting equation should be solved numerically to find ζ_1 as a function of a. Then, s as a function of a is readily found from Equation (27). The yield criterion should be checked in the elastic region using the solution in Equation (23). The calculation should be stopped when the yield condition is satisfied at one point of the elastic region. Denote the corresponding value of a as a_1.

In Case (ii), Equation (11) becomes

$$m = \frac{\sqrt{3}}{a} \int_{-1}^{\zeta_1} \left(\frac{\sigma_\eta}{\sigma_0}\right) d\zeta + \frac{\sqrt{3}}{a} \int_{\zeta_1}^{0} \left(\frac{\sigma_\eta}{\sigma_0}\right) d\zeta. \tag{29}$$

In the first integrand, σ_η/σ_0 should be eliminated by means of Equation (23) and in the second by means of Equation (25).

Case (iii). In this case, there are two plastic regions, $-1 \le \zeta \le \zeta_2$ and $\zeta_1 \le \zeta \le 0$, and one elastic region, $\zeta_1 \le \zeta \le \zeta_2$. At the beginning of this stage of the process, $a = a_1$ and $\zeta_2 = -1$ or $a = a_2$ and $\zeta_1 = 0$. Let σ_{n1} be the value of σ_ζ at $\zeta = \zeta_1$ and σ_{n2} be the value of σ_ζ at $\zeta = \zeta_2$. Then, the elastic solution in Equation (15) can be rewritten as

$$\frac{\sigma_\zeta}{\sigma_0} = \frac{a}{3k} \int_{\zeta_1}^{\zeta} \frac{g(\chi)\ln[4(\chi a + s)]}{(\chi a + s)} d\chi + \frac{\sigma_{n1}}{\sigma_0}, \quad \frac{\sigma_\eta}{\sigma_0} = \frac{\sigma_\zeta}{\sigma_0} + \frac{2}{3k} g(\zeta)\ln[4(\zeta a + s)]. \tag{30}$$

It follows from this solution that

$$\frac{\sigma_{n2}}{\sigma_0} = \frac{a}{3k} \int_{\zeta_1}^{\zeta_2} \frac{g(\chi)\ln[4(\chi a + s)]}{(\chi a + s)} d\chi + \frac{\sigma_{n1}}{\sigma_0}. \tag{31}$$

The solution in Equation (18) is valid in the plastic region $-1 \le \zeta \le \zeta_2$ and the solution in Equation (25) in the plastic region $\zeta_1 \le \zeta \le 0$. Then,

$$\frac{\sigma_{n2}}{\sigma_0} = -\frac{a}{\sqrt{3}} \int_{-1}^{\zeta_2} \frac{\beta(\chi)}{(\chi a + s)} d\chi, \tag{32}$$

and

$$\frac{\sigma_{n1}}{\sigma_0} = \frac{a}{\sqrt{3}} \int_{0}^{\zeta_1} \frac{\beta(\chi)}{(\chi a + s)} d\chi. \tag{33}$$

Equations (20) and (27) are valid. Therefore,

$$\exp\left[\frac{\sqrt{3}k\beta(\zeta_1)}{g(\zeta_1)}\right] - 4\zeta_1 a = \exp\left[-\frac{\sqrt{3}k\beta(\zeta_2)}{g(\zeta_2)}\right] - 4\zeta_2 a, \tag{34}$$

and

$$a = \frac{1}{4(\zeta_2 - \zeta_1)}\left\{\exp\left[-\frac{\sqrt{3}k\beta(\zeta_2)}{g(\zeta_2)}\right] - \exp\left[\frac{\sqrt{3}k\beta(\zeta_1)}{g(\zeta_1)}\right]\right\}. \tag{35}$$

Equations (31)–(33) combine to give

$$\int_{-1}^{\zeta_2} \frac{\beta(\chi)}{(\chi a + s)} d\chi + \frac{1}{\sqrt{3}k}\int_{\zeta_1}^{\zeta_2} \frac{g(\chi)\ln[4(\chi a + s)]}{(\chi a + s)} d\chi + \int_{0}^{\zeta_1} \frac{\beta(\chi)}{(\chi a + s)} d\chi = 0. \tag{36}$$

Eliminating in this equation s by means of Equation (20) or Equation (27) and then a by means of Equation (35) supplies the equation to find ζ_1 as a function of ζ_2 (or ζ_2 as a function of ζ_1). Then, a as a function of ζ_1 (or ζ_2) is found from Equation (35) and s as a function of ζ_1 (or ζ_2) from Equation (20) or (27). The distribution of the stresses is determined from Equation (30) with the use of Equations (32) and (33) in the elastic region, from Equation (18) in the region $-1 \le \zeta \le \zeta_2$ and from Equation (25) in the region $\zeta_1 \le \zeta \le 0$.

In Case (iii), Equation (11) becomes

$$m = \frac{\sqrt{3}}{a}\int_{-1}^{\zeta_2} \frac{\sigma_\eta}{\sigma_0} d\zeta + \frac{\sqrt{3}}{a}\int_{\zeta_2}^{\zeta_1} \frac{\sigma_\eta}{\sigma_0} d\zeta + \frac{\sqrt{3}}{a}\int_{\zeta_1}^{0} \frac{\sigma_\eta}{\sigma_0} d\zeta. \tag{37}$$

In the first integrand, σ_η/σ_0 should be eliminated by means of Equation (18), in the second by means of Equation (30) and the third by means of Equation (25). As usual, it is necessary to verify that the yield criterion is not violated in the elastic region.

4. Unloading

It is assumed that unloading is purely elastic. This assumption should be verified a posteriori. At this stage of the process, the strains can be considered as infinitesimal. Let a_f and s_f be the values of a and s, respectively, at the end of loading. These values are known from the solution given in the previous section. Using Equation (10), the values of r_{CD} and r_{AB} at the end of loading, r_{CD}^f and r_{AB}^f, are determined as

$$\frac{r_{\text{CD}}^f}{H} = \sqrt{\frac{s_f}{a_f^2} - \frac{1}{a_f}} \equiv R_f, \quad \frac{r_{\text{AB}}^f}{H} = \frac{\sqrt{s_f}}{a_f} = r_f. \tag{38}$$

It is convenient to introduce a polar coordinate system (r, θ) with the origin at $x = -H\sqrt{s_f}/a_f$ and $y = 0$ (point O_1 in Figure 1). The coordinate curves of this coordinate system coincide with the coordinate curves of the (ζ, η)-coordinate system. Therefore, $\sigma_\zeta = \sigma_r$ and $\sigma_\eta = \sigma_\theta$ where σ_r and σ_θ are the normal stresses in the polar coordinate system. Moreover, $r = R_f H$ at $\zeta = 0$ and $r = r_f H$ at $\zeta = -1$. The equilibrium equation for the increment of the stresses, $\Delta\sigma_\zeta$ and $\Delta\sigma_\eta$, in the polar coordinate system can be written as

$$\frac{\partial(\Delta\sigma_\zeta)}{\partial\rho} = \frac{\Delta\sigma_\eta - \Delta\sigma_\zeta}{\rho}, \tag{39}$$

where $\rho = r/H$. Since $\sigma_\zeta = 0$ at $\zeta = 0$ and $\zeta = -1$ at any stage of the process, the increment of this stress should satisfy the conditions

$$\Delta\sigma_\zeta = 0, \tag{40}$$

for $\zeta = 0$ and $\zeta = -1$.

The displacement components from the configuration corresponding to the end of loading in the polar coordinate system are supposed to be

$$u_r = H\left(\frac{U_0 R_f^2}{\rho} - \frac{\rho V_0}{2}\right) \quad \text{and} \quad u_\theta = H\rho\theta V_0, \tag{41}$$

where U_0 and V_0 are dimensionless constants. Using Equation (41), the increment of the normal strains in the polar coordinate system is determined as

$$\Delta\varepsilon_r = -\frac{V_0}{2} - \frac{U_0 R_f^2}{\rho^2}, \quad \Delta\varepsilon_\theta = \frac{V_0}{2} + \frac{U_0 R_f^2}{\rho^2}. \tag{42}$$

The increment of the deviatoric stresses is found from Equation (42) and the Hooke's law (Equation (9)) where the stresses and strains should be replaced with the corresponding increments. Then,

$$\Delta\tau_r = -G_0 g(\zeta)\left(V_0 + 2U_0\frac{R_f^2}{\rho^2}\right), \quad \Delta\tau_\theta = G_0 g(\zeta)\left(V_0 + 2U_0\frac{R_f^2}{\rho^2}\right). \tag{43}$$

Using this solution, the right hand side of Equation (36) can be rewritten as

$$\frac{\Delta\sigma_\eta - \Delta\sigma_\zeta}{\rho} = \frac{\Delta\sigma_\theta - \Delta\sigma_r}{\rho} = \frac{\Delta\tau_\theta - \Delta\tau_r}{\rho} = 2G_0 g(\zeta)\left(V_0 + 2U_0\frac{R_f^2}{\rho^2}\right). \tag{44}$$

The Lagrangian coordinate ζ at the end of loading is expressed in terms of ρ as [18]

$$\zeta = \frac{\left(\rho^2 a_f - s_f\right)}{a_f}. \tag{45}$$

Using this equation, it is possible to eliminate ζ in Equation (44). Then, substituting Equation (44) into Equation (39) and integrating gives

$$\frac{\Delta\sigma_\zeta}{\sigma_0} = \frac{2}{3k}\int_{r_f}^{\rho} \frac{g(\zeta)}{\chi}\left(V_0 + 2U_0\frac{R_f^2}{\chi^2}\right)d\chi. \tag{46}$$

It is evident that this solution satisfies the boundary condition in Equation (40) at $\zeta = -1$ (or $\rho = r_f$). The other boundary conditions in Equations (40) and (46) combine to yield

$$V_0\int_{r_f}^{R_f} \frac{g(\zeta)}{\rho}d\rho + 2U_0 R_f^2\int_{r_f}^{R_f} \frac{g(\zeta)}{\rho^3}d\rho = 0. \tag{47}$$

Solving this equation for V_0 results in

$$V_0 = -2U_0 R_f^2\int_{r_f}^{R_f} \frac{g(\zeta)}{\rho^3}d\rho\left[\int_{r_f}^{R_f} \frac{g(\zeta)}{\rho}d\rho\right]^{-1}. \tag{48}$$

Using Equations (43) and (46), it is possible to represent the distribution of $\Delta\sigma_\eta$ as

$$\frac{\Delta\sigma_\eta}{\sigma_0} = \frac{\Delta\sigma_\zeta}{\sigma_0} - \frac{2\Delta\tau_r}{\sigma_0} = \frac{2}{3k}\int_{r_f}^{\rho} \frac{g(\zeta)}{\chi}\left(V_0 + 2U_0\frac{R_f^2}{\chi^2}\right)d\chi + \frac{2}{3k}g(\zeta)\left(V_0 + 2U_0\frac{R_f^2}{\rho^2}\right). \quad (49)$$

The constant V_0 can be eliminated in Equations (46) and (49) by means of Equation (48). It is then obvious that both $\Delta\sigma_\zeta$ and $\Delta\sigma_\eta$ are proportional to U_0. The distribution of the residual stresses follows from Equations (46) and (49) in the form

$$\frac{\sigma_\zeta^{res}}{\sigma_0} = \frac{\sigma_\zeta^f}{\sigma_0} + \frac{2}{3k}\int_{r_f}^{\rho} \frac{g(\zeta)}{\mu}\left(V_0 + 2U_0\frac{R_f^2}{\chi^2}\right)d\chi,$$
$$\frac{\sigma_\eta^{res}}{\sigma_0} = \frac{\sigma_\eta^f}{\sigma_0} + \frac{2}{3k}\int_{r_f}^{\rho} \frac{g(\zeta)}{\chi}\left(V_0 + 2U_0\frac{R_f^2}{\chi^2}\right)d\chi + \frac{2}{3k}g(\zeta)\left(V_0 + 2U_0\frac{R_f^2}{\rho^2}\right). \quad (50)$$

As before, ζ should be eliminated by means of Equation (45) and V_0 by means of Equation (48). The constant U_0 remains to be found. To this end, it is necessary to use the condition that the bending moment vanishes at the end of unloading. Using Equations (11) and (45), this condition can be represented as

$$\int_{R_f}^{r_f}\left(\frac{\sigma_\eta^{res}}{\sigma_0}\right)\rho d\rho = 0. \quad (51)$$

This equation should be solved for U_0 numerically. Then, Equation (50) supplies the distribution of the residual stresses. To verify that the solution given in Section 4 is valid, this distribution should be substituted into the yield criterion in Equation (7) where σ_ζ and σ_η should be replaced with σ_ζ^{res} and σ_η^{res}, respectively. The left-hand side of Equation (7) should be less than or equal to $\left(2/\sqrt{3}\right)\sigma_0\beta(\zeta)$ in the range $-1 \le \zeta \le 0$.

5. Numerical Examples

Several numerical examples are presented in this section, based on the analytical solutions developed in the previous sections. Our chosen modulus gradient function is $g(\zeta) = 1 + (G_1/G_0 - 1)(-\zeta)^N$, and yield stress gradient function $\beta(\zeta) = 1 + (\sigma_1/\sigma_0 - 1)(-\zeta)^N$. The power law exponent N controls the functional distribution of material properties along the thickness coordinate ζ. The power law distributions in modulus and yield stress with the same N have been proposed in the literature [26,27]. The material parameters used in our numerical calculations are listed in Table 1.

Table 1. Material parameters used in the numerical examples.

	G_0, GPa	G_1, GPa	σ_0, GPa	σ_1, GPa	N
Homogeneous	30	30	1	1	0.0001
FGM Case (i)	30	10	1	0.1	1
FGM Case (i)	30	10	1	0.1	3
FGM Case (ii)	10	30	0.1	1	1
FGM Case (ii)	10	30	0.1	1	3

5.1. Homogeneous Sheet under Bending

When the homogenous sheet is under bending, both edges will simultaneously develop plastic zones. Figure 3a shows the movement of the two elastic-plastic boundaries toward the centerline of the sheet, as deformation magnitude increases. The deformation magnitude is measured by parameter a. The applied bending moment is a function of a, as shown in Figure 3b. As an illustration of the developed analytical solutions in previous sections, Figure 4 shows the stress distributions along the sheet under two different deformation magnitudes. As can be seen, the plastic zones increase with a for σ_η, while σ_ζ remains in elastic regime. The reason for σ_η is not perfectly horizontal in the plastic zone is because our numerical codes do not allow N set equal to zero, hence a very small N is chosen, as shown in Table 1. After unloading, Figure 5a,b shows the residual stress distributions under two different as. Larger a increases the magnitude of residual stresses after unloading. Moreover, the residual stress σ_ζ is zero at the left and right edges, as indicated by the red short-dashed line.

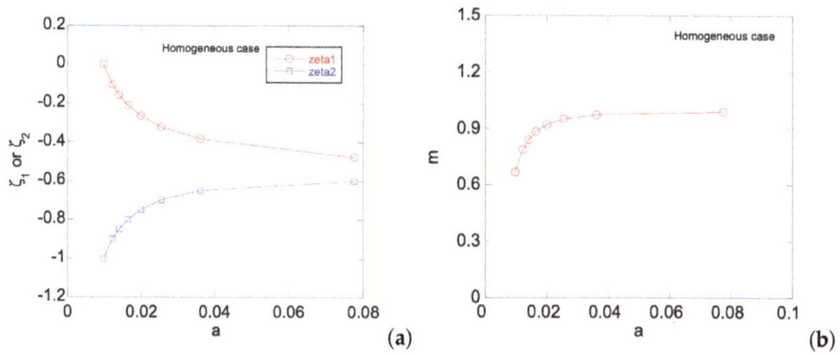

Figure 3. For the homogeneous sheet under bending: (**a**) ζ_1 or ζ_2 vs. a; and (**b**) applied bending moment m vs. a.

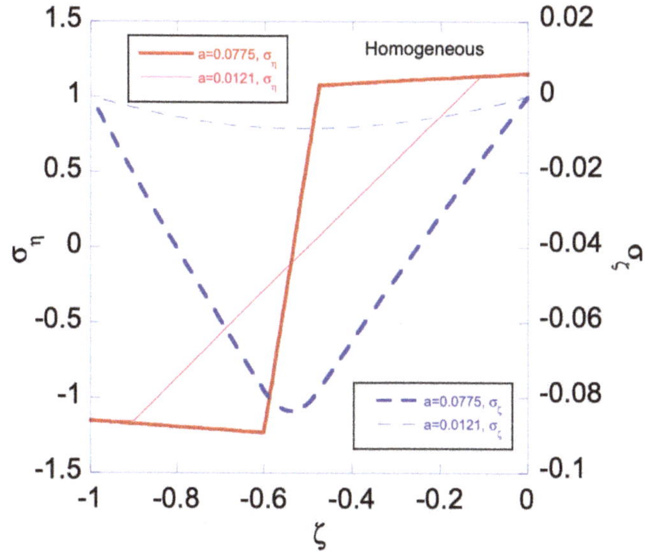

Figure 4. Stress distributions for the bent homogenous sheet with two different loading conditions $a = a_1^h = 0.0121$ and $a = a_2^h = 0.0775$. Two plastic zones, one developed from the left edge and the other from the right edge, increase their size as a increases for σ_η.

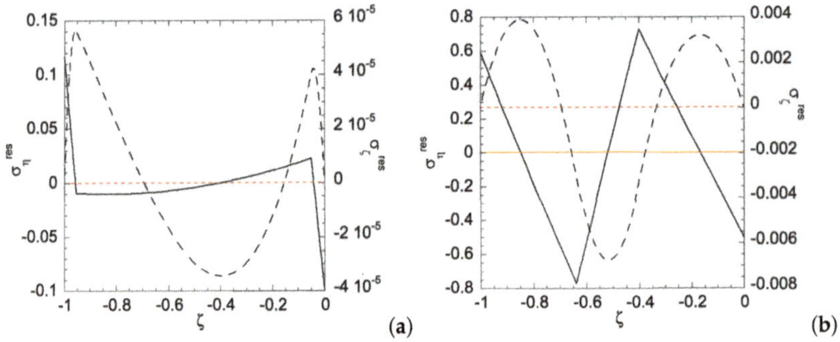

Figure 5. Residual stress distribution along the homogeneous sheet for: (**a**) $a = a^h_{r1} = 0.0096$; and (**b**) $a = a^h_{r1} = 0.0295$. Solid line is for residual σ_η and dashed line for residual σ_ζ. Zeros of σ^{res}_η and σ^{res}_ζ are indicated by orange solid line and red dashed line, respectively.

5.2. *FGM Sheet Belonged to Case (i) under Bending*

In Case (i), the left edge ($\zeta = -1$) of the sheet has smaller yield stress, hence a plastic zone will start on the left edge first. Figure 6 shows the relationship between the applied bending moment and deformation magnitude a with the gradient function exponent $N = 1$ and 3. As can be seen, larger m is required for $N = 3$ than that for $N = 1$, as a increases. Under given deformation magnitudes, Figure 7 shows the stress distributions in the Case (i) FGM under plastic deformation. Larger plastic zone is developed at the left edge as deformation increases. After unloading, residual stress distributions are shown in Figures 8 and 9 for the $N = 1$ and $N = 3$ FGM, respectively. Residual stresses are more predominant at the left edge. In addition, the residual stress σ_ζ is zero at the left and right edges, as indicated by the red short-dashed line.

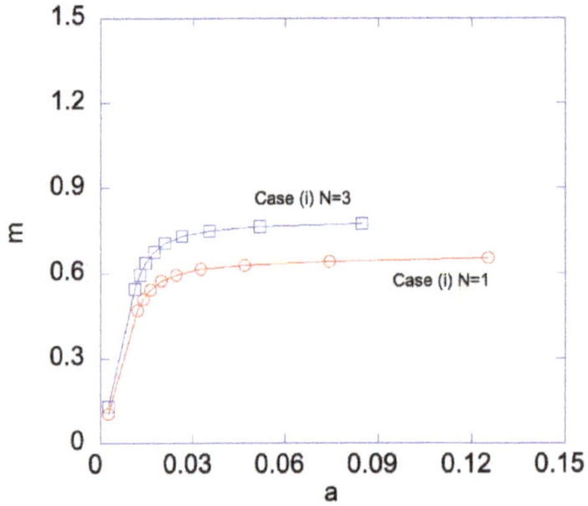

Figure 6. Bending moment m vs. a for Case (i) with $N = 1$ and $N = 3$. With sufficiently large deformation, i.e. large a, Case (iii) is automatically developed, hence both edges are plastically deformed.

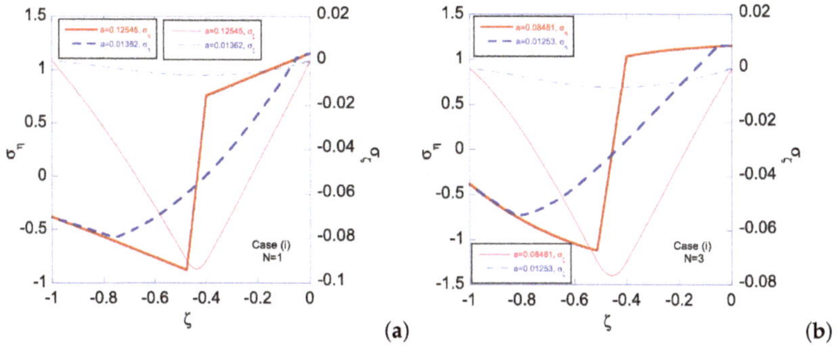

Figure 7. Stress distributions for Case (i) with $N = 1$ with two different deformation magnitudes, i.e., two different as, for: (**a**) $N = 1$; and (**b**) $N = 3$. Two plastic zones, one developed from the left edge and the other from the right edge, increase their size as a increases for σ_η.

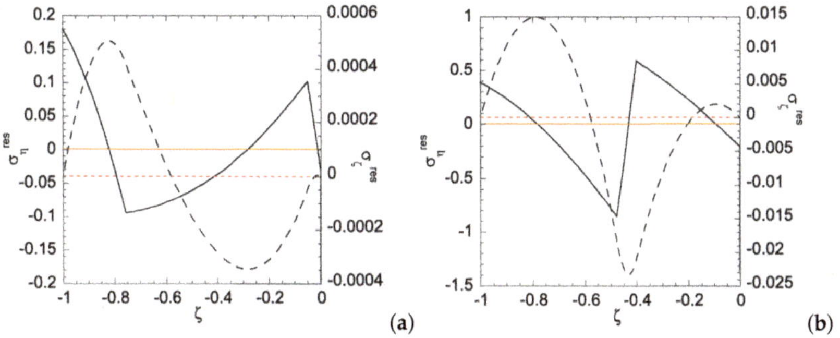

Figure 8. Residual stress distributions for Case (i) with $N = 1$ under deformation: (**a**) $a = a_{r1}^{(i)} = 0.011785065$; and (**b**) $a = a_{r1}^{(i)} = 0.074421129$. Solid line is for residual σ_η and dashed line for residual σ_ζ. Zeros of σ_η^{res} and σ_ζ^{res} are indicated by orange solid line and red dashed line, respectively.

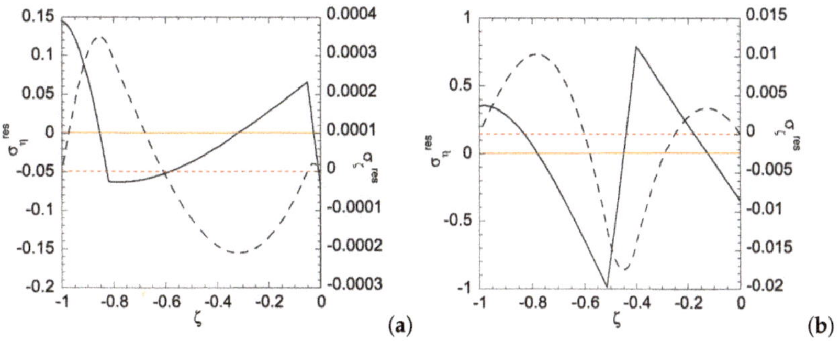

Figure 9. Residual stress distributions for Case (i) with $N = 3$ under deformation: (**a**) $a = a_{r3}^{(i)} = 0.011023776$; and (**b**) $a = a_{r4}^{(i)} = 0.051782217$. Solid line is for residual σ_η and dashed line for residual σ_ζ. Zeros of σ_η^{res} and σ_ζ^{res} are indicated by orange solid line and red dashed line, respectively.

5.3. FGM Sheet Belonged to Case (ii) under Bending

In Case (ii), the right edge ($\zeta = 0$) of the sheet has smaller yield stress, hence a plastic zone will start on the right edge first. Figure 10 shows the relationship between the applied bending moment and deformation magnitude a with the gradient function exponent $N = 1$ and 3. As can be seen, larger m is required for $N = 1$ than that for $N = 3$, as a increases. Under given deformation magnitudes, Figure 11 shows the stress distributions in the Case (ii) FGM under plastic deformation. Larger plastic zone is developed at the right edge as deformation increases. After unloading, residual stress distributions are shown in Figures 12 and 13 for the $N = 1$ and $N = 3$ FGM, respectively. Residual stresses are more predominant at the right edge. The residual stress σ_ζ is zero at the left and right edges, as indicated by the red short-dashed line. The results from the illustrative examples solved here may serve as benchmark solutions for data obtained from numerical or experimental methods.

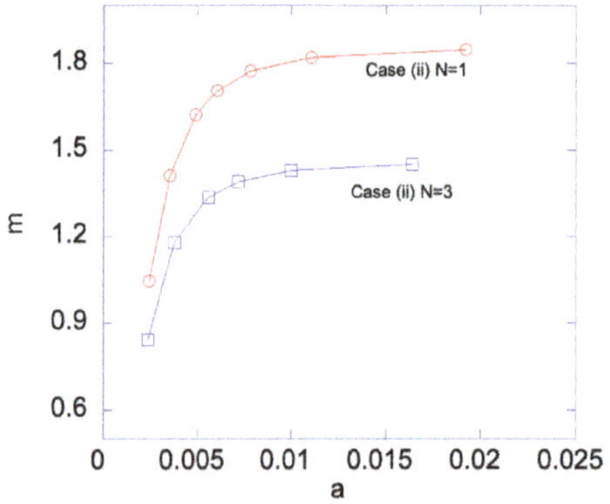

Figure 10. Bending moment m vs. a for Case (ii) with $N = 1$ and $N = 3$. With sufficiently large deformation, i.e., large a, Case (iii) is developed, hence both edges are plastically deformed.

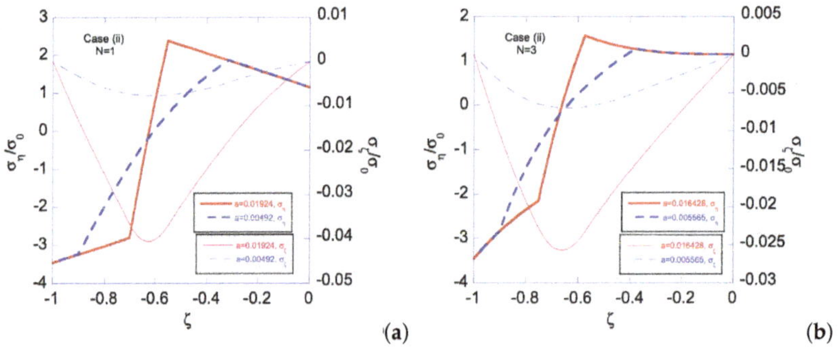

Figure 11. Stress distributions for Case (ii) with two different deformation magnitudes, i.e., two different as, for: (**a**) $N = 1$; and (**b**) $N = 3$. Two plastic zones, one developed from the left edge and the other from the right edge, increase their size as a increases for σ_η.

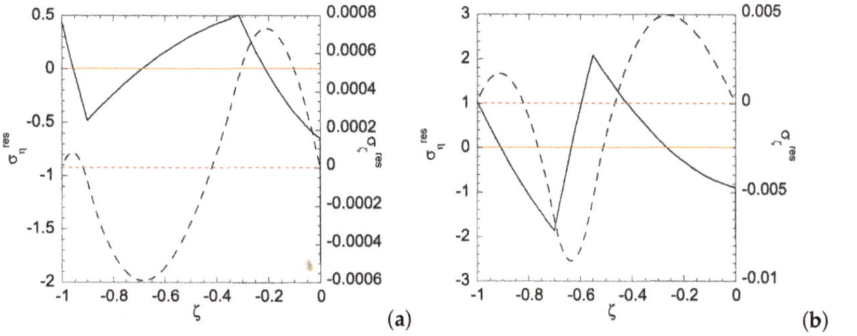

Figure 12. Residual stress distributions for Case (ii) with $N = 1$ under deformation: (**a**) $a = a_{r1}^{(ii)} = 0.004915193$; and (**b**) $a = a_{r1}^{(ii)} = 0.019238382$. Solid curve is for residual σ_η and dashed curve for residual σ_ζ. Zeros of σ_η^{res} and σ_ζ^{res} are indicated by orange solid line and red dashed line, respectively.

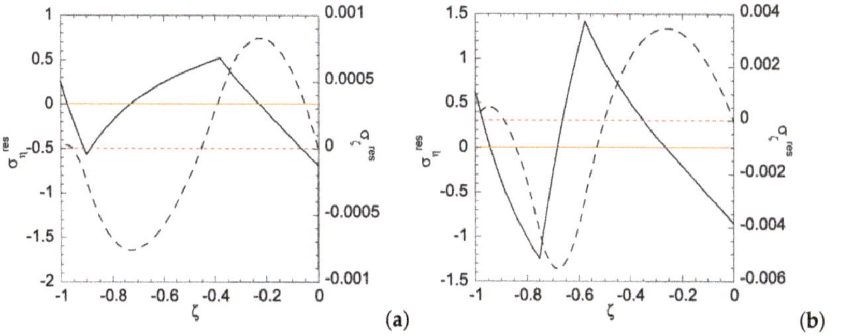

Figure 13. Residual stress distributions for Case (ii) with $N = 3$ under deformation: (**a**) $a = a_{r3}^{(ii)} = 0.005565273$; and (**b**) $a = a_{r4}^{(ii)} = 0.016428139$. Solid curve is for residual σ_η and dashed curve for residual σ_ζ. Zeros of σ_η^{res} and σ_ζ^{res} are indicated by orange solid line and red dashed line, respectively.

6. Conclusions

Efficient analytical and numerical methods and procedures have been developed and programmed to predict the distribution of stresses in a sheet of incompressible material subject to plane strain pure bending at large strain and then the distribution of residual stresses after unloading. Springback is also predicted. It has been assumed that the sheet is made of functionally graded material. The general theory has been developed for an arbitrary through thickness distribution of material properties assuming that the initiation of plastic yielding occurs at one of the surfaces of the sheet. This assumption can be verified using the purely elastic solution (Equation (15)) and the yield criterion (Equation (7)). It is possible to use the general solutions (Equations (15), (18) and (25)) even if the assumption is not satisfied but constants of integration should be added. Then, these general solutions should be combined to satisfy the boundary conditions and the conditions at elastic/plastic boundaries. An illustrative example is concentrated on power law distributions of material properties. Using the numerical code developed in this work enables the effect of parameters involved in these laws to be predicted effectively. The calculated examples show the analytical solutions derived here can systematically treat the plastic problems of the homogenous or functionally graded sheet. The magnitudes of applied moment may be strongly influenced by the power law exponent as deformation increases, which provides an effective way to design the functionally graded sheets.

The method employed to derive the solution in this paper can be extended to cyclic loading. This new solution may be useful for the interpretation of experimental data from the reverse bending fatigue test (for example, [28]).

Author Contributions: Conceptualization, S.A., Y.-C.W. and L.L.; methodology, S.A. and Y.-C.W.; software, S.A. and Y.-C.W.; validation, S.A., Y.-C.W. and L.L.; formal analysis, S.A. and Y.-C.W.; investigation, S.A., Y.-C.W. and L.L.; resources, S.A., Y.-C.W. and L.L.; data curation, S.A. and Y.-C.W.; writing–original draft preparation, S.A., Y.-C.W. and L.L.; writing–review and editing, S.A., Y.-C.W. and L.L.; visualization, S.A. and Y.-C.W.; supervision, S.A., Y.-C.W. and L.L.; project administration, S.A., Y.-C.W. and L.L.; funding acquisition, S.A. and Y.-C.W.

Funding: S.A. acknowledges support from the Russian Foundation for Basic Research (Project 17-58-560005). This research was funded, in part, by Ministry of Science and Technology, Taiwan, grant number MOST 107-2221-E-006-028. This research was, in part, supported by the Ministry of Education, Taiwan, and the Aim for the Top University Project to the National Cheng Kung University (NCKU).

Acknowledgments: This work was initiated while Y.-C.W. was a visiting researcher at Beihang University, Beijing, China. Y.-C.W. is grateful to the National Center for High-performance Computing, Taiwan, for computer time and facilities.

Conflicts of Interest: The authors declare no conflict of interest.

References

1. Birman V.; Byrd, L.W. Modeling and analysis of functionally graded materials and structures. *Appl. Mech. Rev.* **2007**, *60*, 195–216. [CrossRef]
2. Tutuncu, N.; Ozturk, M. Exact solutions for stresses in functionally graded pressure vessels. *Compos. Part B Eng.* **2001**, *32*, 683–686. [CrossRef]
3. Akis, T. Elastoplastic analysis of functionally graded spherical pressure vessels. *Comp. Mater. Sci.* **2009**, *46*, 545–554. [CrossRef]
4. Sadeghian, M.; Toussi, H.E. Axisymmetric yielding of functionally graded spherical vessel under thermos-mechanical loading. *Comp. Mater. Sci.* **2011**, *50*, 975–981. [CrossRef]
5. Li, X.-Y.; Li, P.-D.; Kang, G.-Z. Axisymmetric thermo-elasticity field in a functionally graded circular plate of transversely isotropic material. *Math. Mech. Solids* **2013**, *18*, 464–475. [CrossRef]
6. Horgan, C.O.; Chan, A.M. The pressurized hollow cylinder or disk problem for functionally graded isotropic linearly elastic materials. *J. Elast.* **1999**, *55*, 43–59.:1007625401963. [CrossRef]
7. Tutuncu, N. Stresses in thick-walled FGM cylinders with exponentially-varying properties. *Eng. Struct.* **2007**, *29*, 2032–2035. [CrossRef]
8. You, L.H.; Wang, J.X.; Tang, B.P. Deformations and stresses in annular disks made of functionally graded materials subjected to internal and/or external pressure. *Meccanica* **2009**, *44*, 283–292. [CrossRef]
9. Jabbari, M.; Sohrabpour, S.; Eslami, M.R. Mechanical and thermal stresses in a functionally graded hollow cylinder due to radially symmetric loads. *Int. J. Press. Vessels Pip.* **2002**, *79*, 493–497. [CrossRef]
10. Eraslan, A.N.; Akis, T. Plane strain analytical solutions for a functionally graded elastic–plastic pressurized tube. *Int. J. Press. Vessels Pip.* **2006**, *83*, 635–644. [CrossRef]
11. Horgan, C.O.; Chan, A.M. The stress response of functionally graded isotropic linearly elastic rotating disks. *J. Elast.* **1999**, *55*, 219–230.:1007644331856. [CrossRef]
12. Peng X.-L.; Li X.-F. Effects of gradient on stress distribution in rotating functionally graded solid disks. *J. Mech. Sci. Technol.* **2012**, *26*, 1483–1492. [CrossRef]
13. Bayat, M.; Saleem, M.; Sahari, B.B.; Hamouda, A.M.S.; Mahdi, E. Analysis of functionally graded rotating disks with variable thickness. *Mech. Res. Commun.* **2008**, *35*, 283–309. [CrossRef]
14. Peng, X.-L.; Li, X.-F. Elastic analysis of rotating functionally graded polar orthotropic disks. *Int. J. Mech. Sci.* **2012**, *60*, 84–91. [CrossRef]
15. Dai, H.-L.; Dai, T.; Zheng, H.-Y. Stresses distributions in a rotating functionally graded piezoelectric hollow cylinder. *Meccanica* **2012**, *47*, 423–436. [CrossRef]
16. Callioglu, H.; Sayer, M.; Demir, E. Elastic–plastic stress analysis of rotating functionally graded discs. *Thin-Walled Struct.* **2015**, *94*, 38–44. [CrossRef]
17. Elishakoff, I.; Pentaras, D.; Gentilini, C. *Mechanics of Functionally Graded Material Structures*; World Scientific: Singapore, 2016; ISBN 978-981-4656-58-0.

18. Alexandrov, S.; Kim, J.-H.; Chung, K.; Kang, T.-J. An alternative approach to analysis of plane-strain pure bending at large strains. *J. Strain Anal. Eng. Des.* **2006**, *41*, 397–410. [CrossRef]
19. Hill, R. *The Mathematical Theory of Plasticity*; Clarendon Press: Oxford, UK, 1950; ISBN 0-19-850367-9.
20. Alexandrov, S.; Hwang, Y.-M. The bending moment and springback in pure bending of anisotropic sheets. *Int. J. Solids Struct.* **2009**, *46*, 4361–4368. [CrossRef]
21. Alexandrov, S.; Hwang, Y.-M. Plane strain bending with isotropic strain hardening at large strains. *Trans. ASME J. Appl. Mech.* **2010**, *77*, 064502. [CrossRef]
22. Alexandrov, S.; Gelin, J.-C. Plane strain pure bending of sheets with damage evolution at large strains. *Int. J. Solids Struct.* **2011**, *48*, 1637–1643. [CrossRef]
23. Alexandrov, S.; Hwang, Y.-M. Influence of Bauschinger effect on springback and residual stresses in plane strain pure bending. *Acta Mech.* **2011**, *220*, 47–59. [CrossRef]
24. Verguts, H.; Sowerby, R. The pure plastic bending of laminated sheet metals. *Int. J. Mech. Sci.* **1975**, *17*, 31–51. [CrossRef]
25. Romano, G.; Barretta, R.; Diaco, M. Geometric continuum mechanics. *Meccanica* **2014**, *49*, 111–133. [CrossRef]
26. Nakumura, T.; Wang, T.; Sampath, S. Determination of properties of graded materials by inverse analysis and instrumented indentation. *Acta Mater.* **2000**, *48*, 4293–4306. [CrossRef]
27. Huang, H.; Chen, B.; Han, Q. Investigation on buckling behaviors of elastoplastic functionally graded cylindrical shells subjected to torsional loads. *Comput. Struct.* **2014**, *118*, 234–240. [CrossRef]
28. Faghidian, S.A.; Jozie, A.; Sheykhloo M.J.; Shamsi, A. A novel method for analysis of fatigue life measurements based on modified Shepard method. *Int. J. Fatigue* **2014**, *68*, 144–149. [CrossRef]

materials

MDPI

Article

Stress Concentration and Damage Factor Due to Central Elliptical Hole in Functionally Graded Panels Subjected to Uniform Tensile Traction

Wenshuai Wang [1], Hongting Yuan [1], Xing Li [1,*] and Pengpeng Shi [2,*]

[1] School of Mathematics and Statistics, Ningxia University, Yinchuan 750021, China; wws@nxu.edu.cn (W.W.); yhtgfky@163.com (H.Y.)

[2] State Key Laboratory for Strength and Vibration of Mechanical Structures, Shaanxi Engineering Research Center of NDT and Structural Integrity Evaluation, School of Aerospace, Xi'an Jiaotong University, Xi'an 710049, China

* Correspondence: li_x@nxu.edu.cn (X.L.); shipengpeng@xjtu.edu.cn (P.S.)

Received: 28 November 2018; Accepted: 27 January 2019; Published: 30 January 2019

Abstract: Functionally graded material (FGM) can optimize the mechanical properties of composites by designing the spatial variation of material properties. In this paper, the stress distribution of functionally graded panel with a central elliptical hole under uniaxial tensile load is analyzed. Based on the inhomogeneity variation and three different gradient directions, the effects of the inhomogeneity on the stress concentration factor and damage factor are discussed. The study results show that when Young's modulus increases with the distance from the hole, the stress concentration factor decreases compared with that of homogeneous material, and the optimal design of r-FGM is better than that of x-FGM and y-FGM when the tensile load. In addition, when the associated variation of ultimate stress is considered, the choice of scheme to reduce the failure index is related to the strength-modulus exponent ratio. When the strength-modulus exponent ratio is small, the failure index changes with the index of power-law, which means there is an optimal FGM design. But when the strength-modulus exponent ratio is large, the optimal design modulus design is to select a uniform material that maximizes the modulus at each point. These research results have a certain reference value for further in-depth understanding of the inhomogeneous design for FGM.

Keywords: functionally graded materials; inhomogeneous composite materials; material design; stress concentration factor; failure and damage; elliptical hole; finite element method

1. Introduction

Functionally graded materials (FGM) are a class of composite materials that have smooth and continuous changes in material properties, thus reducing the stress concentrations in the conventional composite materials [1]. Since the concept of FGM was proposed by researchers in the late 1980s [2], extensive research works have been carried out on it. Many review papers have systematically introduced and forecasted the different progress of FGM researches. Recently, Zhang et al. [3] introduced the development of the emerging additive manufacturing research on FGM. Xu et al. [4] reviewed the state of the art of energy absorption of FGM, and discussed the effects of the graded properties on the crashworthiness. Cramer et al. [5] proposed a review of functionally graded thermoelectric generators, which is considered to be an effective solution for the temperature bandwidth, current output range, and lifetime. Petit et al. [6] introduced the rationale for using FGM in the biomedical field, and reviewed the three main types of graded materials (eg., composition, porosity and microstructural graded ceramics). The mechanical problems of the FGM have also drawn much attention. The progress of the resistance of FGM to contact deformation and damage is reviewed by Suresh [7]. Birman [8] outlined the steps of thermoelastic analysis of FGM, from their micromechanical characterization to the structural response. The fracture studies for FGM continuum can be found in the survey by Shanmugavel et al. [9]. In addition, Jha et al. [10] published another detailed overview focused on the thermoelastic statics, vibration and stability analysis of FGM plates. Some comprehensive reviews of the developments, applications, various mathematical idealizations of materials, temperature profiles, modeling techniques and solutions methods for the thermal analysis of FGM plates are presented by Swaminathan et al [11,12]. Some scholars applied the FGM concept to the elastostatic problems and obtained some exact solutions of orthotropic inhomogeneous Saint-Venant beams and isotropic Kirchhoff plates [13,14]. In addition, some scholars applied the concept of FGM to the study of mechanical behavior of nanomaterials by using nonlocal model or gradient elasticity model [15,16].

Due to the specific functional requirements, the influences of inhomogeneous variation on the FGM properties are studied, which has an important reference value for the preparation, performance and use of FGM. Since holes or inclusions are common defects in materials, the stress concentration of FGM has been extensively studied by finite element method and analytical method. Mohammadi et al. [17] used the Frobenius series solution to analyze the effect of inhomogeneous stiffness and Poisson's ratio on the stress concentration factor around the circular holes of infinite plates. Based on the complex function method and the conformal mapping technique, Yang and Gao [18] solved the stress concentration problem of FGM infinite plates with elliptic holes. Dave and Sharma [19] also used the complex variable function method to solve the problem of the FGM plate with rectangular holes. Based on the variable separation method, the analytical solutions of stress and strain distribution around the circular elastic inclusion and elliptical nano-fiber inclusion are obtained by Shi [20,21]. Goyat, et al [22] used the extended finite element method to analyze the stress concentration of the FGM layer in an infinite plate with a pair of circular holes under different loads. Based on the first-order shear deformation theory and Von-Karman hypothesis, Mehrparvar and Ghannadpour [23] analyzed the non-linear behavior of FGM plates with square and rectangular notches. In addition, Shi et al. [24–27] used the integral equation method to study the influence of the existence of the central circular hole on the interface fracture behavior of the FGM composite cylindrical structure.

To reduce the stress concentration factor, Sburlati [28] studied the effect of an inhomogeneous annular made of FGM on the stress distribution around a hole in a homogeneous plate. Aiming to reduce the stress concentration factor around the notch, Gouasmi, et al. [29] used the finite element method to study the performance of the FGM layer near the notch of the ceramic plate. Sburlati, et al. [30] analyzed the effect of FGM layers on the stress concentration factor in a homogeneous plate with holes based on the finite element method. Hsu and Chien [31] combined the finite element method and image processing technology to evaluate the influence of electronic discharge machining parameters on the surface quality of the plate with holes, which can quickly evaluate the stress concentration factor. Based on the finite element method and U-transform method, Yang, et al. [32] analyzed and studied the three-dimensional stress concentration of rectangular holes. Kubair and Bhanu-Chandar [33] investigated the FGM with the elastic modulus of power law and exponential variation, and simulated the FGM plates with circular holes under uniaxial tension by the multi-parameter finite element method. Nie, at al. [34] analyzed the stress concentration of FGM plates with Young's modulus of radial variation and Poisson's ratio under uniaxial tension. Kim and Paulino [35] analyzed the effects of elastic modulus and Poisson's ratio on the properties of isotropic and orthotropic FGM plates by isoparametric gradient finite element method. In addition, with consideration of the associated variation of ultimate stress, the combined optimization using both moduli and ultimate stress is studied by Huang et al. [36], and the optimization for the full spatial variation is completed by Chen et al. [37].

In this paper, the stress distribution of FGM panels with a central elliptical hole under uniaxial tension load is analyzed, and the effects of the inhomogeneous properties on the stress concentration factor (SCF), failure index and damage factor are discussed. In Section 2, the problem description is given. Two inhomogeneous variations and three different gradient directions are proposed here. In Section 3, the stress problems due to a central elliptical hole for FGM with different forms of elastic modulus and different gradient direction are calculated, and the influences of the inhomogeneous characteristics of FGM on stress concentration, failure index and damage factor are analyzed in detail. The conclusion for this paper is given in Section 4.

2. Problem Description

2.1. Problem Description

We consider an isotropic and linearly elastic FGM panel with a central elliptical hole subjected to a uniform tensile traction, as shown in Figure 1a. The length and width of the rectangular panel are L and W, respectively, and the semi-major/semi-minor axes of the ellipse hole are a and b, respectively. In this paper, a finite-size rectangular panel is selected, and the left and right end are subjected to a tensile load $\sigma_0 = P/(tW)$, where P is the value of the force, t is the thickness of the panel. Here, we use cylindrical coordinates (r, θ) and Cartesian coordinates (x, y) with origin at the hole center to describe this problem.

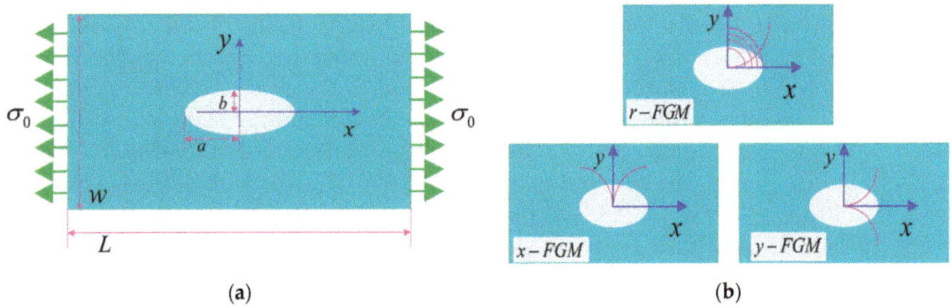

Figure 1. Schematic sketch of boundary value problem (**a**) a rectangle panel with an elliptical hole subjected to a uniform tensile traction; (**b**) three different gradient directions. The origin of the Cartesian and polar coordinates coincides with the center of the elliptical hole. In the case of the r-, x- and y-FGM symmetric property variations are shown.

2.2. Inhomogeneity Variation

Previous studies show that the effect of varying Poisson ratio on the stress distribution is negligible [34]. In this paper, we assume Poisson's ratio to be constant and set to be $v = 0.25$.

In the study of mechanical problems of FGM, the problem is often analyzed by assuming that the material parameters satisfy a certain function form, which can simplify the complexity of the problem. In order to discuss the effect of the inhomogeneity of the FGM, we assume that the nodal values of Young's modulus satisfies the power-law inhomogeneous variation:

$$E(\phi) = E_{ref}\left[1 + \gamma(\phi/Lg)^c\right], \tag{1}$$

where E_{ref} is the reference value of Young's modulus, γ is the modulus ratio, c is the index of power-law variation, Lg is the inhomogeneity length scale, ϕ is a simple function of (x, y).

In order to study the effect of different gradient directions, we assume the following forms for the function ϕ.

$$\phi = \begin{cases} \sqrt{x^2+y^2} \\ x \quad \text{or} \quad |x| \\ y \quad \text{or} \quad |y| \end{cases}, \tag{2}$$

and three different gradient directions of FGM are shown in Figure 1b.

For the power-law inhomogeneous variation, the gradient variation reflected by Equation (1) can be divided into the following two cases according to the different values of parameter c.

Case 1: When $c > 0$, for a finite panel problem, Lg can be set to half length of the rectangular panel. The material parameters at the center of the circle are satisfied:

$$E_0 = E(0) = E_{ref}, \tag{3}$$

and:

$$E(\phi) = E_0\left[1 + \gamma(\phi/Lg)^c\right] \tag{4}$$

This gradient variation Equation (4) is consistent with the power-law gradient variation given in [33].

Case 2: When $c < 0$, Lg can be set to be the semi-minor axis of the elliptical hole. For an infinite panel problem, the material parameters at infinity point satisfies:

$$E_\infty = E(\infty) = E_{ref}, \tag{5}$$

and:

$$E(\phi) = E_\infty\left[1 + \gamma(\phi/a)^c\right], \tag{6}$$

when $\phi = r$, this proposed gradient variation is consistent with that given in [34].

The above analysis shows that the gradient variation given in Refs. [33,34] can be unified by the gradient variant expressed by Equation (1) in this paper.

2.3. Stress Concentration and Damage Factor

Here, we calculate the stress concentration of FGM with a central elliptical hole under uniaxial tension with different inhomogeneous parameters. Stress concentration factor K is defined as $K = \sigma_{\max}/\sigma_{nom}$, where $\sigma_{\max}$ is the maximum value of stress component along the x direction in a panel, and $\sigma_{nom} = \frac{P}{t(W-2r)} = \frac{\sigma_0 W}{t(W-2r)}$ is the reference of stress value.

For some materials, its elastic modulus and strength change with varying porosity and material density [36]. Among them, the strength and elastic modulus of the material satisfy the following relationship.

$$\sigma_{allow} = C_0 E^\delta, \tag{7}$$

where factor δ is called the strength-modulus exponent ratio, and σ_{allow} is the limit strength or maximum allowable stress at that point.

By considering the associated variation of ultimate stress, a failure index Φ that accounts for both strength and stress is used for design purposes. Referring to the results of Ref. [36], the failure index can be defined as follows:

$$\Phi = \max\{\psi(x,y)\}, (x,y) \in \Omega, \tag{8}$$

where:

$$\psi = \frac{\max(|\sigma_1|, |\sigma_2|)}{\sigma_0\left(E/E_{ref}\right)^\delta}, \tag{9}$$

where σ_1 and σ_2 are the principal stresses at an arbitrary point in the panel, and $\sigma_0 = P/(tW)$ is the value of the tensile stress.

3. Results and Discussions

In this paper, the finite element method is introduced for the mechanical analysis of FGM. In order to describe the numerical simulations clearer, the parameter values used for in the following numerical simulations are given in Table 1.

Table 1. The values of simulation parameters.

	Values			
Simulations Parameters	**Width of Rectangle Panel *W***	**Length of Rectangle Panel *L***	**Semi-Major Axis of Elliptical Hole *a***	**Semi-Minor Axis of Elliptical Hole *b***
Figure 2	200 mm	L/W changes from 1 to 5	a/W changes from 0.05 to 0.4	$b = a$
Figure 3	200 mm	200 mm	Case for no hole, $b = a = 0$ mm	
Figure 4	200 mm	200 mm	10 mm	$b = a$
Figures 5–12	200 mm	300 mm	60 mm	40 mm

3.1. Verification

Here, the comparisons between the proposed results and other results obtained in the previous researches are given to verify the correctness of the calculation program used in this paper.

3.1.1. Verification 1: Analysis of Homogeneous Rectangular Panel with a Circular Hole

The stress concentration factors of the homogeneous rectangular panel with a central circular hole are analyzed. The stress concentration factors of a central single circular hole in finite width and infinite length panel can be calculated for tension load by the following theoretical formula [38]:

$$K = 3 - 3.14 \times 2a/W + 3.667 \times (2a/W)^2 - 1.527 \times (2a/W)^3. \tag{10}$$

Figure 2a shows the comparison between the results from finite element method and the analytical results. It can be seen that as the length of the rectangular panel increases, the result of stress concentration factors gradually decrease and tend to be stable. When L/W exceeds 3, the results from finite element method for the finite-length rectangular panel are equal to the analytical results for the infinite length rectangular panel. In particular, the numerical solution of the rectangular panel with $L/W = 5$ and the square panel with $L/W = 1$ are given in Figure 2b. It can be seen that the numerical solution of the rectangular panel with $L/W = 5$ is consistent with the analytical result of the infinite length rectangular panel. The results of the square panel decrease first and then increase as the width increases, which has the difference between the results of the finite-size square panel and the results of the infinite rectangular panels. In general, this comparison can prove the correctness of the present calculation program.

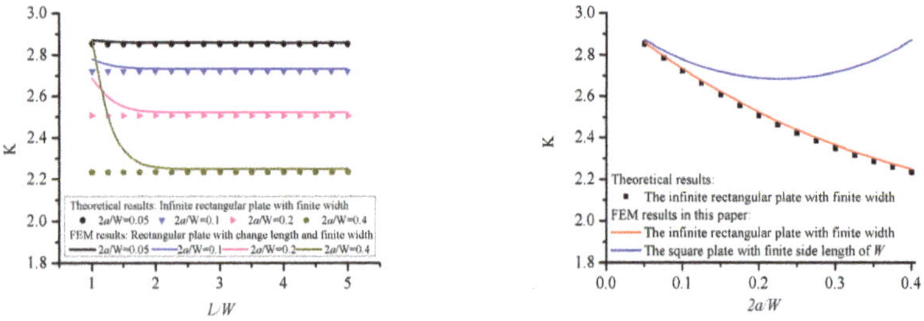

(a) (b)

Figure 2. Variation of stress concentration factors for a central circular hole in a rectangular panel (**a**) rectangular panels with different lengths, (**b**) rectangular and square panels.

3.1.2. Stress Analysis of FGM Panel without Hole

Based on the triangular element, the stress concentration problem of an FGM square panel, without a hole, is solved under uniaxial tension along the y-direction, and then the normal stress distribution is calculated. As shown in Figure 3, the stress results are normalized by referring to the value of external tension load σ_0 along the x-direction. The normalized normal stress for the two configurations is shown in Figure 3. Firstly, the problems raised by Ref. [35] is recalculated and the results are shown in Figure 3a. Here, a softening material means that Young's modulus is progressively decreasing away from the origin of the coordinates, and the hardening material means that Young's modulus gradually increases away from the origin of the coordinates. In this problem, the origin is located at the left or right end of the panel, Young's modulus adopts the exponentially variation and makes it varies along the x direction. The expression of Young's modulus is $E(x) = E_0 \exp(x/L_g)$ and the size of the panel satisfies $w/Lg = \pm 2.08$. When the origin is located at the center of the panel, the problem proposed in Ref. [33] is resolved in Figure 3b. The expression of Young's modulus is $E(x) = E_0 \exp(|x|/Lg)$ and the size of the panel satisfies $w/Lg = \pm 2.08$. As can be seen from Figure 3, the normal stress of the homogeneous panel without hole hardly changes with the change of the position of the x. For FGM panel, even if there is no circular hole defect, there is an inhomogeneous stress distribution in the panel. In Figure 3a, as Young's modulus varies monotonously along x, the stress first increases and then decreases, or first decreases and then increases. Young's modulus discussed in Figure 3b is symmetrical with respect to x = 0, so the distribution of normal stress is symmetrical along the x-axis and the maximum/minimum normal stress appears in the center of the panel. Moreover, the stress variation shown in Figure 3 varies smoothly and satisfies the global equilibrium in an integral sense as $\int (\sigma_{22}/w)dx = \sigma_0$. In general, the results in this paper are consistent with those in Refs [33,35], which proves the feasibility of the present calculation program.

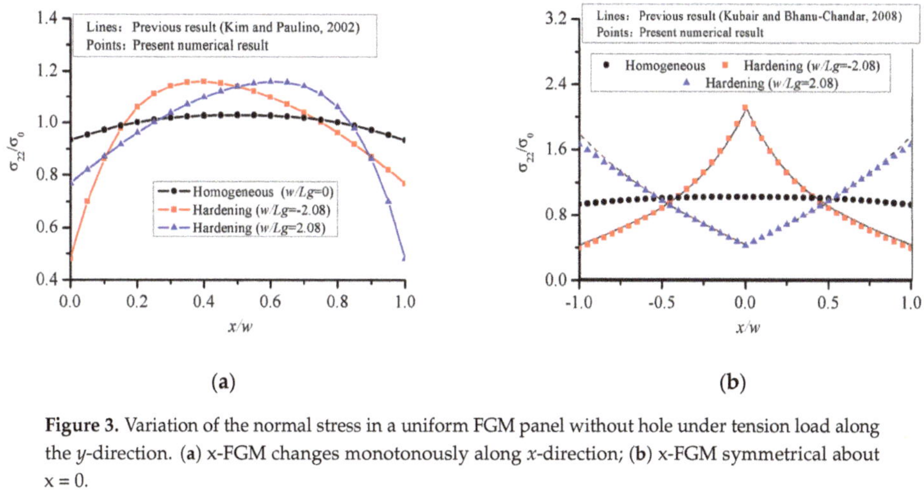

Figure 3. Variation of the normal stress in a uniform FGM panel without hole under tension load along the y-direction. (**a**) x-FGM changes monotonously along x-direction; (**b**) x-FGM symmetrical about x = 0.

3.1.3. Stress Analysis of FGM Panel with a Circular Hole

The stress distribution near the hole of FGM panel with a circular hole under tension load along the x-direction are recalculated which is given in Ref. [34]. Young's modulus in this analysis is $E(r) = E_{ref}[1 + \gamma_1 (r/a)^c]$, Poisson ratio is $v(r) = v_{ref}[1 + \gamma_2 (r/a)^c]$, and $c = -5$. Figure 4 shows that when the value of γ_1 is positive, the hoop stress on the hole surface reaches the maximum value at the point of $x = a$. Correspondingly, when the value of γ_1 is negative, the hoop stress on the hole surface reaches the minimum value at the point of $x = a$. However, the values of γ_2 have little influence on the stress results. In addition, the hoop stress decreases gradually as it moves away from the circular hole. When the value of y/a is close to 5, the stress reaches a stable value. As shown in Figure 4, the new calculation results in this paper are in good agreement with those given in Ref. [34], which confirms the reliability of the present calculation program.

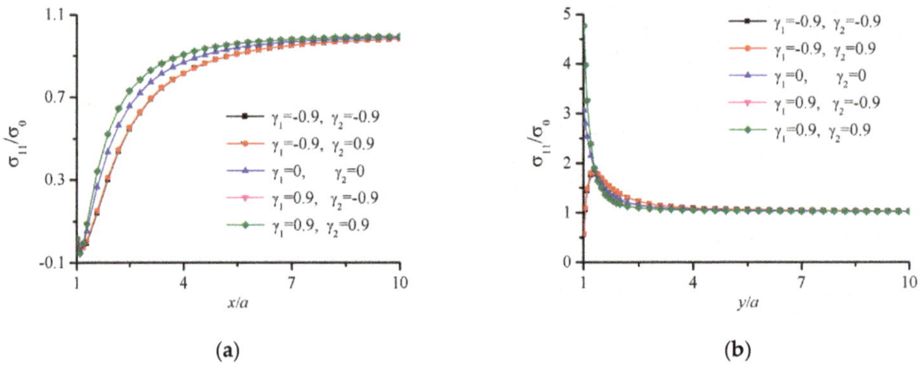

Figure 4. Variation of the (**a**) radial stress on the line y = 0, (**b**) the hoop stress on the line x = 0 in a uniform FGM panel with a circular hole under tension load along the *x*-direction.

3.2. *Stress Concentration Factor*

3.2.1. The Power-Law Inhomogeneous Variation When c > 0

Here, we calculate the stress concentration of FGM with a circular hole under uniaxial tension. Young's modulus varies in the power-law form as $E(\phi) = E_0\left[1 + \gamma(\phi/Lg)^c\right]$, where $c > 0$. By changing the values of c and γ, the variation trend of stress concentration factor is obtained. Figure 5a depicts the variations of Young's modulus E/E_{ref} with ϕ/Lg under different gradient control parameters where $\gamma = -0.5, -0.25, 0, 0.5, 1$ and $c = 1, 2, 3$. When $\gamma = 0$, it satisfies $E = E_0$, which corresponds a homogeneous panel. When $\gamma > 0$, E/E_{ref} gradually increases with the increase of ϕ/Lg. When $\gamma < 0$, E/E_{ref} gradually decreases with the increase of ϕ/Lg. Figure 5b–d gives the stress concentration factor when the elastic modulus changes along the directions of r, x and y, respectively. As shown in Figure 5b–d, when $\gamma > 0$, the stress concentration factor K decreases first and then increases with the increase of c, and when $\gamma < 0$, the stress concentration factor K increases first and then decreases with the increase of c. In addition, it can be seen that when $\gamma > 0$, the stress concentration factor can be reduced compared with that of homogeneous materials. Since the corresponding stress concentration factor becomes minimum when $\gamma = 1$, the dimensionless Von Mises stress distribution are given in Figure 6. It can be seen that the maximum value of the dimensionless Mises stress first decreases and then increases with the increase of c. The means there exists an optimal value of power law index because the stress distribution does not change monotonously with the increasing power law index.

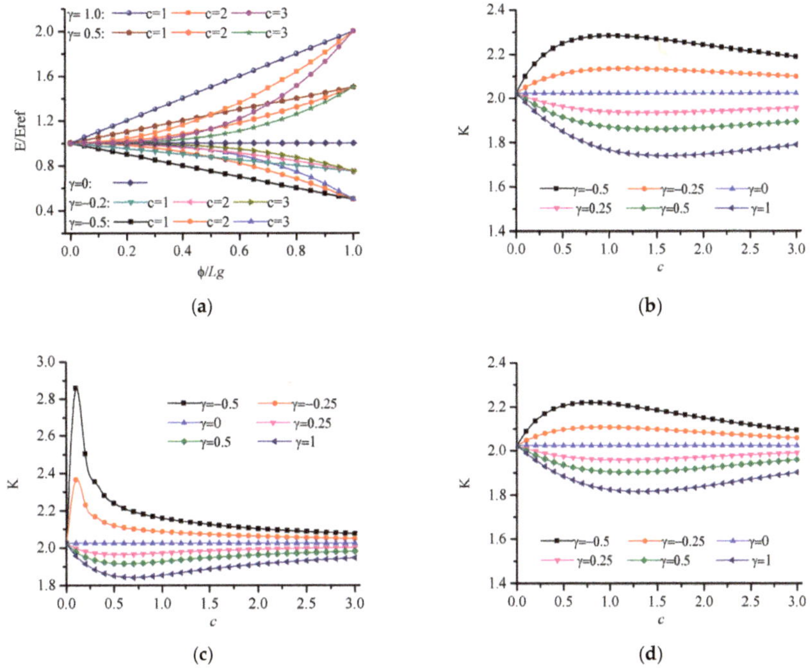

Figure 5. Variation of stress concentration factor K for power-law inhomogeneous variation when c > 0. (**a**) variation of Young modulus; (**b**) results for r-FGM; (**c**) results for x-FGM; (**d**) results for y-FGM.

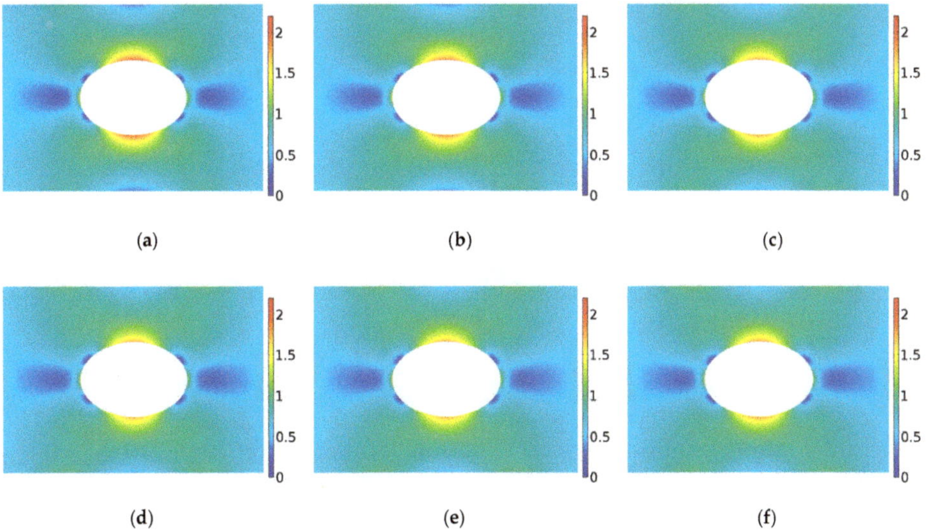

Figure 6. Variation of dimensionless Von Mises stress distribution for r-FGM when $\gamma = 1$; The power-law inhomogeneous variation when c < 0. $\gamma = -0.5$. (**a**) $\gamma = 1$, c = 0.1; (**b**) $\gamma = 1$, c = 0.5; (**c**) $\gamma = 1$, c = 1; (**d**) $\gamma = 1$, c = 1.5; (**e**) $\gamma = 1$, c = 2.0; (**f**) $\gamma = 1$, c = 3.

3.2.2. The Power-Law Inhomogeneous Variation When c < 0

The stress concentration factor of FGM are given in Figure 7 when Young's modulus varies in form of power-law $E(\phi) = E_0\left[1 + \gamma(\phi/a)^c\right]$, where $c < 0$. Figure 7a depicts the variation of Young's modulus E/E_{ref} with ϕ/a under different gradient control parameters where $\gamma = -0.5, 0, 1.0$ and $c = 1, 2, 3$. When $\gamma = 0$, it corresponds a homogeneous panel. When $\gamma > 0$, with the increase of ϕ/a, E/E_{ref} gradually decreases and finally tends to 1. When $\gamma < 0$, with the increase of ϕ/a, E/E_{ref} gradually increases and finally tends to 1. It can be seen that when ϕ/a is large enough, the γ and c have little influence on E/E_{ref}. Figure 7b shows the curve of stress concentration factor K for the r-FGM. As shown in Figure 7b, when the value of γ is positive, K increases significantly with the increase of the absolute value of c. When the value of γ is negative, K first decreases and then increases with the increase of the absolute value of c. The analysis shows that when $\gamma < 0$, the stress concentration factor can be reduced compared with that of homogeneous materials. Figure 8 shows the dimensionless Von Mises stress distribution when $\gamma = -0.5$ which corresponds to the smallest optimal value of the stress concentration factor. It can be seen that the maximum value of the dimensionless Von Mises stress shows a significant decrease with the increase of c.

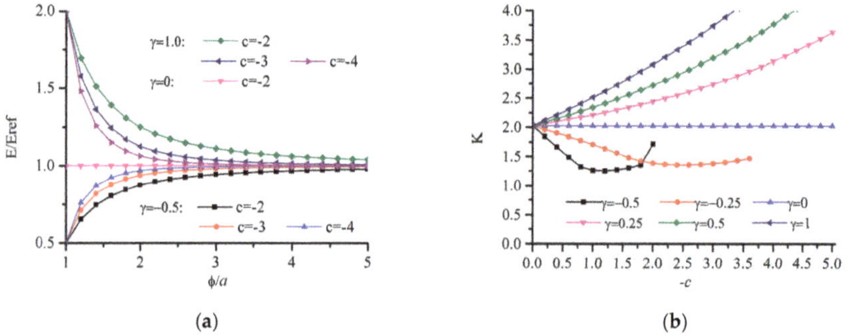

Figure 7. Variation of stress concentration factor K for power-law inhomogeneous variation of r-FGM when c < 0; (**a**) variation of Young's modulus; (**b**) results for r-FGM.

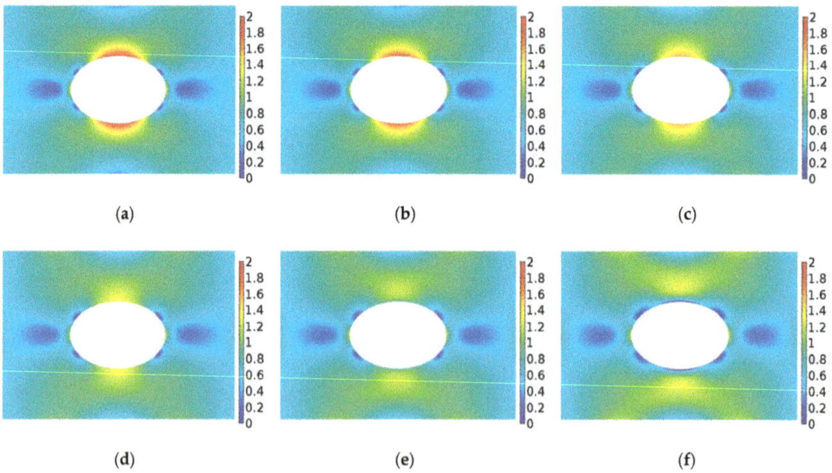

Figure 8. Variation of dimensionless Von Mises stress distribution for r-FGM when $\gamma = -0.5$. (**a**) $\gamma = -0.5$, c = 0.0; (**b**) $\gamma = -0.5$, c = −0.2; (**c**) $\gamma = -0.5$, c = −0.5; (**d**) $\gamma = -0.5$, c = −0.8; (**e**) $\gamma = -0.5$, c = −1.2; (**f**) $\gamma = -0.5$, c = −1.8.

3.3. *Failure Index and Damage Factor*

3.3.1. The Power-Law Inhomogeneous Variation When c > 0

Figure 9 shows the calculation results of the failure index of r-FGM with hole under uniaxial tension with different strength-modulus exponent ratio when Young's modulus varies in the power-law form as $E(\phi) = E_0\left[1 + \gamma(\phi/Lg)^c\right]$, where $c > 0$, and $\gamma = -0.5, -0.3, -0.1, 0.2, 0.5, 1.0$. When this parameters are selected, the modulus of each point of the FGM satisfies $E(\phi) \in [E_0(1+\gamma), E_0]$ when $\gamma < 0$, and $E(\phi) \in [E_0, E_0(1+\gamma)]$ when $\gamma > 0$. As can be seen from Figure 9, the failure index Φ, which is the maximum value of the damage factor, always decreases with the increasing modulus ratio γ. In addition, when $\delta < 0.2$, for the case of $\gamma = 1$, the failure index Φ decreases rapidly and then tends to stabilize as the index of power-law increases. However, when $\delta > 0.3$, for the situation of $\gamma = 1$, the failure index Φ shows a monotonously increasing trend as the index of power-law increases. Figure 10 shows the trend of the dimensional damage factor ψ with the index of power-law under optimal condition of $\gamma = 1$. It can be clearly seen that when $\delta = 0$, the area of the FGM panel susceptible to damage (corresponding to the area shown by red in Figure 10a increases first and then decreases as the index of power-law increases, which is consistent with the result of $\gamma = 1$ in Figure 9a. And the value of the dimensional damage factor is minimized when c = 1.5, which means that the optimal anti-failure performance of FGM is achieved. When $\delta = 0.5$, the damage factor increases slightly with the increasing power-law index, which is consistent with the result of $\gamma = 1$ in Figure 9f.

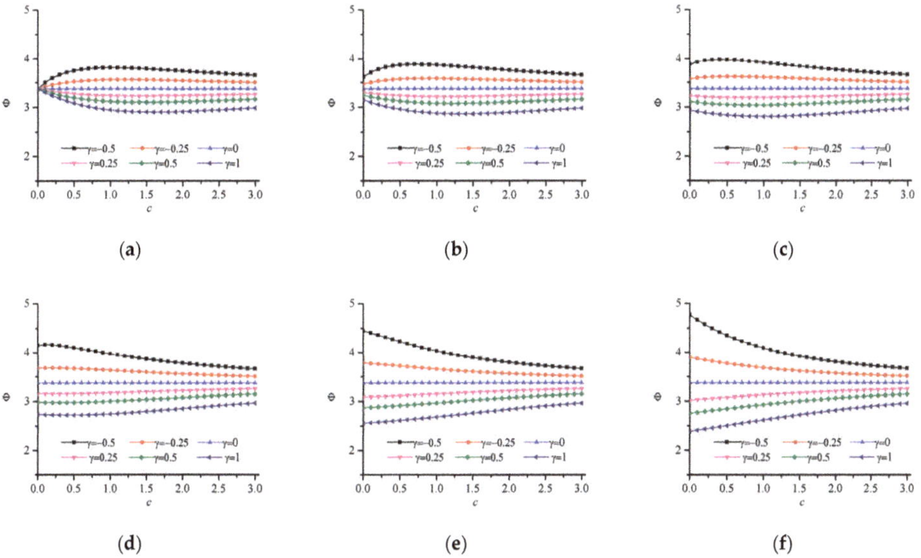

Figure 9. Variation of the failure index with different strength-modulus exponent ratio for the power-law inhomogeneous variation when c > 0. **(a)** $\delta = 0$; **(b)** $\delta = 0.1$; **(c)** $\delta = 0.2$; **(d)** $\delta = 0.3$; **(e)** $\delta = 0.4$; **(f)** $\delta = 0.5$.

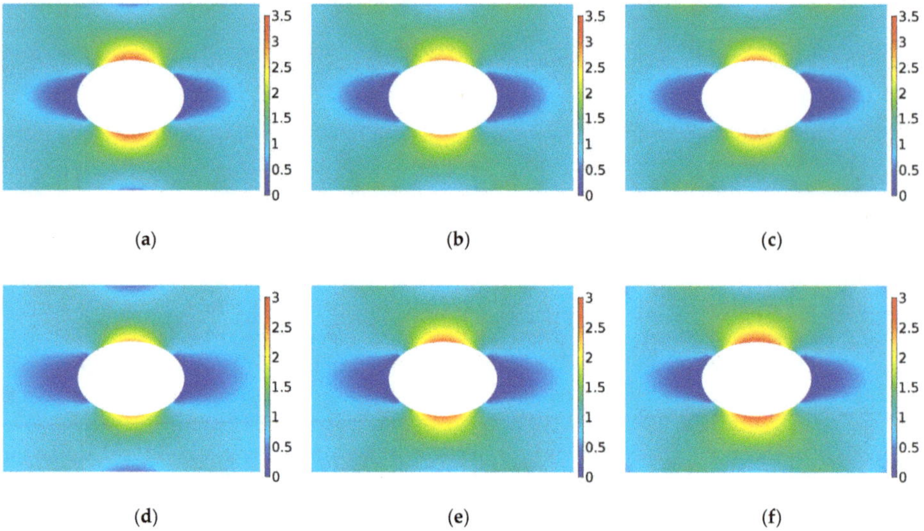

Figure 10. Variation of the damage factor ψ with power-law index under optimal condition of $\gamma = 1$. (**a**) δ = 0.0, γ = 1, c = 0.0; (**b**) δ = 0.0, γ = 1, c = 1.5; (**c**) δ = 0.0, γ = 1, c = 3.0; (**d**) δ = 0.5, γ = 1, c = 0.0; (**e**) δ = 0.5, γ = 1, c = 1.5; (**f**) δ = 0.5, γ = 1, c = 3.0.

Similarly, the variations of the failure index with different strength-modulus exponent ratio for the x-FGM and y-FGM are given in Figure 11. Here we still choose $\gamma = -0.5, -0.3, -0.1, 0.2, 0.5, 1.0$, which makes the material modulus at each point satisfy $\max\{E(\phi)\}/\min\{E(\phi)\} < 2$. From Figure 11, it can be seen that, for the elliptical hole, the trend curves of r-FGM and y-FGM are basically the same. In addition, it can be seen that when $\delta = 0$, the failure index changes with the index of power-law, there is an optimal functional gradient design function. When $\delta > 0.2$, the failure index increases with the index of power-law, which means that the optimal design modulus design is to select a uniform material that maximizes the modulus at each point. This phenomenon is because the maximum allowable stress of the material is a function of strength-modulus exponent ratio and modulus. When the strength-modulus exponent ratio is small, the change of the material modulus has little effect on the limit strength. The optimal design is to reduce the absolute stress at each point by adjusting the material modulus distribution. When the strength-modulus exponent ratio is large, increasing the material modulus causes the corresponding limit strength to increase rapidly, and then the damage factor at each point can be rapidly reduced. So, when $\delta > 0.2$ the solution is to select a uniform material that maximizes the modulus at each point.

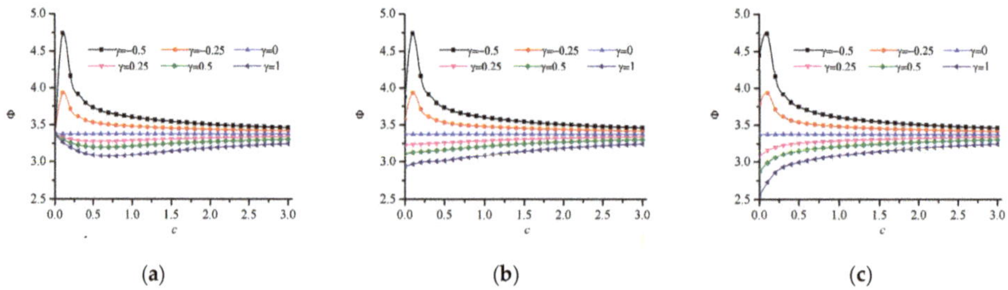

Figure 11. *Cont.*

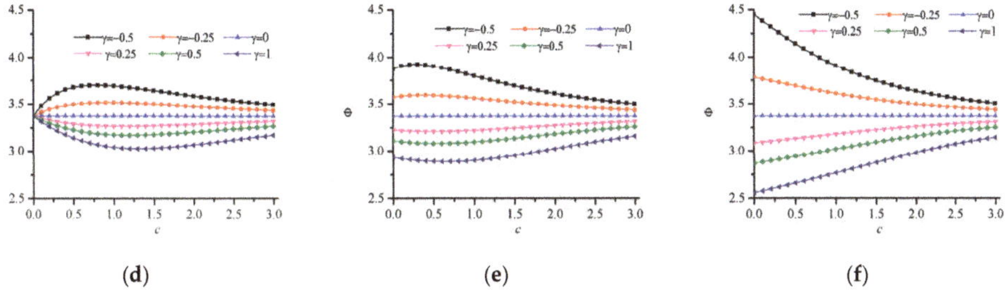

Figure 11. Variation of the failure index with different strength-modulus exponent ratio for the power-law inhomogeneous variation of x-FGM and y-FGM when c > 0. x-FGM: **(a)** $\delta = 0.0$; **(b)** $\delta = 0.2$; **(c)** $\delta = 0.4$; y-FGM: **(d)** $\delta = 0.0$; **(e)** $\delta = 0.2$; **(f)** $\delta = 0.4$.

3.3.2. The Power-Law Inhomogeneous Variation When c < 0

Here, the failure index of FGM is calculated when Young's modulus varies in form of power-law as $E(\phi) = E_0\left[1+\gamma(\phi/a)^c\right]$, where $c < 0$. Here we still choose the parameter $\gamma = -0.5, -0.3, -0.1, 0.2, 0.5, 1.0$, because this makes the material modulus at each point satisfy $\max\{E(\phi)\}/\min\{E(\phi)\} < 2$. From Figure 12a–d, when $\delta < 0.3$ the failure index reaches a minimum value when $\gamma = -0.5$, and the material damage resistance is maximized. However, from Figure 12e,f, when $\delta > 0.3$, the failure index reaches a minimum at $\gamma = 1$, where the material damage resistance is maximized. This optimization result can still be interpreted as the result of competition between the reduced stress value and the increase of the limit strength value. As shown in Figure 7a, this optimization result still shows that when $\delta < 0.3$, the optimal design is the modulus increasing with distance from the hole, and when $\delta > 0.3$, the optimal design is to maximize the modulus at each point.

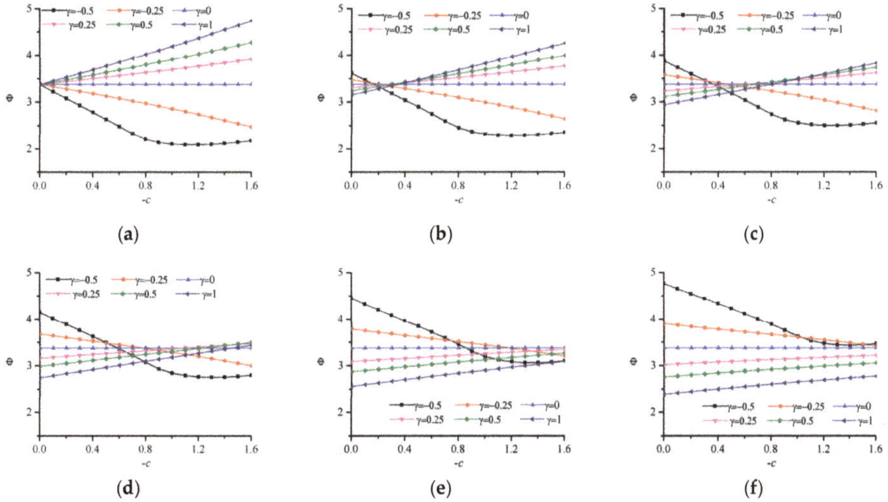

Figure 12. Variation of the failure index with different strength-modulus exponent ratio for the power-law inhomogeneous variation of r-FGM when c < 0. **(a)** $\delta = 0$; **(b)** $\delta = 0.1$; **(c)** $\delta = 0.2$; **(d)** $\delta = 0.3$; **(e)** $\delta = 0.4$; **(f)** $\delta = 0.5$.

4. Conclusions

In this paper, the effect of the inhomogeneous variation and gradient directions on stress concentration caused by a central elliptical hole in FGM panel under uniaxial tension load is analyzed.

The effects of inhomogeneous characteristic control parameters, such as modulus ratio, the index of power-law variation is considered. The conclusions can be given as follows: (1) When the index of power-law variation is positive, the stress concentration factor of FGM can be reduced compared with that of homogeneous materials. (2) When the tensile load is along the x axis, the optimal designs of r-FGM significantly better than that of x-FGM and y-FGM. (3) When the associated variation of ultimate stress is considered, the choice of scheme to reduce the failure index is related to the strength-modulus exponent ratio. When the strength-modulus exponent ratio is small, the failure index changes with the index of power-law, which means there is an optimal FGM design. But when the strength-modulus exponent ratio is large, the optimal design modulus design is to select a uniform material that maximizes the modulus at each point. These research results have a certain reference value for further in-depth understanding of the inhomogeneous design for FGM.

Author Contributions: Conceptualization and Methodology, X.L. and P.S.; Software and Validation: W.W., H.Y. and P.S.; Funding acquisition: W.W., X.L. and P.S.; Writing—original draft, W.W., H.Y. and P.S.; Writing—review & editing, X.L. and P.S.

Funding: This research was funded by the Natural Science Foundation of China (Grant Nos. 11802225 and 11561055), the China Postdoctoral Science Foundation (Grant No. 2017M623154) and the Graduate Innovation Project of Ningxia University (GIP2018067).

Conflicts of Interest: The authors declare no conflict of interest.

References

1. Tornabene, F. Free vibration analysis of functionally graded conical, cylindrical shell and annular plate structures with a four-parameter power-law distribution. *Comput. Methods Appl. Mech. Eng.* **2009**, *198*, 2911–2935. [CrossRef]
2. Koizumi, M.; Niino, M. Overview of FGM research in Japan. *Mrs Bull.* **1995**, *20*, 19–21. [CrossRef]
3. Zhang, B.B.; Jaiswal, P.; Rai, R.; Nelaturi, S. Additive manufacturing of functionally graded material objects: A review. *J. Comput. Inf. Sci. Eng.* **2018**, *18*, 223–235. [CrossRef]
4. Xu, F.X.; Zhang, X.; Zhang, H. A review on functionally graded structures and materials for energy absorption. *Eng. Struct.* **2018**, *171*, 309–325. [CrossRef]
5. Cramer, C.L.; Hsin, W.; Kaka, M. Performance of functionally graded thermoelectric materials and devices: A review. *J. Electron. Mater.* **2018**, *47*, 5122–5132. [CrossRef]
6. Petit, C.; Palmero, P. Functionally Graded Ceramics for biomedical application: Concept, manufacturing, properties. *Int. J. Appl. Ceram. Technol.* **2018**, *15*, 820–840. [CrossRef]
7. Suresh, S. Graded materials for resistance to contact deformation and damage. *Science* **2001**, *292*, 2447–2451. [CrossRef]
8. Birman, V. Modeling and Analysis of Functionally Graded Materials and Structures. *Appl. Mech. Rev.* **2007**, *60*, 195–216. [CrossRef]
9. Shanmugavel, P.; Bhaskar, G.B.; Chandrasekaran, M.; Mani, P.S.; Srinivasan, S.P. An overview of fracture analysis in functionally graded materials. *Eur. J. Sci. Res.* **2012**, *68*, 412–439.
10. Jha, D.K.; Kant, T.; Singh, R.K. A critical review of recent research on functionally graded plates. *Compos. Struct.* **2013**, *96*, 833–849. [CrossRef]
11. Swaminathan, K.; Naveenkumar, D.T.; Zenkour, A.M.; Carrera, E. Stress, vibration and buckling analyses of FGM plates—A state-of-the-art review. *Compos. Struct.* **2015**, *120*, 10–31. [CrossRef]
12. Swaminathan, K.; Sangeetha, D.M. Thermal analysis of FGM plates—A critical review of various modeling techniques and solution methods. *Compos. Struct.* **2017**, *160*, 43–60. [CrossRef]
13. Barretta, R. Analogies between Kirchhoff plates and Saint-Venant beams under flexure. *Acta Mech.* **2014**, *225*, 2075–2083. [CrossRef]
14. Barretta, R.; Luciano, R. Analogies between Kirchhoff plates and functionally graded Saint-Venant beams under torsion. *Contin. Mech. Thermodyn.* **2015**, *27*, 499–505. [CrossRef]
15. Barretta, R.; Marotti de Sciarra, F. A nonlocal model for carbon nanotubes under axial loads. *Adv. Mater. Sci. Eng.* **2013**, *2013*, 360935. [CrossRef]

16. Barretta, R.; Čanađija, M.; Luciano, R.; de Sciarra, F.M. Stress-driven modeling of nonlocal thermoelastic behavior of nanobeams. *Int. J. Eng. Sci.* **2018**, *126*, 53–67. [CrossRef]
17. Mohammadi, M.; Dryden, J.R.; Jiang, L. Stress concentration around a hole in a radially inhomogeneous plate. *Int. J. Solids Struct.* **2011**, *48*, 483–491. [CrossRef]
18. Yang, Q.; Gao, C.F. Reduction of the stress concentration around an elliptic hole by using a functionally graded layer. *Acta Mech.* **2016**, *227*, 2427–2437. [CrossRef]
19. Dave, J.M.; Sharma, D.S. Stress field around rectangular hole in functionally graded plate. *Int. J. Mech. Sci.* **2018**, *136*, 360–370. [CrossRef]
20. Shi, P.P. Stress field of a radially functionally graded panel with a circular elastic inclusion under static anti-plane shear loading. *J. Mech. Sci. Technol.* **2015**, *29*, 1163–1173. [CrossRef]
21. Shi, P.P. Imperfect interface effect for nanocomposites accounting for fiber section shape under antiplane shear. *Appl. Math. Model.* **2017**, *43*, 393–408. [CrossRef]
22. Goyat, V.; Verma, S.; Garg, R.K. Reduction in stress concentration around a pair of circular holes with functionally graded material layer. *Acta Mech.* **2018**, *229*, 1045–1060. [CrossRef]
23. Mehrparvar, M.; Ghannadpour, S.A.M. Plate assembly technique for nonlinear analysis of relatively thick functionally graded plates containing rectangular holes subjected to in-plane compressive load. *Compos. Struct.* **2018**, *202*, 867–880. [CrossRef]
24. Shi, P.P.; Zheng, X.J. The Yoffe-type moving tubular interface crack in a hollow composite cylinder with finite length. *Int. J. Mech. Sci.* **2015**, *98*, 29–38. [CrossRef]
25. Shi, P.P.; Sun, S.; Li, X. Arc-shaped interfacial crack in a non-homogeneous electro-elastic hollow cylinder with orthotropic dielectric layer. *Meccanica* **2013**, *48*, 415–426. [CrossRef]
26. Shi, P.P.; Sun, S.; Li, X. The cyclically symmetric distributed cracks on the arc-shaped interface between a functionally graded magneto-electro-elastic layer and an orthotropic elastic substrate under static anti-plane shear load. *Eng. Fract. Mech.* **2013**, *105*, 238–249. [CrossRef]
27. Shi, P.P. Effect of boundary conditions on the interfacial fracture behavior: Circular arc antiplane crack model for an annular sector bilayer plate. *Theor. Appl. Fract. Mech.* **2016**, *82*, 136–151. [CrossRef]
28. Sburlati, R. Stress concentration factor due to a functionally graded ring around a hole in an isotropic plate. *Int. J. Solids Struct.* **2013**, *50*, 3649–3658. [CrossRef]
29. Gouasmi, S.; Megueni, A.; Bouchikhi, A.S.; Zouggar, K.; Sahli, A. On the reduction of stress concentration factor around a notch using a functionally graded layer. *Mater. Res.* **2015**, *18*, 971–977. [CrossRef]
30. Sburlati, R.; Atashipour, S.R.; Atashipour, S.A. Reduction of the stress concentration factor in a homogeneous panel with hole by using a functionally graded layer. *Compos. Part B Eng.* **2014**, *61*, 99–109. [CrossRef]
31. Hsu, W.H.; Chien, W.T. Effect of electrical discharge machining on stress concentration in titanium alloy holes. *Materials* **2016**, *9*, 957. [CrossRef]
32. Yang, Y.; Cheng, Y.; Zhu, W. Stress concentration around a rectangular cuboid hole in a three-dimensional elastic body under tension loading. *Arch. Appl. Mech.* **2018**, *88*, 1229–1241. [CrossRef]
33. Kubair, D.V.; Bhanu-Chandar, B. Stress concentration factor due to a circular hole in functionally graded panels under uniaxial tension. *Int. J. Mech. Sci.* **2008**, *50*, 732–742. [CrossRef]
34. Nie, G.J.; Zhong, Z.; Batra, R.C. Material tailoring for reducing stress concentration factor at a circular hole in a functionally graded material (FGM) panel. *Compos. Struct.* **2018**, *205*, 49–57. [CrossRef]
35. Kim, J.H.; Paulino, G.H. Isoparametric graded finite elements for nonhomogeneous isotropic and orthotropic materials. *J. Appl. Mech.* **2002**, *69*, 502–514. [CrossRef]
36. Huang, J.; Venkataraman, S.; Rapoff, A.J.; Haftka, R.T. Optimization of axisymmetric elastic modulus distributions around a hole for increased strength. *Struct. Multidiscip. Optim.* **2003**, *25*, 225–236. [CrossRef]
37. Chen, Z.; Li, W.; Negahban, M.; Saiter, J.M.; Delpouve, N.; Tan, L.; Li, Z. Approaching the upper bound of load capacity: Functional grading with interpenetrating polymer networks. *Mater. Des.* **2018**, *137*, 152–163. [CrossRef]
38. Young, W.C.; Budynas, R.G.; Sadegh, A.M. *Roark's Formulas for Stress and Strain*, 8th ed.; McGraw-Hill Companies: New York, NY, USA, 2012.

materials

MDPI

Article

Neuro-Fuzzy Modelling of the Metallic Surface Characterization on Linear Dry Contact between Plastic Material Reinforced with SGF and Alloyed Steel

Victor Vlădăreanu, Lucian Căpitanu and Luige Vlădăreanu *

Institute of Solid Mechanics of the Romanian Academy, C-tin Mille 15, 010141 Bucharest, Romania; victor.vladareanu@vipro.edu.ro or vladareanuv@gmail.com (V.V.); lucian.capitanu@yahoo.com (L.C.); luigiv2007@gmail.com (L.V.)

* Correspondence: luige.vladareanu@vipro.edu.ro; Tel.: +40-21-315-7478

Received: 23 May 2018; Accepted: 5 July 2018; Published: 10 July 2018

Abstract: This paper presents the modelling of wear data resulting from linear dry contact using artificial neural networks (ANN) and adaptive neuro-fuzzy inference systems (ANFIS) with the aim of constructing predictor models for the depth and volume of the wear scar, with great impact in the characterization of new industrial processes utilizing existing materials. The dataset is the result of laboratory testing, presenting both numerical and categorical variables whose inclusion into the model allows for a number of possibilities. The width of the wear scar was measured on a microscope, and its depth was calculated. A multitude of experimental tests was performed with normal loads and different speeds, which led to some conclusive results, but in some cases, with relatively high variance. Various options for the automatic generation of fuzzy inference systems were also approached (genfis2). The innovative approach was compared with a baseline model featuring multivariate linear regression optimized using gradient descent, drawing on previous experimentation on the same dataset. The models developed can be implemented in future research and in practical applications under similar conditions, aiming to optimize performance by applying Computer Science. The obtained results lead to highly accurate prediction models which are further integrated into various metallic surface characterizations in the wear process for tribological and robotics research in new industrial processes using short glass fiber reinforced polymers.

Keywords: ANFIS; fuzzy logic; clustering; neural networks; robotics and contact wear

---

1. Introduction

The paper presents new intelligent analytical methods for the characterization of wear in thermoplastic materials armed with short glass fibers (SGF) and steel in a dry contact wear scenario, applied to new industrial processes using existing materials.

Wear phenomena are very complex within injection or extrusion machine cylinders. The tribological hostile environment, high temperature, and corrosive chemical compounds increase this complexity. In the case of processing thermoplastic materials with short glass fiber fillings, complexity is further increased because of their significant abrasion.

The mechanical interaction in the form of wear, which always appears between two or multiple bodies when there are relative speeds, sliding, rolling, pivoting, and so on, defines friction wear. The process complexity is determined by the wear indicators, among which are the linear wear intensity, volumetric wear intensity, gravimetric wear intensity, wear coefficient, wear sensibility coefficient, and apparent energy density.

Previous studies on friction couples with linear contact on thermoplastic material armed with short glass fibers (SGF) and steel in a dry friction context have shown in experiments that even under normal, relatively small loads, large contact pressures and therefore very high contact temperatures may appear, which are close or may even exceed the transformation temperature of the plastic material.

These lead to the need to approach modelling as a dependency between the various variables of interest involved in the friction process and metallic surface characterization in the wear process using advanced statistical and optimization algorithms on a dataset obtained from hardware simulation. The subject draws from growing interest from the research community with the advent of highly advanced, intelligent classification; optimization and regression algorithms; and the wide impact of metallic surface characterization applications in the wear process, emphasizing the abrasive, adhesive, and corrosive wear.

Zhang et al. [1] analyzed the artificial neural network prediction of erosion wear of the polymer. Three independent sets of measurement data were used and the characteristic properties of erosive wear of these polymers to prepare and test the neural networks were explored. For the first two examples of materials, the angle of impact of solid particle erosion and some characteristic properties were selected as the input variables of the ANN.

Similar directions were investigated in Panda et al. and Flepp [2,3] regarding the potential of supervised or unsupervised learning and modelling the results of friction, while surveys by Ripa and Frangu [4] dealt with the various possibilities for undertaking this task. Finally, the paper builds upon previous work done by authors Rus et al. [5]. The importance of the subject matter is illustrated by its use in building artificial joints and prosthesis—Căpitanu et al., Al-Zubaidi et al. [6,7], among others. Căpitanu et al. [6] presented an analytical qualitative–quantitative correlation of friction and wear processes of steel surfaces in linear dry contact with SGF that reveals the nonlinearity of tribological processes in this case.

Wear processes, similar to fabrication processes, involve very complex and nonlinear phenomena. Consequently, analytical models are difficult or impossible to come by. However, improvements in the performance and reliability of mechanical equipment and production instruments require precise modelling and prediction of the wear phenomenon. Artificial neural networks (ANNs) and the related methods investigated in this paper, such as neuro-fuzzy inference systems, possess many desirable properties for modelling systems and processes: the ability to approximate universal functions, learning from experimental data, high tolerance for lacking or noisy data, and good capacity for generalization.

Artificial neural networks are the driving force behind the current advances in artificial intelligence, with useful applications in virtually every computational field. They rely on successively improving a network of weights attached to hidden units called neurons. Neural networks work by solving for the best dynamic weights of a hidden layer of neurons, which determine the strength with which these are fired [8]. While solving for a linear or polynomial regression model provides an explicit relationship between dependent and independent variables, it may be that an implicit representation model such as a neural network would yield better results.

Shukla [9] reported an overview of the applications of artificial neural networks in the processing domain. The property of learning and nonlinear behavior makes them useful for complex nonlinear process modelling, better than analytical methods. They are useful in some specific points in the field: Processes and wear particles, manufacturing, friction parameters, and defects in mechanical structures. Rao et al. [10] worked with rolling element bearings, which are widely used in almost all global industries. Their proactive strategy was to minimize the imminent failures in real time and at minimal cost. Innovative developments have been recorded in the technology of artificial neural networks (ANN). Chen and Savage [11] described an approach to fuzzy networks for a recognition system of multilevel surface roughness (FN-M-ISRR), whose aim was to predict the surface roughness (Ra) under multiple cutting condition, determined by the material of the tool, the tool size, etc.

The same nonlinearity was found by Căpitanu et al. [12] in the behavior of the UHMWPE tibia insert of the total knee prostheses, and Vlădăreanu et al. [13] applied ANN to a versatile intelligent portable robot control platform based on cyber physical systems principles.

The baseline model considered for expressing a dependency between the various variables of interest involved in the friction process and metallic surface characterization in the wear process is a multivariate linear regression. The first step in a multiple variable regression model is to normalize the features and then run a batch gradient descent algorithm on the data, where each iteration minimizes the cost function by simultaneously updating all variable coefficients [14]. The linear regression model also includes regularization factors to prevent over-fitting. The regularization component is included in the cost function and provides a penalty for the data being fitted too closely using polynomial variables.

The resulting dataset is then used to train a linear regression model for each of the two considered dependent variables: Wear depth and wear volume. This model assumes the dependent variables to be a linear combination of the considered independent variables, speed and pressure, and an intercept term, which does not vary with the independent variables. The intercept term is added only for the linear regression problem since the neural network algorithm will do the same on its own.

The optimization problem is then to find the best coefficients that minimize the cost function, which gives a measurement of the difference between the empirical values of the two dependent variables and their estimates obtained through linear regression. Gradient descent is the algorithm used to iteratively arrive at the best possible set of variable coefficients. The learning rate for this version of gradient descent is set to 1.

For each dependent variable, the prediction is a dot-product of the independent variable values and its respective coefficient vector. The linear regression coefficients show the relative influence a certain independent variable has on the prediction of a dependent variable. The intercept term is a baseline starting point for the prediction models, being an aggregation of all the other factors not considered and the inherent randomness of the model.

Fuzzy logic and fuzzy inference systems extend regular logic systems by assigning a degree of membership to elements within sets, which proves to be a very useful ability for modelling complex, unknown, or dynamic systems. Fuzzy Logic has long been used in academia and in industry and is one of the more palpable staples of artificial intelligence in the world today. Fuzzy logic controllers have proven to be robust and relatively easy to design [15]. They seem to suffer from no one major flaw while providing a number of important benefits such as expert knowledge emulation. There are various algorithms for the optimization of fuzzy inference systems' parameters such as genetic algorithms and neural networks. Neuro-fuzzy modelling (ANFIS) attempts to model the behavior of a given system for which arrays of input and output values are provided by creating a fuzzy inference system to produce similar results. The fuzzy inference system is then learned (i.e., its parameters are optimized) using an artificial neural network algorithm. This is a very convenient tool for simulating systems whose mathematical formulae are unknown or very complex. The overall concept is explained in further detail in [15,16], which are part of the authors' previous work on the topic.

ANFIS implementations in the context of wear prediction deals mainly with fault prevention and monitoring. Zuperl and Cus [17] construct a tool monitoring system using a merged neural decision network and wear predictor for tool maintenance, while Lo [18] uses ANFIS and the grey system method for tool failure detection in single point turning operations. As relates to contact wear, Aliman et al. [19] investigate wear rate on a coated aluminum alloy. Shabani et al. [20] also obtained interesting results by combining ANFIS with particle swarm optimization in manufacturing wear resistant nano-composites.

The provenance of the initial fuzzy inference system is of great significance in designing an ANFIS algorithm. Fuzzy inference systems may be automatically obtained from the available data through a number of algorithms, which mainly attempt to group the available data-points into equivalent fuzzy sets and then deduce relations between them, which turn into a set of fuzzy rules. The resulting fuzzy inference systems (FIS) can be used to seed an ANFIS algorithm, that is, to have it start optimization from the previously obtained FIS. In fact, if no seed is specified, the ANFIS algorithm will itself use one of the methods, namely grid partitioning, to generate its starting FIS. As is traditional with ANFIS, all of the FISs investigated were of the Sugeno type due to complexity and computational constraints. In addition, because of the limited amount of data, the generated fuzzy inference systems themselves are considered as possible solutions for the proposed model, since ANFIS models are very susceptible to over-fitting the data in the present context.

The authors have had previous contributions to this and related topics in [5,6,11,12,15,16], among others. The original contribution of the current paper stems from implementing the proposed models on a recently obtained dataset, evaluating the different results and providing a comparison of the effect of the various algorithms. This entails a comprehensive comparative study on the same dataset, while varying the learning algorithm hyper-parameters and associated options, such as investigating the various methods used for fuzzy inference system generation. Finally, the successful models are to be chosen for future implementation in real world applications in the fields of Tribology and Robotics.

The remainder of this paper is divided as follows. Section 2 will discuss the experimental data, the type of considered variables, how the data is processed for the learning application, the procedures for the automatic generation of fuzzy inference systems from the available data, the neural networks, and the adaptive neuro fuzzy inference systems. Section 3 compares the results of the various model solutions on the test sets. Section 4 discusses these results and Section 5 draws conclusions on this and possible future work.

2. Materials and Methods

The study of injection and extruding processes for thermoplastic materials is a complex process due to the phenomena existing inside injection and extruding machines with a permanent interconnection of the influencing factors.

Starting from the material selection stage, either fine dust, or the quantity of short glass fibers (SGF) used, adding materials such as TiO_2 (titanium dioxide) or graphite fibers, the technological process implies transforming the material from a solid to a plastic/liquid phase, which is achieved at temperatures above 1600 °C, with the material suffering deformation, pressing, and heating depending on the machine and the technological process.

In addition to the material complexity and the preparation for injection, an important part is played by determining a predictive model for the wear process in order to increase wear resistance of the work surfaces of the injection and extruding machines.

These are some of the considerations for the neuro-fuzzy modelling of wear data resulting from linear dry contact using artificial neural networks (ANN) and adaptive neuro-fuzzy inference systems (ANFIS).

Wear performance of the two steel alloys, C120 and Rp3, has been previously studied in the case of linear dry contact with each polymer (polyamide and polycarbonate) reinforced with different percentages of short glass fibers (SGF). The functional diagram of the friction couple is presented in Figure 1, from Căpitanu et al. (2014), where it looks at the linear contact. The friction couple comprises a cylindrical plastic liner and a flat polished steel hardened sample.

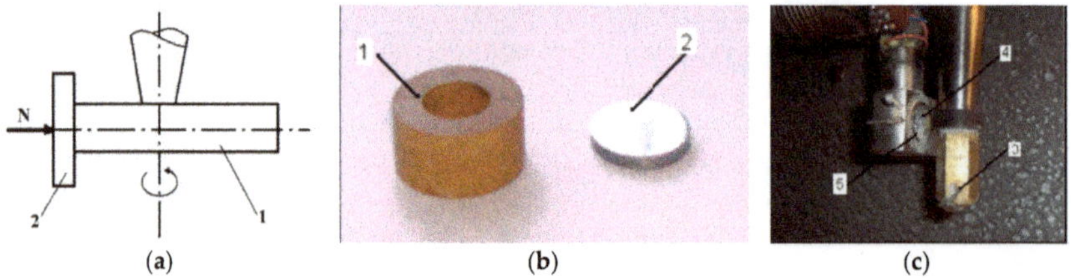

Figure 1. The functional diagram of the friction couple (**a**) the coupling elements (**b**) mounting the coupling to experimental equipment (**c**) movement of the bush against the disc, where 1—cylindrical thimble; 2—steel disc sample; 3—fastening nut; 4—carrier; 5—prism to ensure parallel contact.

The wear scar occurs by penetration of the cylindrical liner, under the influence of the normal load, in the flat sample material. In theory, the holding thimble is considered as rigid and relatively small in view of the backside imprint, so it can be considered as made up of a sum of cylinder areas. This is illustrated schematically in Căpitanu et al. [11].

We considered the following three polymers:

A. Nylonplast AVE Polyamide + 30% glass fibres; $E_{2A} = 40.25$ MPa.
B. Noryl Polyamide + 20% glass fibres; $E_{2B} = 31.76$ MPa.
C. Lexan Polycarbonate + 20% glass fibres; $E_{2C} = 42.08$ MPa.

Numerical values were determined by the elasticity modules (E) listed above, and the deformed liner radii (r_2), imposing p_{max} is provided as $p_{max} < 0.5\ H$, where H is the Brinell hardness for the plastic liner, enough so that it will not be crushed. The approximate depth of the wear scar is calculated with the relation: $h \approx l^2/8r_2$.

The imposed condition allows the following values of the maximum contact pressure of the dry linear couplings contact to be established, in the case of three plastic materials (A, B, C) reinforced with SGF, the five normal loads (contact pressures), indexes 1 to 5 of notations of the pressures that have been subjected to tests, for each of the seven relative sliding speeds used (18.56; 27.85; 37.13; 46.41; 55.70, 111.4; and 153.57 cm/s):

$$p_{A1} = 16.3\text{ MPa};\ p_{A2} = 23.5\text{ MPa};\ p_{A3} = 28.2\text{ MPa};\ p_{A4} = 32.6\text{ MPa};\ p_{A5} = 36.4\text{ MPa}$$

$$p_{B1} = 12.3\text{ MPa};\ p_{B2} = 17.4\text{ MPa};\ p_{B3} = 21.4\text{ MPa};\ p_{B4} = 24.6\text{ MPa};\ p_{B5} = 27.6\text{ MPa}$$

$$p_{C1} = 16.9\text{ MPa};\ p_{C2} = 23.9\text{ MPa:}\ p_{C3} = 29.3\text{ MPa};\ p_{C4} = 33.8\text{ MPa};\ p_{C5} = 37.8\text{ MPa}$$

After inspecting and measuring the wear scars of the metal surfaces, the widths of each wear scar were measured and their volume was calculated (the amount of material lost through wear). Their variation curves were also traced depending on the applied load (contact pressure), the relative speed of sliding contact with the temperature specification of the optical image and the presentation of the scar. This quantitative–qualitative assessment was presented in Căpitanu et al. [11]. All tests took place for 60 min, so that the calculated wear volumes are actually wear rates.

The increased friction coefficient entails increasing the wear rate and contact temperature, but after our data, it has not yet been possible to establish a mathematical relationship between the two sizes, which is widely recognized. This is why a suggestive graphical representation was sought to provide a qualitative correlation between the two sizes that relates them to the contact temperature and based on which to determine a quantitative correlation.

All the variation curves of the output parameter of the frictional system (amount of wear, depth of wear, friction coefficient, contact temperature) depending on input parameters, normal load (contact pressure), the sliding speed while maintaining the steady state surface (roughness R_a), shows a strong nonlinearity due to the behavior of the elastic-plastic polymers tested. In this situation, we tried an approach to model metal surface wear through advanced data fitting algorithms, because of their ability to model very complex and strongly nonlinear phenomena. This was the qualitative–quantitative analytical approach previously achieved. The graphic processing of these results was presented in Căpitanu et al. [5].

The data for metallic surface characterization on linear dry contact between plastic material reinforced with SGF and alloyed steel was obtained through experiments run on friction couples with linear contact using three different types of polymers on two different types of steel variants. Aside from alternating the materials used, the speed and pressure applied to them were varied under the same operating conditions. This was done with regard to the particulars of each material combination and the levels of speed and pressure.

The method used approaches of the study of wear on a metallic surface in the case of dry linear contact, plastic reinforced with SGF on surfaces of alloyed steel, C120 and Rp3, through the method of artificial neural networks. This is necessary because the wear processes involve very complex and powerfully nonlinear phenomena, and analytic models are difficult or impossible to obtain. Furthermore, the multiple inputs (normal load, contact pressure, sliding speed, measured contact temperature, materials properties) and outputs (width and depth of the wear scar, contact temperature) influence each other continually.

The resulting dataset includes the following information, seen in Table 1.

Table 1. Dataset.

Variable	Mode	Type	Range	Unit
Material	Independent	Categorical	N/A	N/A
		Numerical	-	-
Speed	Independent	Numerical	18.56–153.55	cm/s
Pressure	Independent	Numerical	10–50	N
Depth	Dependent	Numerical	0.9–9.1	10^{-4} mm^3
Volume	Dependent	Numerical	0.13–3.48	10^{-4} mm^3

For each of the variables considered, the table describes the following characteristics:

- Mode is whether the variable is treated as a dependent or an independent variable. The contention of any modelling technique is that the dependent variables can be represented as some relation, whether explicit or implicit, of the independent variables. In the present context, the paper investigates the effect of various materials, speeds, and pressures on the depth and volume of the wear scar.
- Range shows the numerical limits of each considered variable.
- Unit displays the unit of measurement for each variable.
- Type is one of three possibilities for the nature of the variable: Numerical, categorical, or ordinal [21]. The latter is not present in the experimental dataset. Speed and pressure are obviously numerical.

For the material type, there are three possible options since a fitting application cannot work with simple labels as inputs. The first is not considered proper, and is only shown as a comparison to the other two. The second and third options lead to separate optimization problems, both of which are considered in parallel for each of the models involved. Figure 2 shows a graphical representation of the options available in the current context, while Table 2 briefly outlines the benefits and drawbacks of each option.

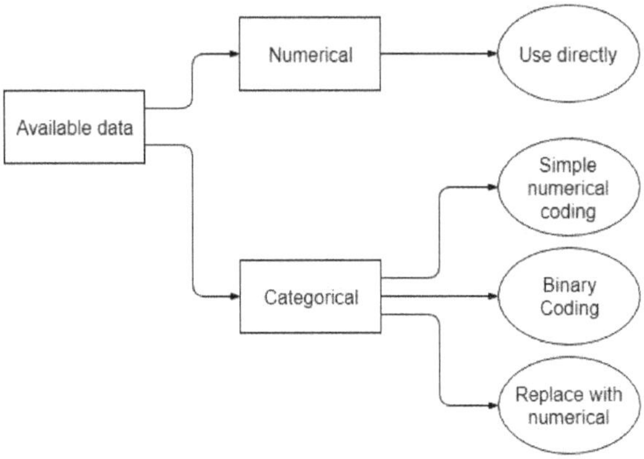

Figure 2. Flowchart for using available data.

Table 2. Comparison of coding options.

Option No.	Coding	Benefits	Disadvantages
1	Simple Numerical Coding	Simplest implementation	Implicitly orders labels, introducing bias.
2	Binary Coding	Easy to implement. Translates naturally to actual context.	Introduces multiple new variables. May cause rank deficiencies when working with matrices.
3	Replace with numerical	Leads to a completely numerical problem. Eliminates the issues in simple and binary coding.	Introduces many new variables. May cause dimensionality problems with small datasets.

The first is coding the categorical variable as a simple numerical variable, with different integer numbers for each of the labels. For example, each of the material combinations used in the experiments can be assigned an integer, transforming it into a discrete numerical variable and allowing it to interact with the rest of the independent numerical variables.

$$\text{AVE} + 30\%\ \text{SGF}; \text{C120 steel}$$

$$\text{AVE} + 30\%\ \text{SGF}; \text{Rp3 steel}$$

$$\text{Lexan} + 20\%\ \text{SGF}; \text{C120 steel}$$

$$\text{Noryl} + 20\%\ \text{SGF}; \text{C120 steel}$$

However, this suffers from implicitly ordering the labels, which would negatively impact the model and possibly introduce hidden biases, since the assigned numbers have no mathematical meaning [21].

The second option is coding the categorical variable into multiple binary variables, one for each of the material labels. This is especially advantageous in the present context, since each value of the categorical variable is logically a combination of two materials, the scratched surface and the object used for scratching. As an aside, this was also possible for the first representation, but it further increased the danger of misrepresentation through the interaction of numerical labels. The second representation is shown in Table 3.

Each of the materials that make up the labels of the categorical variable will be binary variables themselves. The binary representation translates naturally to whether that particular material is present or not. This is the standard representation and is implemented in some of the models. The main drawback is that the scarcity of the resulting matrices may lead to rank deficiency in some of the methods for automatically generating fuzzy inference systems (genfis). Table 4 shows an excerpt of the available date for training, using this coding option.

Table 3. Binary coding of variables.

Material	AVE	Lexan	Noryl	C120	Rp3
AVE + 30% SGF with C120 steel	1	0	0	1	0
AVE + 30% SGF with Rp3 steel	1	0	0	0	1
Lexan + 20% SGF with C120 steel	0	1	0	1	0
Noryl + 20% SGF with C120 steel	0	0	1	1	0

Table 4. Dataset excerpt using binary coding.

Independent								Dependent	
Material	Speed	Pressure	A	C	R	L	N	Depth	Volume
AVE + 30% SGF/C120	18.56	20	1	1	0	0	0	2.4798	0.4404
AVE + 30% SGF/C120	27.85	20	1	1	0	0	0	3.7076	0.5338
AVE + 30% SGF/C120	37.13	30	1	1	0	0	0	5.1336	0.9418
AVE + 30% SGF/C120	46.41	10	1	1	0	0	0	3.8871	0.2714
AVE + 30% SGF/C120	111.4	10	1	1	0	0	0	4.9482	0.283
AVE + 30% SGF/Rp3	18.56	40	1	0	1	0	0	3.9708	1.1247
AVE + 30% SGF/Rp3	27.85	20	1	0	1	0	0	3.4464	0.5164
AVE + 30% SGF/Rp3	37.13	30	1	0	1	0	0	4.2392	0.8627
AVE + 30% SGF/Rp3	46.41	20	1	0	1	0	0	4.5242	0.5833
AVE + 30% SGF/Rp3	46.41	30	1	0	1	0	0	5.0392	0.9377
Lexan + 20% SGF/C120	27.85	40	0	1	0	1	0	4.9169	1.1594
Lexan + 20% SGF/C120	46.41	10	0	1	0	1	0	4.0361	0.2582
Noryl + 20% SGF/C120	46.41	40	0	1	0	0	1	6.3271	2.4946
Noryl + 20% SGF/C120	55.7	20	0	1	0	0	1	4.3885	1.0289
Noryl + 20% SGF/C120	55.7	30	0	1	0	0	1	4.9133	1.6474

The third option is replacing the categorical variable altogether. Instead of using the material labels, the model will introduce, as independent variables, the numerical characteristics of the various materials. To this end, five new variables for the polymers—specific weight, water absorption, elasticity, thermal conductivity and linear dilation—and three for the steel—S_{max}, P_{max} and Ni_{max}—are introduced. The chosen variables have a full complement of values for each of the materials used in the experiment. The eight new variables are shown in Tables 5 and 6.

Table 5. Polyamide characteristics.

Polyamide	Weight	Absorption	Elasticity	Conductivity	Dilation
AVE	1.35	0.8	80	0.34	3.3
Noryl	1.27	0.06	84	0.196	2.5
Lexan	1.35	0.16	86	0.5	2.68

Table 6. Steel characteristics.

Steel	S_{max}	P_{max}	Ni_{max}
Rp3	0.02	0.025	0.4
C120	0.025	0.03	0.35

This gives the dataset eight independent variables that describe the variation in material on both sides of the experiment. The higher dimensionality is a slight disadvantage, but the representation is more natural due to them being actual numerical variables. An excerpt of the dataset is shown in Table 7 (numbers and labels are heavily truncated to allow representation).

In conclusion, the first coding option—simple numerical coding—is not suited to the task, while the second and third options are both used. This will create two distinct learning problems, which will be called the binary coding problem (option 2) and the numerical coding problem (option 3), on which the various algorithms are run. The results are shown and discussed for both cases in the appropriate sections.

Table 7. Dataset excerpt using numerical coding.

Independent											Dependent	
Material	Sp	Pr	Wgt	Abs	Els	Cd	Dil	S	P	Ni	Dpt	Vol
A/C120	18.5	40	1.35	0.8	80	0.34	3.3	0.025	0.03	0.35	5.48	1.30
A/C120	27.8	10	1.35	0.8	80	0.34	3.3	0.025	0.03	0.35	2.91	0.23
A/C120	27.8	40	1.35	0.8	80	0.34	3.3	0.025	0.03	0.35	6.13	1.37
A/C120	37.1	10	1.35	0.8	80	0.34	3.3	0.025	0.03	0.35	3.68	0.29
A/C120	46.4	40	1.35	0.8	80	0.34	3.3	0.025	0.03	0.35	8.54	1.62
A/C120	57.7	10	1.35	0.8	80	0.34	3.3	0.025	0.03	0.35	4.47	0.29
A/C120	111	30	1.35	0.8	80	0.34	3.3	0.025	0.03	0.35	8.00	1.17
A/C120	153	10	1.35	0.8	80	0.34	3.3	0.025	0.03	0.35	5.17	0.31
A/Rp3	18.5	40	1.35	0.8	80	0.34	3.3	0.02	0.025	0.4	3.97	1.12
A/Rp3	27.8	20	1.35	0.8	80	0.34	3.3	0.02	0.025	0.4	3.44	0.51
A/Rp3	37.1	30	1.35	0.8	80	0.34	3.3	0.02	0.025	0.4	4.23	0.86
A/Rp3	46.4	20	1.35	0.8	80	0.34	3.3	0.02	0.025	0.4	4.52	0.58
L/C120	27.8	40	1.35	0.16	86	0.5	2.6	0.025	0.03	0.35	4.91	1.15
L/C120	46.4	10	1.35	0.16	86	0.5	2.6	0.025	0.03	0.35	4.03	0.25
N/C120	46.4	40	1.27	0.06	84	0.19	2.5	0.025	0.03	0.35	6.32	2.49
N/C120	55.7	20	1.27	0.06	84	0.19	2.5	0.025	0.03	0.35	4.38	1.02

An example of plotted data is shown in Figure 3. Due to the increased dimensionality, the data cannot be plotted as a graph dependent on the independent variables. Therefore, a two-dimensional graph based on the index was chosen instead. For the same reason, the data now has no discernible outliers, as can be seen in the figure, so all pre-processing conditions were eliminated.

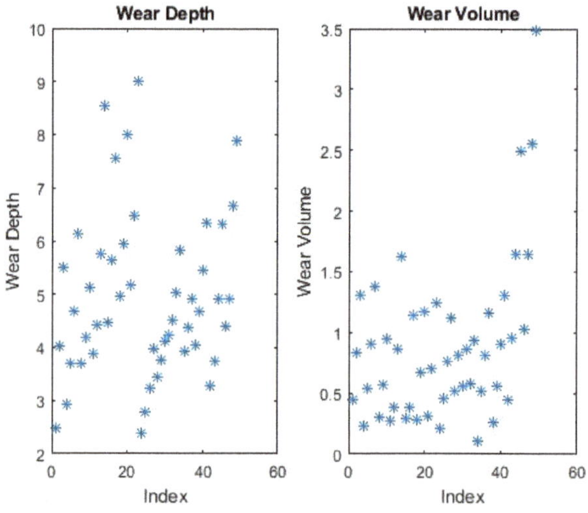

Figure 3. Dependent variables plotted on index for (**a**) wear depth and (**b**) wear volume.

The available data-points were randomly divided into a training set and a test set, with ratios of 70% of the total and 30% of the total, respectively. This is done by constructing a random index of 70% true values on a vector with the same length as the number of lines (i.e., samples) in the dataset. The training and test sets are then easily separated based on this index. The training test is used for optimizing the parameters of the various models including hyper-parameters (i.e., cross-validation). The test set is not available to any algorithm until the model is complete, when it is used to evaluate its performance on a heretofore unseen set of data belonging to the same phenomenon. Therefore, the evaluation of each model will take place both on the training set as well as on the test set.

3. Results

For metallic surface characterization on linear dry contact between the plastic material reinforced with SGF and alloyed steel, a multitude of experimental tests was performed with normal loads and different speeds, which led to the modelling of the metallic surface characterization on linear dry contact between the plastic material reinforced with SGF and alloyed steel using four predictor models presented below.

3.1. Linear Regression

The first model is a first-order multivariate linear regression optimized using batch gradient descent. The results obtained here will be used as a baseline for all other models. The optimization problem is described as $Y = \hat{\theta} * X$, where X is the input data, namely the dependent variables, containing all data-points, plus an intercept term. Y is the output data, alternatively the wear speed or wear volume, that the algorithm is trying to learn, and θ is the matrix of coefficients used to estimate Y from X. The challenge is finding the best coefficients, which minimizes the error between the actual Y and the estimate. This is obtained from $\hat{Y} = \theta_{LR} * X$, or, in extended form:

$$\begin{bmatrix} \hat{y_1} \\ \hat{y_2} \\ \hat{y_3} \\ \vdots \\ \hat{y_n} \end{bmatrix} = \begin{bmatrix} \theta_{11} & \cdots & \theta_{1(m+1)} \\ \vdots & \ddots & \vdots \\ \theta_{n1} & \cdots & \theta_{n(m+1)} \end{bmatrix} \begin{bmatrix} x_1 \\ x_2 \\ \vdots \\ x_m \\ int \end{bmatrix} \tag{1}$$

There will be m features and an intercept term, which helps prevent over-fitting. The total sum of all errors, across all values, is defined as the cost function. It is this cost function that the optimization algorithms attempt to minimize.

Gradient descent is an optimization algorithm where the potential solution is improved each iteration by moving along the feature gradient in the variable space. While it requires that the target function be differentiable and it is somewhat susceptible to local minima, gradient descent provides a stable and computationally inexpensive algorithm for function optimization.

As noted in the previous chapter, both representations of the dataset, using the two options for coding the categorical variable, were investigated in parallel. After running the gradient descent algorithm, the following coefficients were obtained, as shown in Table 8. Running the algorithm takes between 1 and 2 s for each of the dependent variables—the last run was timed internally at 1.174 s.

Table 8. Coefficient values from the linear regression algorithm.

Binary Coding			Numerical Coding		
Features	Thetas		Features	Thetas	
	Depth	Volume		Depth	Volume
Intercept	0.5470	−0.1621	Intercept	0.1632	4.5657
Speed	0.0339	0.0015	Speed	0.0313	0.0035
Pressure	0.0961	0.0391	Pressure	0.1095	0.0450
AVE	0.2615	−0.1706	Weight	0.0252	−0.1091
Lexan	0.3347	0.1097	Absorption	0.0826	−0.3794
Noryl	−0.3257	−0.1250	Elasticity	−0.0004	−0.0373
C120	0.0184	−0.2411	Conductivity	0.0394	−0.4060
Rp3	−0.2729	0.3989	Dilation	0.1059	−0.4695
			Smax	0.0197	0.0072
			Pmax	0.0188	0.0062
			Nimax	−0.1858	−0.0552

The obtained models will now predict the depth and volume of the wear scar as the linear combination of the vector of thetas and the value of a given feature set.

For example, using binary coding, the wear depth will be predicted as

$$D = 0.547 + 0.034Sp + 0.096P + 0.261A + 0.335L - 0.326N + 0.018C - 0.273R \quad (2)$$

Figures 4 and 5 show the training fits for wear and volume, with both coding options, using linear regression.

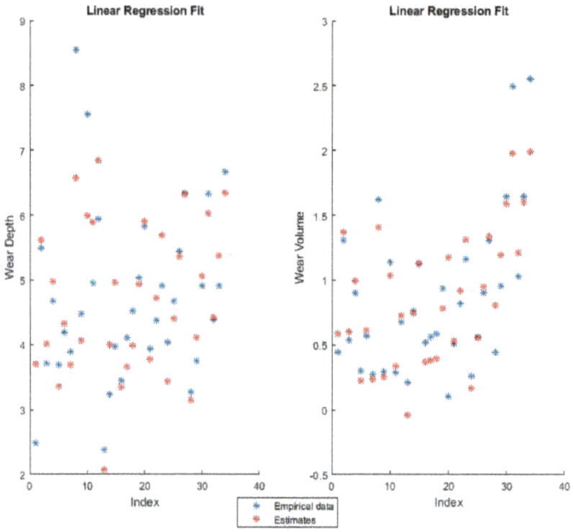

(a) (b)

Figure 4. Linear regression fit for (**a**) wear depth and (**b**) wear volume in binary coding.

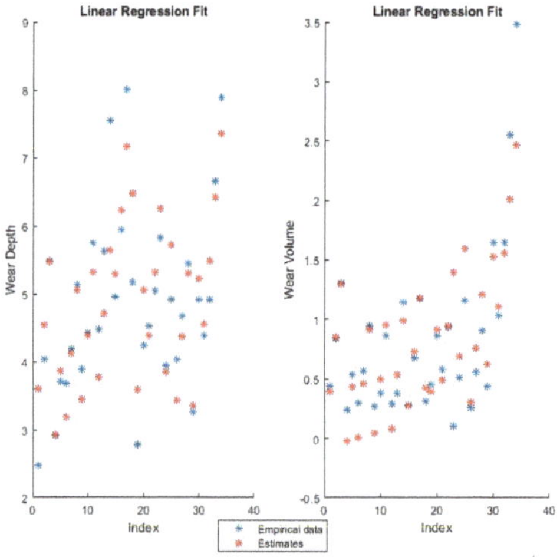

(a) (b)

Figure 5. Linear regression fit for (**a**) wear depth and (**b**) wear volume in numerical coding.

The blue points represent the actual experimental data available for training, while the red points are the estimates that the model would obtain for the same input data.

The linear regression model has a decent performance of fitting the training data, but is obviously at a disadvantage because of the non-linearity present in the dataset. One interesting point of note is that the relative influence of the various features can be directly ascertained from the theta values, which shows both the magnitude and the direction of the dependency.

3.2. Neural Networks

Neural network models are centered on a layer of hidden features (i.e., neurons) which control the prediction. Neural networks have an input layer that matches the considered independent variables and an output layer which matches the dependent variable. The model is optimized by successively tuning the weights associated to these neurons as well as their activation functions. A standard neural network is exemplified in Figure 6 [22].

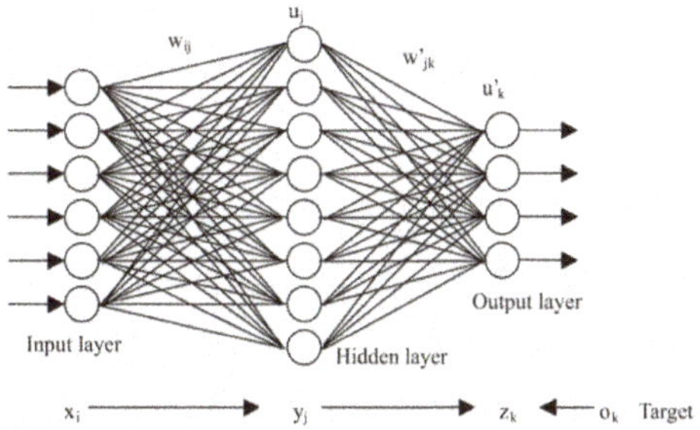

Figure 6. Artificial Neural Network with one hidden layer.

The most important hyper-parameter of a neural network model is the number of neurons in the hidden layer. For the purposes of this application, all models contain 25 hidden neurons. This number was selected after running the algorithm with various levels of neurons and settling on the best performance in terms of the correlation coefficient for both the training and test sets. The training time varies greatly with the choice of training algorithm selected: Some may require only 1–2 s, while some configurations can take up to 30 s. The neural network selected here was timed at 12 s.

The weights of the selected neural networks are far too numerous to display in the paper. For example, to pass from the input layer of 10 neurons (10 features in the numerical coding case) to the hidden layer of 25 neurons, a matrix of 25 lines and 10 columns is required. Figures 7 and 8 show the training fits for wear and volume with both coding options using a neural network model.

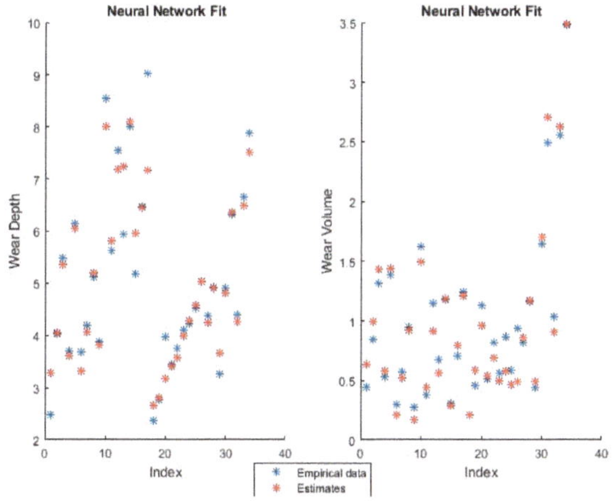

(a) (b)

Figure 7. Neural Network fit for (**a**) wear depth and (**b**) wear volume in binary coding.

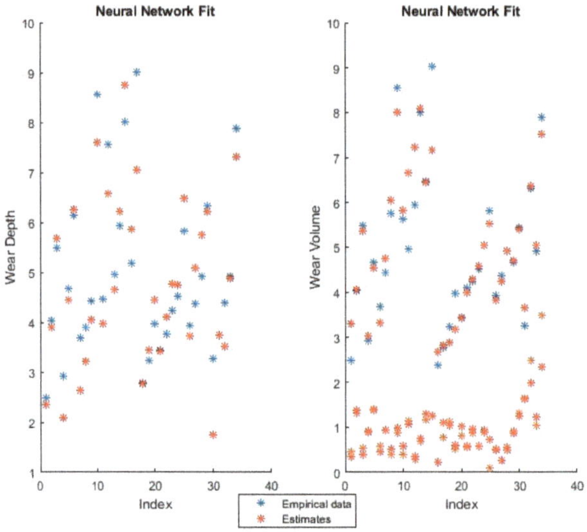

(a) (b)

Figure 8. Neural network fit for (**a**) wear depth and (**b**) wear volume in numerical coding.

As with the linear regression figures, the blue points are the empirical data and the red points the estimates. It should be noted that perfectly correct estimates (red points) will overwrite the empirical data representation (blue points) where they coincide.

Neural networks give a very good fit of the training data. For example, notice how virtually all empirical points in the lower half of the volume fit, as seen in the graphical representation, have been overwritten by their estimates. Such good behavior on the training set, however, always raises suspicions of over-fitting, which will be verified or invalidated when the model is used on the training set.

3.3. Generated Fuzzy Inference System

Fuzzy inference systems can be automatically generated and then used as actual models for predicting the future behavior of a system. The generated FIS can then be deployed either as such, or further optimized using an ANFIS algorithm, which will be discussed in the next section. As these tend to be computationally intensive models, the FIS is of the Sugeno type, which employs linear or constant functions as outputs, as opposed to the Mamdani type FIS, whose outputs are also fuzzy membership functions. This type of FIS has fewer parameters and is therefore somewhat less computationally intensive. As will be discussed in the final results section, there is also the danger of over-fitting since there are so many parameters and comparatively few data-points.

Given a set of raw data, there are three options for obtaining a working FIS: Grid partitioning (genfis1), subtractive clustering (genfis2), and fuzzy c-means clustering (genfis3). There are a number of hyper-parameters to be tuned such as the number of membership functions per input and the shape of each membership functions. These were chosen after some trials to be four membership functions per numerical variable and two per binary variable, when dealing with the binary coding option. The shape of the membership functions was kept as standard bell curves (Gaussian distributions). Generating fuzzy inference systems is performed quickly, usually within a second. The last run yielding internal timings of 0.177 s for genfis1, 0.153 s for genfis2, and 0.248 s for genfis3.

Grid partitioning is really only used as a benchmark, since it is commonly held that its performance is unsatisfactory unless further developed with ANFIS [23–25]. However, because it is such a rudimentary starting point for the ANFIS algorithm, it will actually not work on the numerical coding options, as there are simply too few data-points, too many parameters and too computationally intensive a task. Therefore, ANFIS starting from grid partitioning is only investigated for the binary coding problem. The other two options are, however, valid possible solutions both on their own, as well as after further optimization. Figures 9 and 10 show a sample of the fuzzy inference spaces of the FISs obtained through automatic generation. Since it is impossible to show the n-dimensional graph of the FIS, the first two inputs are chosen as the independent variable axis.

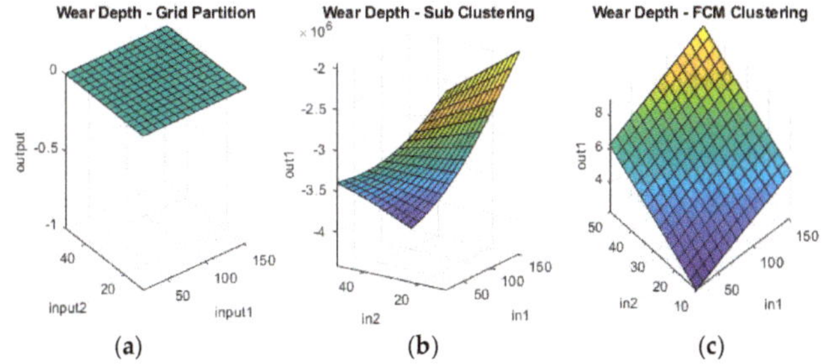

Figure 9. Fuzzy inference systems generated through (**a**) grid partitioning; (**b**) sub-clustering; and (**c**) Fuzzy c-Means (FCM) clustering for wear depth in numerical coding.

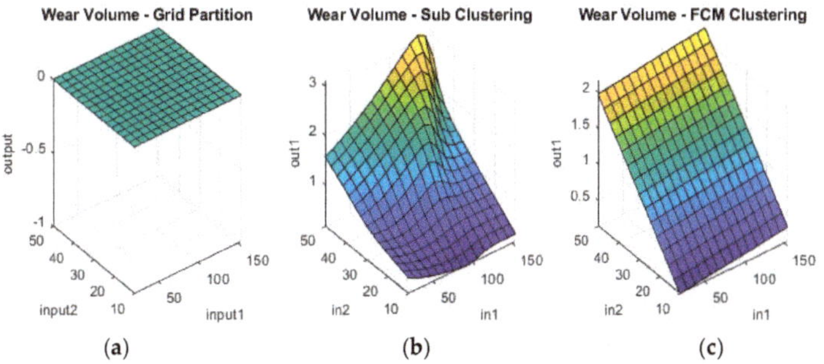

Figure 10. Fuzzy inference systems generated through (**a**) grid partitioning; (**b**) sub-clustering; and (**c**) FCM clustering for wear volume in numerical coding.

As can be seen from Figures 9 and 10, grid partitioning provides a simple starting point for further optimization, while sub-clustering and FCM clustering present very interesting fuzzy inference spaces. Figures 11 and 12 show a selection of the fit obtained using the generated fuzzy inference systems.

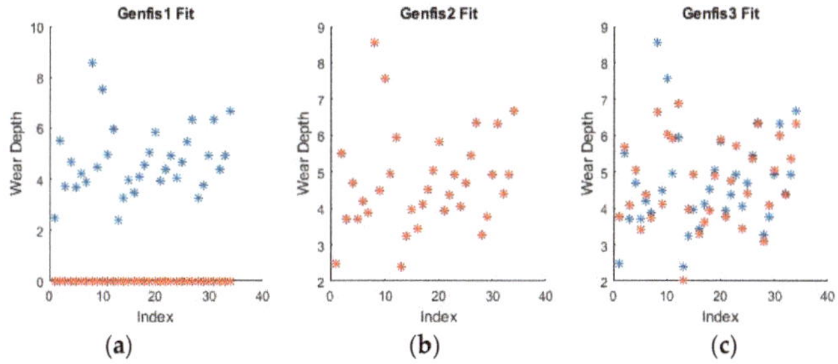

Figure 11. Fuzzy interference systems (FIS) fit generated through (**a**) grid partitioning; (**b**) sub-clustering; and (**c**) FCM clustering for wear depth in binary coding.

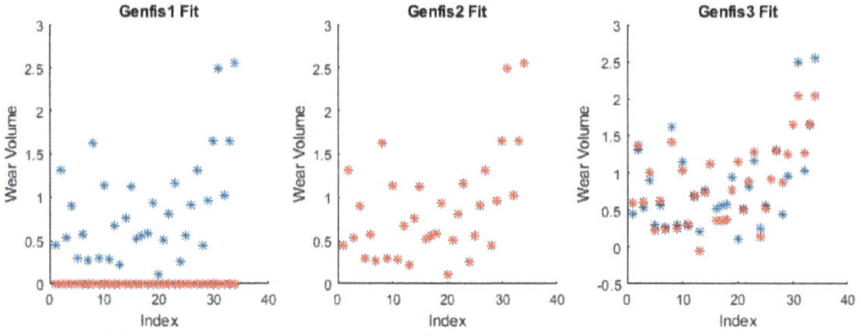

(a) (b) (c)

Figure 12. FIS fit generated through (**a**) grid partitioning; (**b**) sub-clustering; and (**c**) FCM clustering for wear volume in numerical coding.

From the spread of the empirical and estimated data, some early conclusions can be drawn about generated fuzzy inference systems. As already discussed, grid partitioning is simply an empty form for a FIS, which is the starting point of an ANFIS algorithm, as no attempt is made to fit the experimental data. The FIS obtained through subtractive clustering has near perfect performance on the training set, which naturally raises concerns about the possibility of over-fitting the data. These will be addressed when performing on the test set. The third method, fuzzy c-means clustering, obtains good, if not great, performance and is a very good start for further optimization.

3.4. Adaptive Neuro-Fuzzy Inference Systems

ANFIS uses back propagation to determine the premise parameters of each rule. The consequent parameters are then determined using a least mean squares algorithm. An iteration of the learning algorithm consists of two passes: In the forward pass, the premise parameters are fixed and the consequent parameters are optimized through an iterative least squares approach, while going through the rule base system. In the backward pass, the consequent parameters are fixed, while the premise parameters are modified using back propagation. This algorithm continues until the target error is met or the number of iterations exceeds a predetermined threshold. An excellent description of the ANFIS architecture and learning procedure is given by Denai et al. [26].

Adaptive neuro-fuzzy inference systems construct a FIS capable of predicting the values of the dependent variables through further optimization of a generated FIS structure. If no initial FIS is specified, one is created through a genfis1 (grid partitioning) algorithm. With the exception of grid partitioning in the case of numerical coding, as discussed previously, all generated FISs were run through an ANFIS algorithm in the hopes of improving performance. The algorithm is relatively fast, usually lasting a few seconds. It will take longer when starting from a FIS obtained through grid partitioning (genfis1), since it requires more optimization. The inference systems discussed in the paper were timed at 4.987 s for ANFIS 1, 1.019 s for ANFIS 2, and 1.866 s for ANFIS 3.

The algorithm was used with a standard set of hyper-parameters such as 50 generations and the inherited FISs obtained at the previous step. Figures 13 and 14 show a selection of the resulting fuzzy inference spaces.

The fuzzy inference spaces obtained after running the algorithm have a very interesting configuration. The resulting rule base spaces are clearly nonlinear, which should lead to a good ability to fit the present dataset. Figures 15 and 16 show a selection of the resulting performance when fitting the training set data.

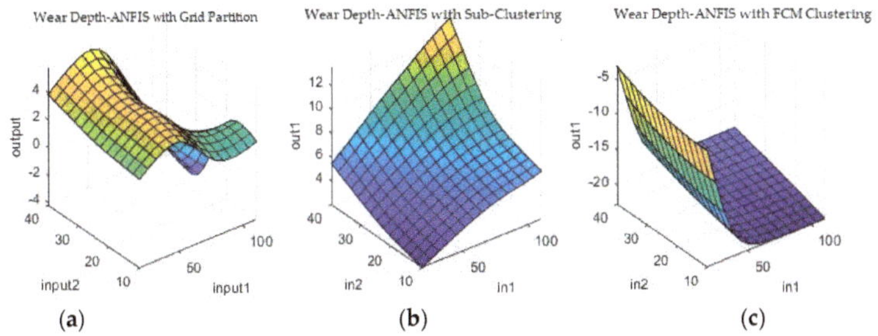

Figure 13. Fuzzy inference systems generated through adaptive neuro-fuzzy inference systems (ANFIS) with (**a**) grid partitioning; (**b**) sub-clustering; and (**c**) FCM clustering for wear depth in binary coding.

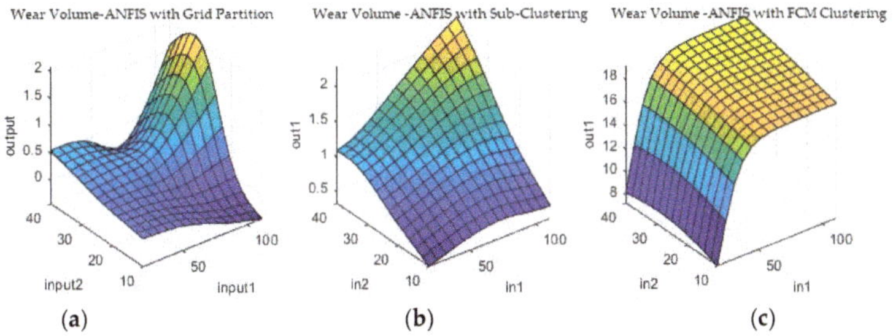

Figure 14. Fuzzy inference systems generated through ANFIS with (**a**) grid partitioning; (**b**) sub-clustering; and (**c**) FCM clustering for wear volume in binary coding.

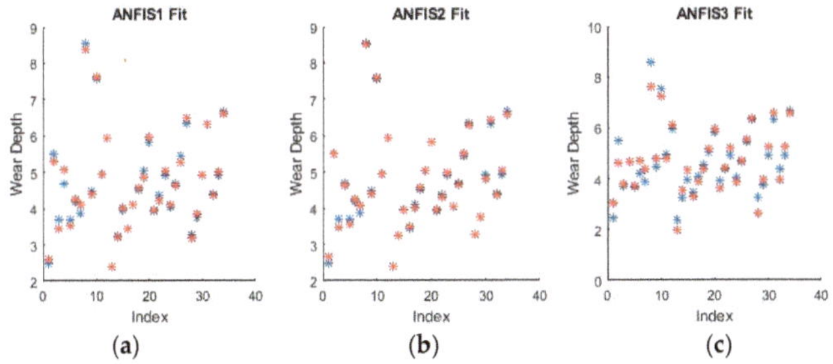

Figure 15. ANFIS fit generated through (**a**) grid partitioning; (**b**) sub-clustering; and (**c**) FCM clustering for wear depth in binary coding.

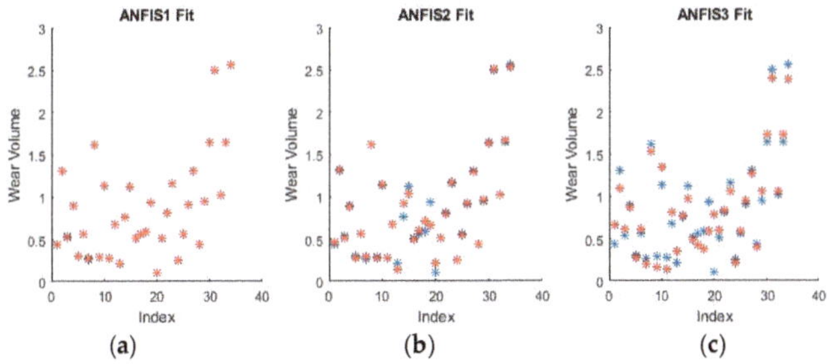

Figure 16. ANFIS fit generated through (**a**) grid partitioning; (**b**) sub-clustering; and (**c**) FCM clustering for wear volume in binary coding.

The performance shown by all ANFISs is very encouraging as the spreads shown above seem to reveal very good behavior from all systems. Pending the requisite investigation on the test sets, the ANFISs appear to be the frontrunners for fitting the experimental dataset. Of particular note is the near perfect performance obtained in fitting the wear volume.

3.5. Model Results

While the figures provide a good overview of the models' behavior, they fail to give an analytical measure of model performance. For this, two metrics were used, both of which gave a numerical measure of the error: The mean square error (MSE) and the mean absolute error (MAE). Since these are metrics for the error of the model, lower values correlate to better performance. The mathematical expressions are shown below:

$$MSE = \frac{1}{n} \sum_{i=\overline{1...n}} (X_i * \theta - Y_i)^2 \quad (3)$$

$$MAE = \frac{1}{n} \sum_{i=\overline{1...n}} |X_i * \theta - Y_i| \quad (4)$$

These two error functions should provide the definitive information on which model best fits the available experimental data. The results are put together into Tables 9 and 10 below, one for each dependent variable.

Table 9. Wear depth fitting metrics.

Model	Binary Coding				Numerical Coding			
	Train Suite		Test Suite		Train Suite		Test Suite	
	MSE	MAE	MSE	MAE	MSE	MAE	MSE	MAE
Linear	0.38528	0.46629	0.64289	0.54639	0.54483	0.53454	0.49382	0.50348
Neural	0.07182	0.18512	0.29991	0.30002	0.52048	0.55583	0.19614	0.33355
Genfis1	26.306	4.92209	27.094	4.93573	29.5935	5.19193	19.64452	4.32411
Genfis2	0.01578	0.07577	0.19928	0.33022	0.02053	0.09137	0.27421	0.41283
Genfis3	0.38405	0.47340	0.56615	0.59409	0.40373	0.47233	0.51991	0.59842
ANFIS1	0.02162	0.10473	2.98506	0.78510				
ANFIS2	0.02265	0.11829	0.30524	0.43454	0.04433	0.15980	0.16904	0.35767
ANFIS3	0.19765	0.34180	0.42079	0.49680	0.11747	0.29467	0.50163	0.62247

Table 10. Wear volume fitting metrics.

Model	Binary Coding				Numerical Coding			
	Train Suite		Test Suite		Train Suite		Test Suite	
	MSE	MAE	MSE	MAE	MSE	MAE	MSE	MAE
Linear	0.06750	0.15916	0.09580	0.17017	0.11225	0.20713	0.08110	0.19386
Neural	0.02062	0.10657	0.01620	0.10758	0.08371	0.17435	0.04756	0.14160
Genfis1	1.09606	0.86541	1.45574	0.93736	1.49774	0.98549	1.37690	0.91272
Genfis2	0.00000	0.00000	1435	10.48023	0.00965	0.03834	1.82954	0.56139
Genfis3	0.06648	0.15842	0.03457	0.11739	0.07857	0.19069	0.04393	0.16326
ANFIS1	0.00000	0.00017	1.63150	0.47196				
ANFIS2	0.00441	0.03373	0.02079	0.11009	0.00965	0.03821	0.00435	0.04772
ANFIS3	0.02919	0.11852	0.05496	0.20210	0.00002	0.00182	0.00043	0.01699

It should first be noted that there is a random element to the above results. No two algorithm runs result in exactly the same models, hence there is some variation in the implementation of the simulation. However, the algorithms have been run multiple times and each set of results supports the general conclusions which are to be drawn from the above tables.

The obtained prediction function provides a good fit of the model, with little error and no over-fitting, as can be concluded from the testing phase of the algorithm. Once trained, the predictive model can be used instead of the actual analytical approach in any application where the dependent variables are needed and the independent variable values are available, within a similar context. This reduces the mathematical complexity of the overall application and could find use in a range of computationally intensive models. It is these functions that will be further improved through the addition of more data as it becomes available, and will be used to predict the future behavior of the experiment.

4. Discussion

The discussion of results revolves around the test scores for the two types of error functions obtained by each model.

In judging the various algorithms, a lower value for MSE and MAE norms are preferable since these represent errors in fitting. For example, in modelling the wear volume with binary coding, the ANFIS2 model is preferable to the linear regression as it achieves lower error grades in both MSE and MSE for the training suite ($0.004 < 0.067$ and $0.033 < 0.159$) as well as the test suite ($0.02 < 0.095$ and $0.11 < 0.17$). Furthermore, it is relatively easy to spot an over-fitting model as it will have good to great performance on the training suite, but poor performance on the test suite. This means it fits the available training data too closely and is ill-equipped to handle new incoming data from the same experiment. As an example, Genfis2 fits the training data for wear volume in binary coding perfectly—the errors are actually zero, but it has very large errors when used on the test data (MSE = 1453 and MAE = 10.48).

As relates to the coding option being used, the variation in results from binary to numerical coding has generally been minimal. While the numbers are obviously different, the same tendencies exist in both representations of the problem.

Linear regression actually performs very well on the test set and in fact overall, suggesting there may be some linear dependencies between the dependent and independent variables after all. It does not provide the level of performance of the neural network or ANFIS models, but as a baseline model, it does very well.

The neural network models have the best overall performance in terms of being the most balanced. While occasionally overtaken by some of the ANFIS or even Genfis algorithms, the neural network models always give very good performance, significantly overtaking the baseline models (i.e., the linear regression models).

The Genfis models achieved mixed results. While sometimes showing superb performance, overtaking even the neural networks or some ANFIS models, they are susceptible to dangerous levels of over-fitting. When that happens, the model is virtually useless as it achieves disproportionate error levels on the test sets—see Genfis2 for wear volume fitting in binary coding. Genfis1 was never meant to be an actual fitting model and of course performs the most poorly, although it does have a few instances when it is workable. However, these should be seen as chance encounters with a low-value dataset, rather than indicative of possible improved performance. As has been mentioned, Genfis1 is only meant to provide a starting point for further optimization. Genfis2 and 3 perform very well, sometimes challenging the performance of NN or ANFIS algorithms, with the aforementioned caveat of possibly disastrous over-fitting. It should also be noted that, in the binary case, Genfis is greatly helped by being able to declare two, membership, function variables, which very nicely model the binary ones.

ANFIS models performed very well throughout all tests and are the main contender of the neural network algorithms, which they oftentimes outperformed. The variation of their results was slightly higher than that of the neural networks, which is their only real disadvantage. However, given the inherent variation of results from different runs of the algorithms, it can be said that their performance is comparable and, in some instances, preferable, to that of the neural networks. Additionally, similar to neural networks and in contrast to some of the Genfis models, they do not seem to be significantly affected by over-fitting, even though they are constructed on rather complex structures.

The elapsed times of the various algorithms are generally comparable, lasting in the range of a few seconds, with the exception of training neural networks under certain configurations. While less training time is always preferable, it is not a major factor for the selection of this model, since the envisioned applications do not require real-time performance. As more experimental data becomes available, the elapsed times will likely increase and future work will deal with improving the speed of all the algorithms discussed in the paper. As a final recommendation and conclusion of the results presented, both neural networks and ANFIS (type 2 and 3) models perform very well on both the training and test sets and could be used for implementation. Any further choice between these two architectures would have more to do with either personal preference or external requirements for an eventual expanded application.

5. Conclusions

The studies on modelling the friction and wear of metallic surfaces for friction couples with linear contact between a thermoplastic material armed with short glass fibers (SGF) and steel are complex and outline the influence of input parameters specific to the tribological system (normal load: contact pressure, the relative sliding speed, the friction type, and the characteristics of the materials in contact) on the output parameters of the tribological system: The wear volume of metallic material and the depth of the wear material.

Neuro-fuzzy modelling of the metallic surface characterization on linear dry contact between plastic material reinforced with SGF and alloyed steel proves that the friction force is not only dependent on the friction coefficient and normal load, as previously thought, but also on the sliding speed and the physical and mechanical properties of the materials, which has significant impact in the characterization of new industrial processes utilizing short glass fiber reinforced polymers.

Using advanced statistical and optimization algorithms on a dataset obtained from the hardware simulation, the results of the research led to modelling a dependency between the various variables of interest involved in the friction process. The subject draws a growing interest from the research community with the advent of highly advanced, intelligent classification, and optimization and regression algorithms.

This research focused on processing an experimental dataset on contact wear, with the aim of obtaining prediction models for the two variables of interest, wear depth, and wear volume. The data was obtained through experiments run on friction couples with linear contact using three different types of polymers on two different types of steel variants. Aside from alternating the materials used, the speed and pressure applied to them were varied under the same operating conditions. The data was pre-processed and coded for use with numerical learning algorithms. Since two viable coding options were found, both resulting numerical problems were treated separately. Various possible data fitting models were then investigated, including a baseline model for reference (linear regression) and different variation of possible considered structures (Genfis and ANFIS models). These models were then judged on their ability to fit the available data, as quantified by two standard error norms, calculated for each model. The end result is a comparison of the performance of the multiple prediction systems and a discussion of the various factors that influence these numerical results.

The chosen models are to be implemented in future work and in any practical implementation in similar conditions for metallic surface characterization within the field of tribology, robotics, and with wide impact metallic surface characterization applications in the wear process, emphasizing the abrasive, adhesive, and corrosive wear. It will be of great interest to investigate the interaction of the current prediction models with real world applications for which there are many options [6,11,12,27].

Another very important direction for further research focuses on gathering more experimental data to allow the development of even more complex enhanced models, either pertaining to new developments in the state of the art, or for retraining the current architectures. Over-fitting, an issue present and discussed throughout the work, should be greatly diminished with training on more available data. This is because most of the models do not over-fit out of design or implementation reasons, but simply because some are too complex (i.e., including too many parameters) for the current dataset.

Author Contributions: Conceptualization, V.V., L.C. and L.V.; Funding acquisition, L.V.; Investigation, V.V., L.C. and L.V.; Methodology, L.V.; Resources, L.C.; Software, V.V. and L.V.; Supervision, L.V.; Validation, L.V. and L.C.; Visualization, L.V., V.V., and L.C.; Writing—original draft, V.V.; Writing—review & editing, L.V. and V.V.

Funding: The paper was funded by the European Commission Marie Skłodowska-Curie SMOOTH project, Smart Robots for Fire-Fighting, H2020-MSCA-RISE-2016-734875, Yanshan University: "Joint Laboratory of Intelligent Rehabilitation Robot" project, KY201501009, Collaborative research agreement between Yanshan University, China and Romanian Academy by IMSAR, RO, and the UEFISCDI Multi-MonD2 Project, Multi-Agent Intelligent Systems Platform for Water Quality Monitoring on the Romanian Danube and Danube Delta, PN-III-P1-1.2-PCCDI2017-0637/33PCCDI/01.03.2018.

Acknowledgments: This work was supported by a grant of the Romanian Ministry of Research and Innovation, CCCDI-UEFISCDI, MultiMonD2 project number PN-III-P1-1.2-PCCDI2017-0637/33PCCDI/01.03.2018, within PNCDI III, and by the European Commission Marie Skłodowska-Curie SMOOTH project, Smart Robots for Fire-Fighting, H2020-MSCA-RISE-2016-734875, Yanshan University: "Joint Laboratory of Intelligent Rehabilitation Robot" project, KY201501009, Collaborative research agreement between Yanshan University, China and Romanian Academy by IMSAR, RO.

Conflicts of Interest: The authors declare no conflict of interest.

References

1. Zhang, Z.; Barkoula, N.M.; Karger-Kocsis, J.; Friedrich, K. Artificial neural network predictions on erosive wear of polymers. *Wear* **2003**, *255*, 708–713. [CrossRef]
2. Panda, S.S.; Singh, A.K.; Chakraborty, D.; Pal, S.K. Drill wear monitoring using back propagation neural network. *J. Mater. Process. Technol.* **2006**, *172*, 283–290. [CrossRef]
3. Flepp, B. Wear Diagnosis of Mechanical Seals with Neural Networks. Ph.D. Thesis, ETH Zurich, Zurich, Switzerland, 1999.
4. Ripa, M.; Frangu, L. A survey of artificial neural networks applications in wear and manufacturing processes. *J. Tribol.* **2004**, *8*, 35–42.
5. Rus, D.; Căpitanu, L.; Badita, L.L. A qualitative correlation between friction coefficient and steel surface wear in linear dry sliding contact to polymers with SGF. *Friction* **2014**, *2*, 47–57. [CrossRef]
6. Căpitanu, L.; Vlădăreanu, L.; Onisoru, J.; Iarovici, A.; Tiganesteanu, C.; Dima, M. Mathematical model and artificial knee joints wear control. *J. Balkan Tribol. Assoc.* **2008**, *14*, 87–101.
7. Willert, H.G.; Semlitsch, M. Reactions of the articular capsule to wear products of artificial joint prostheses. *J. Biomed. Mater. Res.* **1977**, *11*, 157–164. [CrossRef] [PubMed]
8. Hinton, G.E. *Neural Network Architectures for Artificial Intelligence*; American Association for Artificial Intelligence: Menlo Park, CA, USA, 1988.
9. Shukla, R.K. Applications of artificial neural networks in manufacturing: An overview. In Proceedings of the National Conference on Trends and Advances in Mechanical Engineering, Faridabad, Haryana, Indian, 9–10 December 2006.
10. Rao, B.K.N.; Pai, P.S.; Nagabhushana, T.N. Failure diagnosis and prognosis of rolling—Element bearings using artificial neural networks: A critical overview. *J. Phys. Conf. Ser.* **2012**, *364*, 012023. [CrossRef]
11. Chen, J.C.; Savage, M. A Fuzzy-Net-Based Multilevel In-process Surface Roughness Recognition System in Milling Operations. *Int. J. Adv. Manuf. Technol.* **2001**, *17*, 670–676. [CrossRef]
12. Căpitanu, L.; Vlădăreanu, L.; Florescu, V. The Knee Wear Prediction of UHMWPE Tibial Insert Using VIPRO Platform. *J. Mech. Eng. Autom.* **2015**, *5*, 591–600.
13. Vlădăreanu, V.; Dumitrache, I.; Vlădăreanu, L.; Sacală, I.S.; Tonț, G.; Moisescu, M.A. Versatile intelligent portable robot control platform based on cyber physical systems principles. *Stud. Inform. Control* **2015**, *24*, 409–418. [CrossRef]
14. James, G.; Witten, D.; Hastie, T.; Tibshirani, R. *An Introduction to Statistical Learning*; Springer: New York, NY, USA, 2013; Volume 112.

15. Vlădăreanu, V.; Şchiopu, P.; Deng, M.C. Robots Extension Control Using Fuzzy Smoothing. In Proceedings of the International Conference on Advanced Mechatronic Systems (ICAMECHS), Luoyang, China, 25–27 September 2013; pp. 511–516.
16. Vlădăreanu, V.; Şchiopu, P.; Deng, M.C.; Yu, H. Intelligent Extended Control of the Walking Robot Motion. In Proceedings of the International Conference on Advanced Mechatronic Systems (ICAMECHS), Kumamoto, Japan, 10–12 August 2014; pp. 489–495.
17. Župerl, U.; Čuš, F. Merged Neural Decision System and ANFIS Wear Predictor for Supporting Tool Condition Monitoring. *Trans. FAMENA* **2011**, *35*, 13–26.
18. Lo, S.P. The application of an ANFIS and grey system method in turning tool-failure detection. *Int. J. Adv. Manuf. Technol.* **2002**, *19*, 564–572. [CrossRef]
19. Alimam, H.; Hinnawi, M.; Pradhan, P.; Alkassar, Y. ANN & ANFIS Models for Prediction of Abrasive Wear of 3105 Aluminium Alloy with Polyurethane Coating. *Tribol. Ind.* **2016**, *38*, 221–228.
20. Shabani, M.O.; Shamsipour, M.; Mazahery, A.; Pahlevani, Z. Performance of ANFIS Coupled with PSO in Manufacturing Superior Wear Resistant Aluminum Matrix Nano Composites. *Trans. Indian Inst. Met.* **2017**, 1–9. [CrossRef]
21. Types of Statistical Data: Numerical, Categorical, and Ordinal. Available online: http://www.dummies.com/education/math/statistics/types-of-statistical-data-numerical-categorical-and-ordinal/ (accessed on 9 July 2018).
22. Templeton, G. Artificial Neural Networks are Changing the World. What Are They? Available online: https://www.extremetech.com/extreme/215170-artificial-neural-networks-are-changing-the-world-what-are-they (accessed on 22 June 2018).
23. ANFIS in the Fuzzy Logic Toolbox. Available online: http://www.cs.nthu.edu.tw/~jang/anfisfaq.htm (accessed on 5 December 2017).
24. Jang, J.-S. ANFIS: Adaptive-Network-Based Fuzzy Inference Systems. *IEEE Trans. Syst. Man Cybern.* **1993**, *23*, 665–685. [CrossRef]
25. Jang, J.-S.; Sun, C.-T. Neuro-Fuzzy Modeling and Control. *Proc. IEEE* **1995**, *83*, 378–406. [CrossRef]
26. Denaï, M.A.; Palis, F.; Zeghbib, A. Modeling and control of non-linear systems using soft computing techniques. *Appl. Soft Comput.* **2007**, *7*, 728–738.
27. Moisescu, M.A.; Sacală, I.S.; Dumitrache, I.; Caramihai, S. Predictive Maintenance and Robotic System Design. *J. Fundam. Appl. Sci.* **2018**, *10*, 234–239.

materials

MDPI

Article

The Computation of Complex Dispersion and Properties of Evanescent Lamb Wave in Functionally Graded Piezoelectric-Piezomagnetic Plates

Xiaoming Zhang, Zhi Li and Jiangong Yu *

School of Mechanical and Power Engineering, Henan Polytechnic University, Jiaozuo 454000, China; zxmworld11@hpu.edu.cn (X.Z.); jixielizhi@126.com (Z.L.)

* Correspondence: yujiangong75@tom.com; Tel.: +86-13693919651

Received: 24 May 2018; Accepted: 7 July 2018; Published: 10 July 2018

Abstract: Functionally graded piezoelectric-piezomagnetic material (FGPPM), with a gradual variation of the material properties in the desired direction(s), can improve the conversion of energy among mechanical, electric, and magnetic fields. Full dispersion relations and wave mode shapes are vital to understanding dynamic behaviors of structures made of FGPPM. In this paper, an analytic method based on polynomial expansions is proposed to investigate the complex-valued dispersion and the evanescent Lamb wave in FGPPM plates. Comparisons with other related studies are conducted to validate the correctness of the presented method. Characteristics of the guided wave, including propagating modes and evanescent modes, in various FGPPM plates are studied, and three-dimensional full dispersion and attenuation curves are plotted to gain a deeper insight into the nature of the evanescent wave. The influences of the gradient variation on the dispersion and the magneto-electromechanical coupling factor are illustrated. The displacement amplitude and electric potential and magnetic potential distributions are also discussed in detail. The obtained numerical results could be useful to design and optimize different sensors and transducers made of smart piezoelectric and piezomagnetic materials with high performance by adjusting the gradient property.

Keywords: evanescent wave; polynomial approach; functionally graded piezoelectric-piezomagnetic material; dispersion; attenuation

1. Introduction

Due to the excellent coupling behavior among mechanical, electric, and magnetic fields, piezoelectric-piezomagnetic composites (or magneto-electro-elastic material) composed of piezoelectric and piezomagnetic phases have been increasingly applied to different engineering structures, especially to the smart or intelligent systems as intelligent sensors, damage detectors, etc. [1]. It is found that the smart structures made of functionally graded materials (FGM) possess a better structural performance than traditional composite materials. The concept of FGM has been extended to the development of new piezoelectric-piezomagnetic materials appointed functionally graded piezoelectric-piezomagnetic materials, which can realize the smooth transition of the physical constitutive parameters of the piezoelectric and piezomagnetic materials. FGPPMs have been used in some devices to improve their efficiency and other features. Many applications are closely connected with the vibration and wave propagation of FGM and FGPPM [2–5]. Dispersion relations and wave mode shapes are very important for understanding dynamic behaviors of structures. Wave propagation features in FGPPM plates could be also useful in designing and optimizing the high-accuracy sensors and transducers [6,7].

With the remarkable achievements in fabrication of FGPPM during the decades, many investigators have turned attention to the study of wave propagation in such materials. Wang and Rokhlin [8]

presented the differential equations governing the transfer and stiffness matrices for a functionally graded generally anisotropic magneto-electro-elastic medium and calculated the surface wave velocity dispersion. Pan et al. [9] derived an exact solution for the multilayered plate made of functionally graded, anisotropic and linear magneto-electro-elastic materials based on Pseudo-Stroh formalism. Bhangale et al. [10] carried out the free vibration studies on the simply supported functionally graded magneto-electro-elastic plate by semi-analytical finite element method. Wu et al. [11] investigated the wave propagating characteristics in the non-homogeneous magneto-electro-elastic plates by using orthogonal polynomial approach. By employing the power series technique, Cao [12] investigated the Lamb wave propagation in FGPPM plates. Singh and Rokne [13] investigated the SH wave propagating in FGPPM structures. Xiao et al. [14] investigated the dispersion properties of wave propagation in the functionally graded magneto-electro-elastic plate by the Chebyshev spectral element method.

As is reviewed above, so far, studies on the guided wave in FGPPM structures are limited to the propagating waves, but the evanescent waves have not been investigated. Recently, some studies on pseudo surface acoustic waves (PSAW) in piezoelectric half-spaces find that the PSAW modes have higher velocities and lower attenuations, compared to the classical surface acoustic waves [15,16]. Such modes make the piezoelectric device possess higher resolution. Evanescent wave modes also have the similar features. According to the classification of Auld [17], the complete wave modes consist of propagating modes with real wave number and evanescent modes with complex or purely imaginary wave number. Note that evanescent modes represent local modes that would exist at discontinuities and decay with propagating distance (so referred to as evanescent or non-propagating wave). As early as 1955, Lyon [18] obtained the purely imaginary roots of the dispersion equation for an elastic plate. Remarkable is the work done by Mindlin who demonstrated the presence of complex roots of the Rayleigh-Lamb equation [19]. Freedman [20] studied the imaginary valued Lamb mode spectra covering virtually the full range of the Poisson ratio. Quintanilla et al. [21] calculated the full spectrum for guided wave problems in plates and layered cylinders using a spectral collocation method. More recently, Yan and Yuan [22,23] discussed the potential application of evanescent waves in structural health monitoring and investigated the conversion of evanescent SH and Lamb waves into propagating waves using a semi-analytical approach. Chen et al. [24] studied theoretically the real-valued and imaginary-valued SH waves in a piezoelectric plate of cubic crystals. These researches focused on the simple material and purely imaginary modes. In fact, the search of complex roots corresponding to evanescent waves is a difficult task for FGM with material properties of variable coefficients. To the best of the authors' knowledge, the evanescent waves in FGM or piezoelectric-piezomagnetic composite have not been studied before, which is the motivation of this study.

In this paper, an analytic method based on polynomial expansions is proposed to calculate guided waves in FGPPM plates. The presented method can replace the problem of computing a transcendental dispersion equation by a general eigenvalue problem in wave number. The complete solutions of the dispersion equation, including the purely real, purely imaginary and complex solutions, can be obtained. We plot the full dispersion curves in three dimensional (3D) frequency-complex wave number space to gain a better and deeper insight into the characteristics of evanescent waves. Two known cases are given to validate this approach. The characteristics of evanescent guided waves in various FGPPM plates are illustrated. The effects of different graded fields on the dispersion curves and the coupled electromechanical factor are investigated. The displacement amplitude and electric potential and magnetic potential distributions are also discussed in detail.

2. Mathematics and Formulation of the Problem

Consider a FGPPM plate with varying material properties with regard to thickness (the z-axis). The plate described in Cartesian coordinate system (x, y, z), is infinite horizontally but finite in the z direction with a thickness h, occupies the region $0 \leq z \leq h$, as shown in Figure 1. The wave propagates along the x direction, and the upper and bottom surface of the plate are traction free.

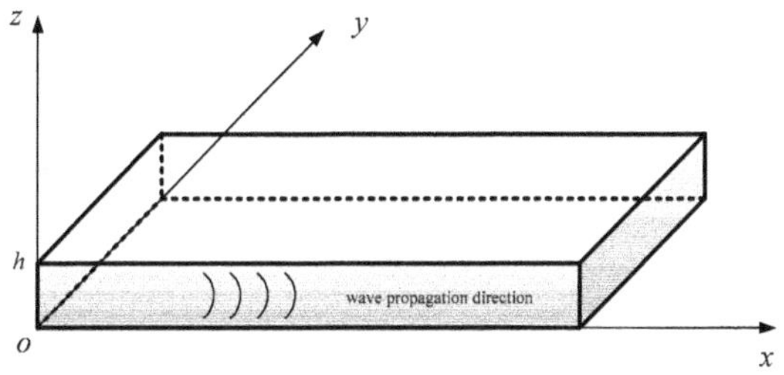

Figure 1. Geometry of the problem.

For the piezoelectric-piezomagnetic medium, the governing field equations (Equation (1)) and the generalized constitutive relations (Equation (2)) can be expressed as [12]:

$$\sigma_{ij,j} = \rho\ddot{u}_i,\ D_{i,i} = 0,\ B_{i,i} = 0 \tag{1}$$

$$\sigma_{ij} = C_{ijkl}\varepsilon_{kl} - e_{kij}E_k - q_{ijk}H_k,\ D_i = e_{ikl}\varepsilon_{kl} + \in_{ik} E_k + g_{ik}H_k,\ B_i = q_{ijk}\varepsilon_{jk} + g_{ik}E_k + \mu_{ik}H_k \tag{2}$$

$$\varepsilon_{ij} = \frac{1}{2}(u_{i,j} + u_{j,i}),\ E_i = -\varphi_{,i},\ H_i = -\psi_{,i} \tag{3}$$

Generalized geometric equations under a rectangular coordinates system are in the above Equations (1)–(3), σ_{ij} is the stress tensor, D_i is the electric displacement and B_i is the magnetic induction C_{ijkl}, e_{kij}, q_{ijk}, $\in_{ik}$, g_{ik} and μ_{ik} are the elastic, piezoelectric, piezomagnetic, dielectric, magnetic, and magnetoelectric parameters of the FGPPM, respectively, while all of them, including the mass density ρ, are functions of z. According to Einstein summation convention, where i, j, k and l = 1, 2, 3 corresponding to x, y, z directions, respectively. ε_{kl}, E_k and H_k are the strain tensor, the electric field, and magnetic field, respectively. u_i ($i = x, y, z$) denotes the mechanical displacement component in the ith direction. φ and ψ are the electric potential and magnetic potential. Comma in subscripts and superposed dot denote spatial and time derivatives, respectively.

For Lamb waves propagating along the x direction, the displacement components, electric potential and magnetic potential can be expressed as

$$u_i(x,z,t) = \exp(\mathrm{i}kx - \mathrm{i}\omega t)U_i(z),\ \varphi(x,z,t) = \exp(\mathrm{i}kx - \mathrm{i}\omega t)X(z), \psi(x,z,t) = \exp(\mathrm{i}kx - \mathrm{i}\omega t)Y(z) \tag{4}$$

where U_i ($i = x, z$) represents the amplitude of the displacements in the ith directions, X and Y represent the amplitude of electric potential and magnetic potential, respectively. k is the wave number, ω is the angular frequency, and i is the imaginary number.

Since the material properties change gradually with thickness and are the functions of z, they can be fitted into the following form:

$$f(z) = f^{(l)}(z/h)^l, l = 0,1,2\ldots,L \tag{5}$$

where f ($f = \rho, C, e, \in, q, g$ and μ) denotes material parameters, l is the order number, $f^{(l)}$ is the coefficient. For homogeneous material, $f(z) = C^{(0)}$, and when $l > 0, f(l)$ is zero.

The following boundary and continuous conditions should be satisfied as follows. For the traction-free boundary condition, it requires that $\sigma_{zz}\big|_{z=0,h} = 0,\ \sigma_{xz}\big|_{z=0,h} = 0,\ \sigma_{yz}\big|_{z=0,h} = 0$. For electric and magnetic open circuit, $D_z\big|_{z=0,h} = 0,\ B_z\big|_{z=0,h} = 0$, and for electric and magnetic shorted circuit, $\varphi\big|_{z=0,h} = 0,\ \psi\big|_{z=0,h} = 0$.

Then take the traction-free and electrical and magnetic open-circuit boundary conditions as an example. Considering the boundary of material, the position-dependent material parameters are given by:

$$f(z) = f(z)\pi(z) \tag{6}$$

where $\pi(z)$ is a rectangular window function defined by $\pi(z) = \begin{cases} 1, & 0 \le z \le h \\ 0, & elsewhere \end{cases}$, whose derivative is a Dirac's delta function, $\delta(z-h) - \delta(z)$. Then the boundary conditions can be automatically incorporated in the constitutive relations [25].

To reduce the number of resolving equations, we substitute Equations (3)–(6) into Equation (2) with following substitution into Equation (1). Consequently, the governing differential equations in terms of the displacement, electric potential and magnetic potential components can be obtained. Here, the case of an orthotropic FGPPM plate with the z direction polarization is given:

$$\begin{aligned} &\left(\tfrac{z}{h}\right)^l \Big[C_{55}^{(l)}U'' + lz^{-1}C_{55}^{(l)}U' + ik\left(C_{13}^{(l)} + C_{55}^{(l)}\right)W' + ik\left(e_{15}^{(l)} + e_{31}^{(l)}\right)X' + ik\left(q_{15}^{(l)} + q_{31}^{(l)}\right)Y' \\ &-k^2C_{11}^{(l)}U + likz^{-1}\left(C_{55}^{(l)}W + e_{15}^{(l)}X + q_{15}^{(l)}Y\right)\Big]\pi(z) + (\delta(z-0) - \delta(z-h))\left(\tfrac{z}{h}\right)^l\Big(C_{55}^{(l)}U' \\ &+ikC_{55}^{(l)}W + ike_{15}^{(l)}X + ikq_{15}^{(l)}Y\Big) = -\tfrac{\rho^{(l)}z^l\omega^2}{h^l}U\pi(z) \end{aligned} \tag{7a}$$

$$\begin{aligned} &\left(\tfrac{z}{h}\right)^l \Big[C_{33}^{(l)}W'' + e_{33}^{(l)}X'' + q_{33}^{(l)}Y'' + ik\left(C_{13}^{(l)} + C_{55}^{(l)}\right)U' + lz^{-1}\left(C_{33}^{(l)}W' + e_{33}^{(l)}X' + q_{33}^{(l)}Y'\right) \\ &+likz^{-1}C_{13}^{(l)}U - k^2\left(C_{55}^{(l)}W + e_{15}^{(l)}X + q_{15}^{(l)}Y\right)\Big]\pi(z) + (\delta(z-0) - \delta(z-h))\left(\tfrac{z}{h}\right)^l\Big(C_{33}^{(l)}W' \\ &+e_{33}^{(l)}X' + q_{33}^{(l)}Y' + ikC_{13}^{(l)}U\Big) = -\tfrac{\rho^{(l)}z^l\omega^2}{h^l}W\pi(z) \end{aligned} \tag{7b}$$

$$\begin{aligned} &\left(\tfrac{z}{h}\right)^l \Big[e_{33}^{(l)}W'' - \epsilon_{33}^{(l)}X'' - g_{33}^{(l)}Y'' + ik\left(e_{15}^{(l)} + e_{31}^{(l)}\right)U' + lz^{-1}\left(e_{33}^{(l)}W' - \epsilon_{33}^{(l)}X' - g_{33}^{(l)}Y'\right) \\ &+likz^{-1}e_{31}^{(l)}U - k^2e_{15}^{(l)}W + k^2\epsilon_{11}^{(l)}X + k^2g_{11}^{(l)}Y\Big]\pi(z) \\ &+(\delta(z-0) - \delta(z-h))\left(\tfrac{z}{h}\right)^l\left(e_{33}^{(l)}W' - \epsilon_{33}^{(l)}X' - g_{33}^{(l)}Y' + ike_{31}^{(l)}U\right) = 0 \end{aligned} \tag{7c}$$

$$\begin{aligned} &\left(\tfrac{z}{h}\right)^l \Big[q_{33}^{(l)}W'' - g_{33}^{(l)}X'' - \mu_{33}^{(l)}Y'' + ik\left(q_{15}^{(l)} + q_{31}^{(l)}\right)U' + likz^{-1}q_{31}^{(l)}U \\ &+lz^{-1}\left(q_{33}^{(l)}W' - g_{33}^{(l)}X' - \mu_{33}^{(l)}Y'\right) - k^2q_{15}^{(l)}W + k^2g_{11}^{(l)}X + k^2\mu_{11}^{(l)}Y\Big]\pi(z) \\ &+(\delta(z-0) - \delta(z-h))\left(\tfrac{z}{h}\right)^l\left(q_{33}^{(l)}W' - g_{33}^{(l)}X' - \mu_{33}^{(l)}Y' + ikq_{31}^{(l)}U\right) = 0 \end{aligned} \tag{7d}$$

where U and W respectively represent the amplitude of vibration in the x and z directions. The superscript (′) is the derivative with respect to z.

The four amplitudes can be expanded into Legendre orthogonal polynomial series as:

$$U(z) = \sum_{m=0}^{\infty} p_m^1 Q_m(z),\ W(z) = \sum_{m=0}^{\infty} p_m^2 Q_m(z),\ X(z) = \sum_{m=0}^{\infty} p_m^3 Q_m(z),\ Y(z) = \sum_{m=0}^{\infty} p_m^4 Q_m(z) \tag{8}$$

where $p_m^\alpha (\alpha = 1,2,3,4)$ are the expansion coefficients, $Q_m(r)$ are an orthonormal set of polynomials in the interval [0,h].

$$Q_m(z) = \sqrt{\frac{2m+1}{h}} P_m\left(\frac{2z-h}{h}\right) \tag{9}$$

where P_m is the Legendre polynomial of order m.

Substituting Equations (8) and (9) into Equation (7), then multiplying both sides of the modified Equation (7) by the complex conjugate $Q_j^*(z)$ with j running from 0 to M, integrating over z from 0 to h, taking advantage of the orthonormality of the polynomial, yields:

$$
\begin{gathered}
k^2\begin{bmatrix} {}^lA_{11}^{j,m} & {}^lA_{12}^{j,m} & {}^lA_{13}^{j,m} & {}^lA_{14}^{j,m} \\ {}^lA_{21}^{j,m} & {}^lA_{22}^{j,m} & {}^lA_{23}^{j,m} & {}^lA_{24}^{j,m} \\ {}^lA_{31}^{j,m} & {}^lA_{32}^{j,m} & {}^lA_{33}^{j,m} & {}^lA_{34}^{j,m} \\ {}^lA_{41}^{j,m} & {}^lA_{42}^{j,m} & {}^lA_{43}^{j,m} & {}^lA_{44}^{j,m} \end{bmatrix}\begin{Bmatrix} p_m^1 \\ p_m^2 \\ p_m^3 \\ p_m^4 \end{Bmatrix} + k\begin{bmatrix} {}^lB_{11}^{j,m} & {}^lB_{12}^{j,m} & {}^lB_{13}^{j,m} & {}^lB_{14}^{j,m} \\ {}^lB_{21}^{j,m} & {}^lB_{22}^{j,m} & {}^lB_{23}^{j,m} & {}^lB_{24}^{j,m} \\ {}^lB_{31}^{j,m} & {}^lB_{32}^{j,m} & {}^lB_{33}^{j,m} & {}^lB_{34}^{j,m} \\ {}^lB_{41}^{j,m} & {}^lB_{42}^{j,m} & {}^lB_{43}^{j,m} & {}^lB_{44}^{j,m} \end{bmatrix}\begin{Bmatrix} p_m^1 \\ p_m^2 \\ p_m^3 \\ p_m^4 \end{Bmatrix} \\
+\begin{bmatrix} {}^lC_{11}^{j,m} & {}^lC_{12}^{j,m} & {}^lC_{13}^{j,m} & {}^lC_{14}^{j,m} \\ {}^lC_{21}^{j,m} & {}^lC_{22}^{j,m} & {}^lC_{23}^{j,m} & {}^lC_{24}^{j,m} \\ {}^lC_{31}^{j,m} & {}^lC_{32}^{j,m} & {}^lC_{33}^{j,m} & {}^lC_{34}^{j,m} \\ {}^lC_{41}^{j,m} & {}^lC_{42}^{j,m} & {}^lC_{43}^{j,m} & {}^lC_{44}^{j,m} \end{bmatrix}\begin{Bmatrix} p_m^1 \\ p_m^2 \\ p_m^3 \\ p_m^4 \end{Bmatrix} = -\omega^2\begin{bmatrix} {}^lM_m^j & 0 & 0 & 0 \\ 0 & {}^lM_m^j & 0 & 0 \\ 0 & 0 & 0 & 0 \\ 0 & 0 & 0 & 0 \end{bmatrix}\begin{Bmatrix} p_m^1 \\ p_m^2 \\ p_m^3 \\ p_m^4 \end{Bmatrix}
\end{gathered} \tag{10}
$$

or is abbreviated as

$$k^2\mathbf{A}\cdot\mathbf{p} + k^1\mathbf{B}\cdot\mathbf{p} + \mathbf{C}\cdot\mathbf{p} = -\omega^2\mathbf{M}\cdot\mathbf{p} \tag{11}$$

where **A**, **B**, **C** and **M** are matrices of order $4(M+1)\cdot(M+1)$, $\mathbf{p} = \left[p_m^1\, p_m^2\, p_m^3\, p_m^4\right]^T$, the elements of the matrices are as following,

$$
\begin{aligned}
{}^lA_{11}^{j,m} &= -\tfrac{1}{h^l}C_{11}^{(l)}\beta(m,l,0,j) {}^lA_{22}^{j,m} = -\tfrac{1}{h^l}C_{55}^{(l)}\beta(m,l,0,j) {}^lA_{23}^{j,m} = -\tfrac{1}{h^l}e_{15}^{(l)}\beta(m,l,0,j) \\
{}^lA_{24}^{j,m} &= -\tfrac{1}{h^l}q_{15}^{(l)}\beta(m,l,0,j) {}^lA_{32}^{j,m} = -\tfrac{1}{h^l}e_{15}^{(l)}\beta(m,l,0,j) {}^lA_{33}^{j,m} = \tfrac{1}{h^l}\in_{11}^{(l)}\beta(m,l,0,j) \\
{}^lA_{34}^{j,m} &= \tfrac{1}{h^l}g_{11}^{(l)}\beta(m,l,0,j) {}^lA_{42}^{j,m} = -\tfrac{1}{h^l}q_{15}^{(l)}\beta(m,l,0,j) {}^lA_{43}^{j,m} = \tfrac{1}{h^l}g_{11}^{(l)}\beta(m,l,0,j) \\
{}^lA_{44}^{j,m} &= \tfrac{1}{h^l}\mu_{11}^{(l)}\beta(m,l,0,j) {}^lA_{12}^{j,m} = {}^lA_{21}^{j,m} = 0 {}^lA_{13}^{j,m} = {}^lA_{31}^{j,m} = 0 {}^lA_{14}^{j,m} = {}^lA_{41}^{j,m} = 0; \\
{}^lB_{12}^{j,m} &= \tfrac{1}{h^l}\{\mathbf{i}\left(C_{13}^{(l)} + C_{55}^{(l)}\right)\beta(m,l,1,j) + liC_{55}^{(l)}\beta(m,l-1,0,j) + iC_{55}^{(l)}\gamma(m,l,0,j)\}, \\
{}^lB_{13}^{j,m} &= \tfrac{1}{h^l}\{\mathbf{i}\left(e_{15}^{(l)} + e_{31}^{(l)}\right)\beta(m,l,1,j) + lie_{15}^{(l)}\beta(m,l-1,0,j) + ie_{15}^{(l)}\gamma(m,l,0,j)\} \\
{}^lB_{14}^{j,m} &= \tfrac{1}{h^l}\{\mathbf{i}\left(q_{15}^{(l)} + q_{31}^{(l)}\right)\beta(m,l,1,j) + liq_{15}^{(l)}\beta(m,l-1,0,j) + iq_{15}^{(l)}\gamma(m,l,0,j)\} \\
{}^lB_{21}^{j,m} &= \tfrac{1}{h^l}\{i\left(C_{13}^{(l)} + C_{55}^{(l)}\right)\beta(m,l,1,j) + liC_{13}^{(l)}\beta(m,l-1,0,j) + iC_{13}^{(l)}\gamma(m,l,0,j)\} \\
{}^lB_{31}^{j,m} &= \tfrac{1}{h^l}\{i\left(e_{15}^{(l)} + e_{31}^{(l)}\right)\beta(m,l,1,j) + lie_{31}^{(l)}\beta(m,l-1,0,j) + ie_{31}^{(l)}\gamma(m,l,0,j)\} \\
{}^lB_{41}^{j,m} &= \tfrac{1}{h^l}\{i\left(q_{15}^{(l)} + q_{31}^{(l)}\right)\beta(m,l,1,j) + liq_{31}^{(l)}\beta(m,l-1,0,j) + iq_{31}^{(l)}\gamma(m,l,0,j)\}, \\
{}^lB_{11}^{j,m} &= {}^lB_{22}^{j,m} = {}^lB_{33}^{j,m} = {}^lB_{44}^{j,m} = 0 {}^lB_{23}^{j,m} = {}^lB_{32}^{j,m} = 0 {}^lB_{24}^{j,m} = {}^lB_{42}^{j,m} = 0 {}^lB_{34}^{j,m} = {}^lB_{43}^{j,m} = 0; \\
{}^lC_{11}^{j,m} &= \tfrac{1}{h^l}\left\{C_{55}^{(l)}\beta(m,l,2,j) + lC_{55}^{(l)}\beta(m,l-1,1,j) + C_{55}^{(l)}\gamma(m,l,1,j)\right\} \\
{}^lC_{22}^{j,m} &= \tfrac{1}{h^l}\left\{C_{33}^{(l)}\beta(m,l,2,j) + lC_{33}^{(l)}\beta(m,l-1,1,j) + C_{33}^{(l)}\gamma(m,l,1,j)\right\} \\
{}^lC_{23}^{j,m} &= \tfrac{1}{h^l}\left\{e_{33}^{(l)}\beta(m,l,2,j) + le_{33}^{(l)}\beta(m,l-1,1,j) + e_{33}^{(l)}\gamma(m,l,1,j)\right\} \\
{}^lC_{24}^{j,m} &= \tfrac{1}{h^l}\left\{q_{33}^{(l)}\beta(m,l,2,j) + lq_{33}^{(l)}\beta(m,l-1,1,j) + q_{33}^{(l)}\gamma(m,l,1,j)\right\}, \\
{}^lC_{32}^{j,m} &= \tfrac{1}{h^l}\left\{e_{33}^{(l)}\beta(m,l,2,j) + le_{33}^{(l)}\beta(m,l-1,1,j) + e_{33}^{(l)}\gamma(m,l,1,j)\right\} \\
{}^lC_{33}^{j,m} &= \tfrac{1}{h^l}\{-\in_{33}^{(l)}\beta(m,l,2,j) - l\in_{33}^{(l)}\beta(m,l-1,1,j) - \in_{33}^{(l)}\gamma(m,l,1,j)\} \\
{}^lC_{34}^{j,m} &= \tfrac{1}{h^l}\{-g_{33}^{(l)}\beta(m,l,2,j) - lg_{33}^{(l)}\beta(m,l-1,1,j) - g_{33}^{(l)}\gamma(m,l,1,j)\} \\
{}^lC_{42}^{j,m} &= \tfrac{1}{h^l}\left\{q_{33}^{(l)}\beta(m,l,2,j) + lq_{33}^{(l)}\beta(m,l-1,1,j) + q_{33}^{(l)}\gamma(m,l,1,j)\right\} \\
{}^lC_{43}^{j,m} &= \tfrac{1}{h^l}\{-\mathbf{g}_{33}^{(l)}\beta(m,l,2,j) - lg_{33}^{(l)}\beta(m,l-1,1,j) - g_{33}^{(l)}\gamma(m,l,1,j)\} \\
{}^lC_{44}^{j,m} &= \tfrac{1}{h^l}\{-\mu_{33}^{(l)}\beta(m,l,2,j) - l\mu_{33}^{(l)}\beta(m,l-1,1,j) - \mu_{33}^{(l)}\gamma(m,l,1,j)\}, \\
{}^lC_{12}^{j,m} &= {}^lC_{21}^{j,m} = 0 {}^lC_{13}^{j,m} = {}^lC_{31}^{j,m} = 0 {}^lC_{14}^{j,m} = {}^lC_{41}^{j,m} = 0 {}^lM_m^j = \tfrac{1}{h^l}\rho^{(l)}\beta(m,l,0,j);
\end{aligned}
$$

with $\beta(m,l,n,j) = \int_0^h Q_j^*(z)z^l\frac{\partial^n Q_m(z)}{\partial z^n}dz, \gamma(m,l,n,j) = \int_0^h Q_j^*(z)z^l\frac{\partial \pi(z)}{\partial z}\frac{\partial^n Q_m(z)}{\partial z^n}dz.$

The objective is to find wave numbers k that satisfy the Equation (11). It is simple and useful for propagating wave, by specifying real k and then solving for ω. But if interest is the evanescent wave, the approach is useless because k is complex and the solving of Equation (11) involves a multivariable search. In order to overcome this difficulty, we develop a new solution procedure as shown below.

We introduce two new vectors:

$$\mathbf{q} = k \cdot \mathbf{p},\ \mathbf{N} = -\omega^2\mathbf{M}. \tag{12}$$

Substitution Equation (12) into Equation (11), and then multiplying both sides of the modified Equation (11) by inverse matrix $\mathbf{A}^{-1}$, yields

$$\mathbf{A}^{-1}(\mathbf{N}-\mathbf{C})\mathbf{p} - (\mathbf{A}^{-1}\mathbf{B})\mathbf{q} = k \cdot \mathbf{q}. \tag{13}$$

Combining Equation (13) and the above vector $\mathbf{q} = k \cdot \mathbf{p}$, we obtain

$$\begin{bmatrix} \mathbf{Z} & \mathbf{I}_{4(\mathbf{M}+1)} \\ \mathbf{A}^{-1}(\mathbf{N}-\mathbf{C}) & -\mathbf{A}^{-1}\mathbf{B} \end{bmatrix} \begin{bmatrix} \mathbf{p} \\ \mathbf{q} \end{bmatrix} = k \begin{bmatrix} \mathbf{p} \\ \mathbf{q} \end{bmatrix}. \tag{14}$$

where $\mathbf{I}$ is the identity matrix and $\mathbf{Z}$ is a zero matrix.

If we define $\mathbf{R} = [\mathbf{pq}]^T$, then Equation (14) can be written as

$$\begin{bmatrix} \mathbf{Z} & \mathbf{I}_{4(M+1)} \\ \mathbf{A}^{-1}(\mathbf{N}-\mathbf{C}) & -\mathbf{A}^{-1}\mathbf{B} \end{bmatrix} \mathbf{R} = k\mathbf{R}. \tag{15}$$

Up to this stage, the problem is reduced to a typical eigenvalue problem, which can be easily solved using an eigensolver routine that yields the complex eigenvalues k. All the developments performed in this paper were implemented in Mathematica software (version 8.0, Wolfram company, Champaign, IL, USA). The calculation technique in the short-circuit case is similar to that which is used in the open-circuit case. The deduction process is not shown to save space.

3. Numerical Results and Discussion

Based on the previous formulations, the computer program in terms of the presented method has been written using Mathematica software to calculate the dispersion and phase velocity curves for the FGPPM plate composed of $CoFe_2O_4$ (top) and $BaTiO_3$ (bottom), h = 1 mm. The material parameters are from literature [26] and are listed in Table 1.

Table 1. Material parameters of two piezoelectric-piezomagnetic materials.

Materials	Property								
	C_{11}	C_{12}	C_{13}	C_{22}	C_{23}	C_{33}	C_{44}	C_{55}	C_{66}
Ba_2TiO_3	166	77	78	166	78	162	43	43	44.6
$CoFe_2O_4$	286	173	170	286	170	269	45.3	45.3	46.5
	Property								
	e_{15}	e_{24}	e_{31}	e_{32}	e_{33}	$\in_{11}$	$\in_{22}$	$\in_{33}$	ρ
Ba_2TiO_3	11.6	11.6	−4.4	−4.3	18.6	196	201	28	5.8
$CoFe_2O_4$	0	0	0	0	0	0.8	0.8	0.93	5.3
	Property								
	q_{15}	q_{24}	q_{31}	q_{32}	q_{33}	μ_{11}	μ_{22}	μ_{33}	
Ba_2TiO_3	0	0	0	0	0	5	5	10	
$CoFe_2O_4$	550	550	580.3	580.3	699.7	−590	−590	157	

Units: C_{ij} (10^9 N/m^2), $\in_{ij}$ (10^{-10} F/m), e_{ij} (C/m^2), q_{ij} (N/Am), μ_{ij} (10^{-6} Ns^2/C^2), ρ (10^3 kg/m^3).

We use the Voigt-type model, as described in the literature [27], to calculate the effective material property of the FGPPM plate:

$$F(z) = F_B V_B(z) + F_C V_C(z),\ V_B(z) + V_C(z) = 1 \tag{16}$$

where F_B and F_C respectively represent the material property of the Ba_2TiO_3 and $CoFe_2O_4$ materials, and V_B and V_C are volume fraction.

Equation (16) can be rewritten as

$$F(z) = F_B + (F_C - F_B)V_C(z) \tag{17}$$

Similar to Equation (5), $V_C(z)$ can be expressed as a power expansion, Here we consider four different gradient fields, $V_C(z) = (z/h)^n$, n = 1, 2 and 3, namely linear, quadratic and cubic graded fields, and sinusoidal graded field $V_C(z) = \sin(0.5\pi z/h)$.

3.1. Approach Validation and Convergence of the Problem

To check the validity and the efficiency of our approach, we make a comparison between our results and the literature results. Because there is no investigation on the evanescent waves in FGPPM so far, we compute the full spectrum of Lamb wave in a steel plate and make a comparison with the available results in literature [17] from a spectral collocation method. The calculating parameters are ρ = 7932 kg/m^3, C_{11} = 281.757 GPa, C_{12} = 113.161 GPa, C_{44} = 84.298 GPa, and h = 10 mm. The non-dimensional frequency and wave number are defined as $\Omega = (\omega h\sqrt{\rho/C44})/\pi, \Psi = kh/\pi$, respectively. The resulting dispersion curves are given in Figure 2. It clearly shows that the numerical results obtained by the present polynomial approach agree well with those obtained by the spectral collocation method, which validates our approach and program.

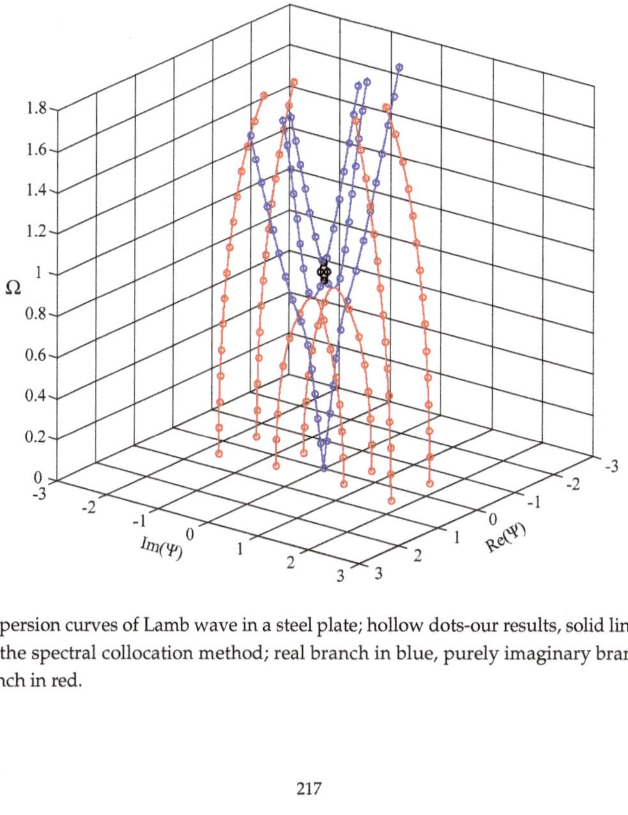

Figure 2. Dispersion curves of Lamb wave in a steel plate; hollow dots-our results, solid lines-literature results from the spectral collocation method; real branch in blue, purely imaginary branch in black, complex branch in red.

The material in the above verification example is isotropic. We calculate the dispersion curves of Lamb wave in an orthotropic plate and make a comparison with the available results in literature [28] from the reverberation-ray matrix method, which serves as a further validation of our approach. The material is PZT-4, and the material parameters are listed in Table 2. Figure 3 shows the obtained frequency spectra. Here again, the agreement is quite good between our results and those from the reverberation ray matrix method.

Table 2. Material parameters.

Material	Property								
	C_{11}	C_{12}	C_{13}	C_{22}	C_{23}	C_{33}	C_{44}	C_{55}	C_{66}
PZT-4	139	78	74	139	74	115	25.6	25.6	30.5
	Property								
	e_{15}	e_{24}	e_{31}	e_{32}	e_{33}	$\in_{11}$	$\in_{22}$	$\in_{33}$	ρ
PZT-4	12.7	12.7	−5.2	−5.2	15.1	65	65	56	7.5

Units: C_{ij} (10^9 N/m^2), $\in_{ij}$ (10^{-10} F/m), e_{ij} (C/m^2), ρ (10^3 kg/m^3).

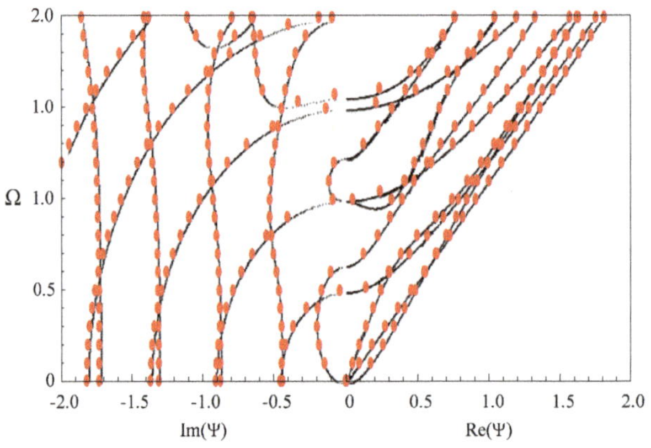

Figure 3. Dispersion curves of Lamb wave in a PZT-4 plate; red dots—our results, black dotted lines—literature results from the reverberation-ray matrix method.

Then we discuss the convergence of the present polynomial approach. We present dispersion curves of propagating Lamb-like wave in a linear FGPPM plate with electric and magnetic open circuit and $h = 1$ mm, when the truncation order M takes 7, 8, 9 and 15, respectively, as shown in Figure 4. It can be seen that more and more order modes converge as M increases. When $M = 7$, the first three modes are convergent. The first four when $M = 8$, and the first seven when $M = 9$. So, we can think that at least the first $(M - 1)/2$ modes are convergent. Similarly, this can be concluded for the purely imaginary modes, and we don't present the dispersion curves of purely imaginary branches for saving space. For evanescent Lamb-like waves, we tabulate the results in Table 3 since graph is not convenient for comparison. These numerical results also show that the complex solutions are convergent as M increases. When $M = 10$ and $M = 11$, the first three modes are convergent. The first four when $M = 12$, the first five when $M = 13$, and the first six when $M = 14$. Obviously, the real solution is easier to converge than the complex one. From these results, good convergence of the present approach can be observed. We take $M = 30$ in this paper.

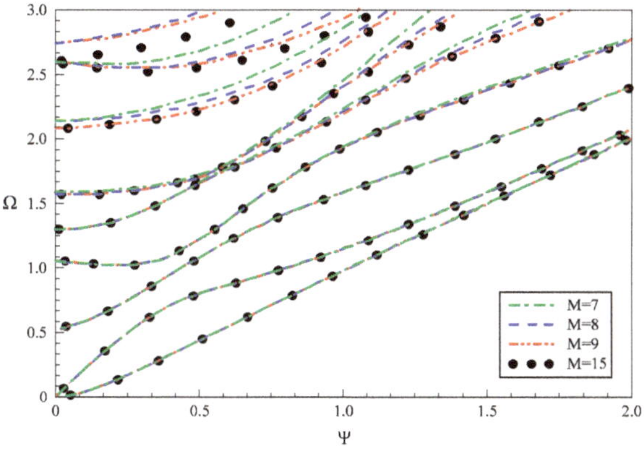

Figure 4. Dispersion curves of propagating Lamb-like wave in a linear functionally graded piezoelectric-piezomagnetic material (FGPPM) plate with various "*M*".

Table 3. Convergence of complex wave numbers of the first six modes (Ω = 0.1).

M	Mode					
	1	2	3	4	5	6
10	0.31308 +0.67653i	0.32055 +1.23591i ™	0.16593 +1.87713i ™	0.33292 +2.58928i ™	0.34536 +3.14395i ™	0.24222 +3.74226i ™
11	0.31308 +0.67653i	0.32055 +1.23591i ™	0.16593 +1.87713i ™	0.33184 +2.59041i	0.34722 +3.14407i ™	0.24481 +3.73460i ™
12	0.31308 +0.67653i	0.32055 +1.23591i ™	0.16593 +1.87713i ™	0.33063 +2.59061i	0.35203 +3.14418i ™	0.25047 +3.72524i
13	0.31308 +0.67653i	0.32055 +1.23591i ™	0.16593 +1.87713i ™	0.33063 +2.59061i	0.35211 +3.14436i ™	0.25504 +3.72496i ™
14	0.31308 +0.67653i	0.32055 +1.23591i ™	0.16593 +1.87713i ™	0.33063 +2.59061i	0.35211 +3.14436i ™	0.25561 +3.72471i ™
20	0.31308 +0.67653i	0.32055 +1.23591i ™	0.16593 +1.87713i ™	0.33063 +2.59061i	0.35211 +3.14436i ™	0.25561 +3.72471i ™

3.2. Full Dispersion Curves of Lamb Wave

Propagating waves have received a lot of attention, and here we put the emphasis on evanescent waves. We plot the full dispersion curves in 3D frequency-complex wave number space for a clearer visualization of the solutions and a better understanding of the nature of the modes, when necessary, with a different color for clarity. Figure 5a plots the full dispersion curves of Lamb wave for a linear FGPPM plate with electric and magnetic open circuit. Since the eigenvalues are computed for one ω at a time, the dispersion curves are constructed of unconnected dots and the points near the cut-off frequencies become sparse. We can observe that purely real and purely imaginary solutions appear in pairs of opposite signs and the complex ones appear in quadruples of complex conjugates and opposite signs. Purely real wave numbers correspond to the propagating wave, and purely imaginary and complex wave numbers correspond to the evanescent wave. For a given frequency, a certain small number of real branches exist together with an infinite number of complex and purely imaginary branches (mostly imaginary with few complexes in the given range). For clarity, Figure 5b shows one quadrant dispersion curves in a small range. For complex branches, most of them start from 0 frequency and end at the minima of the purely real branches. Occasionally, one connecting two purely imaginary branches appears. The real part of the complex branches is usually small. For purely imaginary branches, most of them start from 0 frequency and end at cut-off frequencies with increasing frequency, and some with small wave numbers start from one cut-off frequency and terminate the other one.

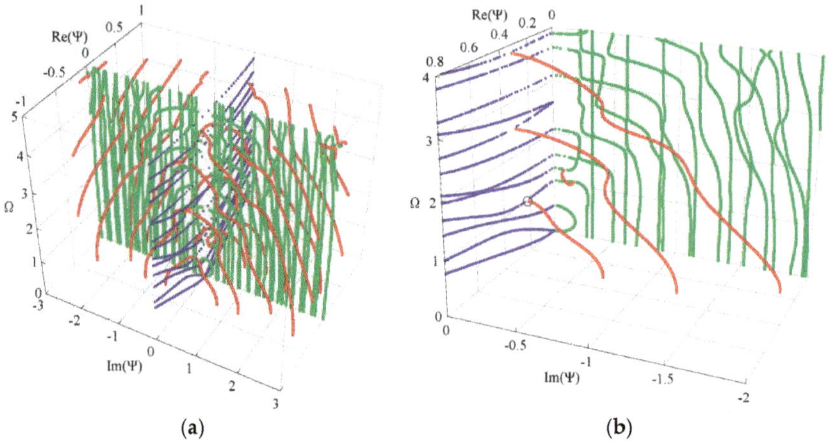

Figure 5. 3D dispersion curves of Lamb wave: (**a**) four quadrants, (**b**) one quadrant; blue—real solutions, green—imaginary solutions, red—complex solutions.

Figure 6 shows the phase velocity dispersion and attenuation curves of the first three propagating and complex branches. The dimensionless phase velocity and frequency and attenuation are defined by $\text{Vp} = \omega/(\text{Re}(k) \cdot \sqrt{C_{55}/\rho})$, $\text{fh} = \omega h/(2\pi\sqrt{C_{55}/\rho})$ and $\text{Im}(\text{kh})$. We can find from these curves that the phase velocity of a propagating mode is decreased and gradually tends to a steady value with increasing frequency, but the velocity of an evanescent mode becomes bigger as well as the attenuation decreases. At high frequency, the evanescent mode has a very small attenuation, and its phase velocity is noticeably bigger than that of a propagating mode. For example, at *fh* = 2–3, the phase velocity of the second evanescent mode is about 8, but that of the propagating mode is below 2. Also, the wave dispersion is quite weak in this frequency range.

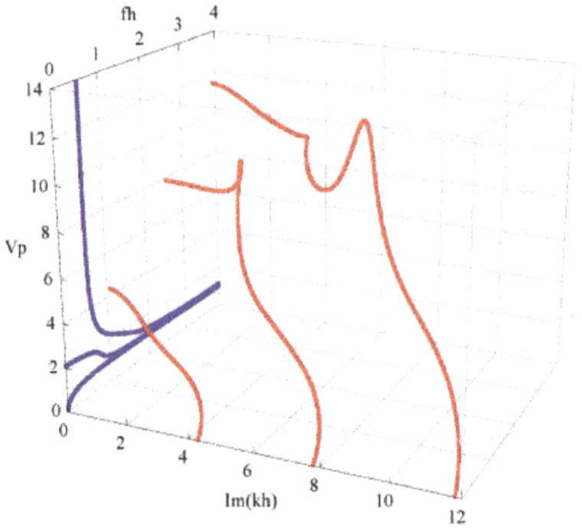

Figure 6. The phase velocity dispersion and attenuation curves; propagating wave in blue, evanescent wave in red.

3.3. *Influences of Graded Field on Dispersion Curves and the Electromechanical Coupling Factor*

Considering two graded shapes, cubic and sinusoidal graded fields. Figure 7 shows their dispersion curves of Lamb wave. The results show that the effect of the graded field on dispersion characteristics of Lamb wave is significant, including the propagating modes and evanescent modes. Comparison between Figures 7 and 5b, we can notice that the imaginary part of the complex branches for the sinusoidal graded case, at $\Omega = 0$ plane, is bigger than that for the linear and cubic cases. Interestingly, for the cubic graded cases, the complex branch connecting two purely imaginary branches disappears and turns into a different one connecting a purely imaginary branch and a real branch. For clarity, Figure 8 shows the frequency spectra and phase velocity spectra of Lamb propagating wave for the three graded fields. Obviously, the effect of the graded field is little on the low mode, but becomes significant with increasing the mode order and wave number. The phase velocity for the sinusoidal graded field is bigger than that for the linear graded field, while the linear bigger than the cubic. The reason lies in that the different graded fields result in different material volume distributions, and the wave velocity depends on the material properties. Figure 9 gives the variation curves of the three gradient fields in the z direction. The $CoFe_2O_4$ content for the sinusoidal graded field is the highest, and the wave velocity of Ba_2TiO_3 is slower than that of $CoFe_2O_4$.

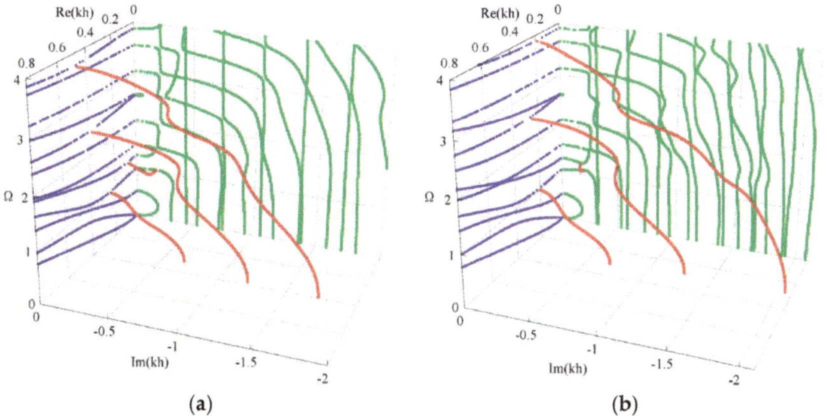

Figure 7. 3D dispersion curves of Lamb wave: (**a**) cubic graded field, (**b**) sinusoidal graded field. blue—real branches, green—purely imaginary branches, red—complex branches.

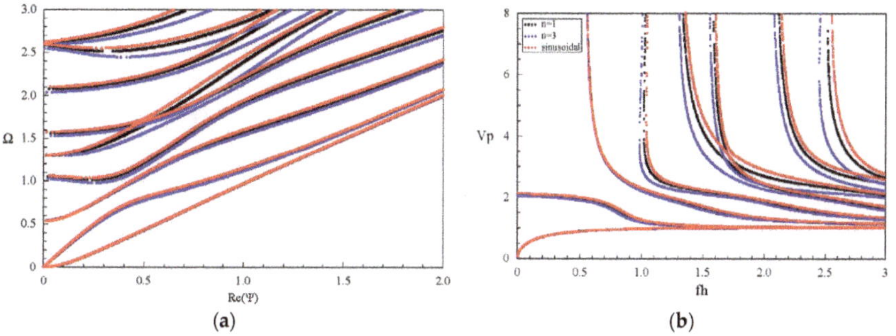

Figure 8. Dispersion curves of propagating Lamb wave for FGPPM plates with different graded fields; (**a**) Frequency spectra; (**b**) Phase velocity spectra.

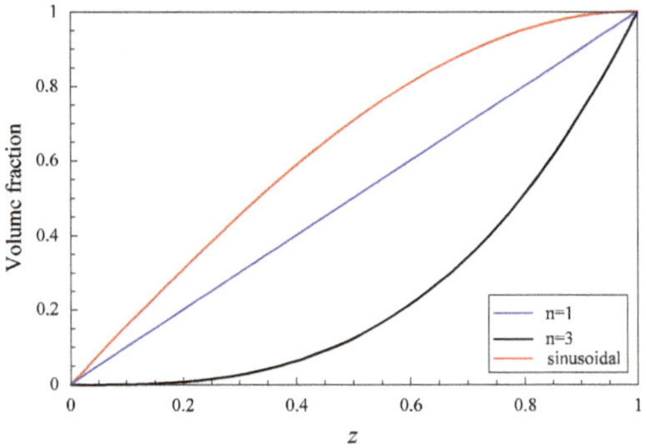

Figure 9. Variation curves of the three graded functions.

The magneto-electromechanical coupling factor K^2 is an important parameter for designing acoustic wave devices. A high magneto-electromechanical coupling factor is expected in engineering applications. It is defined as [29]

$$K^2 = \frac{2|Voc - Vsc|}{Voc} \tag{18}$$

where *Voc* and *Vsc* are the phase velocities for the electric and magnetic open circuit and short circuit, respectively.

To illustrate the effect of graded field on the K^2, we calculate the K^2 for S0 modes of four different FGPPM plates, as shown in Figure 10. We can find that the K^2 reaches a maximum at a certain wave number and tend to the same little value with increasing wave number, which implies the influence of the graded field on the energy propagation of Lamb wave in high-frequency zone is insignificant. It reaches a maximum from 4.4% for the sinusoidal graded field to 9.5% for the cubic graded field. They are located near $kh = 2$ and $kh = 1.5$ respectively. The K^2 for the cubic graded field is always bigger than that of the other three graded cases. Also the maximum of K^2 shifts to the smaller wave number when the graded power exponent is increasing.

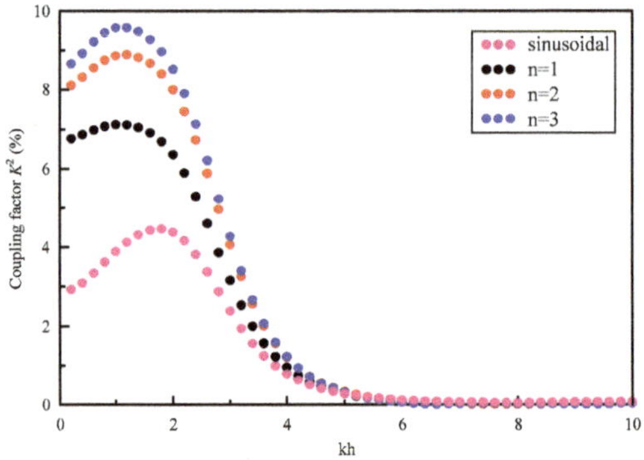

Figure 10. Effect of graded field on magneto-electromechanical coupling factor for S0 mode.

3.4. Wave Structure Analysis

The distributions of displacement and electric potential and magnetic potential fields can be obtained according to Equations (4) and (8). Considering a special position where the complex branch firstly collapses onto the real branch at about $\Omega = 1.0$, as marked with a circle in Figure 5b. Figures 11 and 12 present the distributions of the physical quantities in the *z* and *x* directions when

$\Omega = 1.01115$, $\Psi = 0.23216 - 0.04612i$, and $\Omega = 1.01911$, $\Psi = 0.17938$, respectively. As seen in these figures, the real branch propagates without any attenuation, and the complex branch exhibits an oscillatory distribution and propagates a very long distance, about a few tens of thicknesses of the plate. The displacement u_z and electric potential and magnetic potential distributions change along the z direction in a nearly anti-symmetric manner. The displacement u_x exhibits a nearly symmetric manner. The distribution of displacement u_z of the complex branch is very similar to that of the real branch, implying the evanescent wave mode converts into the propagating wave mode.

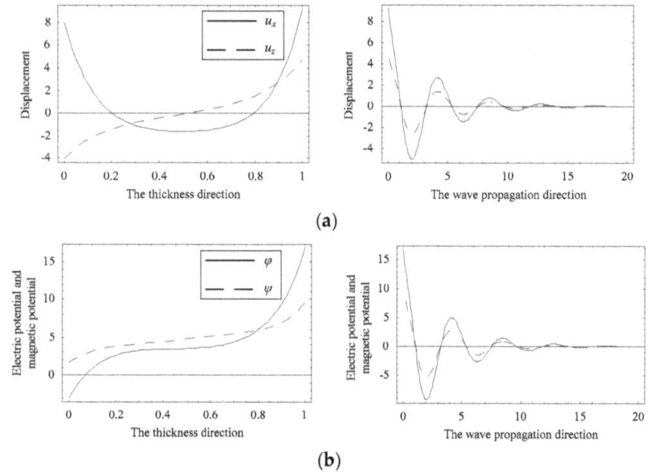

Figure 11. Distributions of the physical quantities when $\Omega = 1.01115$, $\Psi = 0.23216 - 0.04612i$. (**a**) displacement distribution, (**b**) electric potential and magnetic potential distribution.

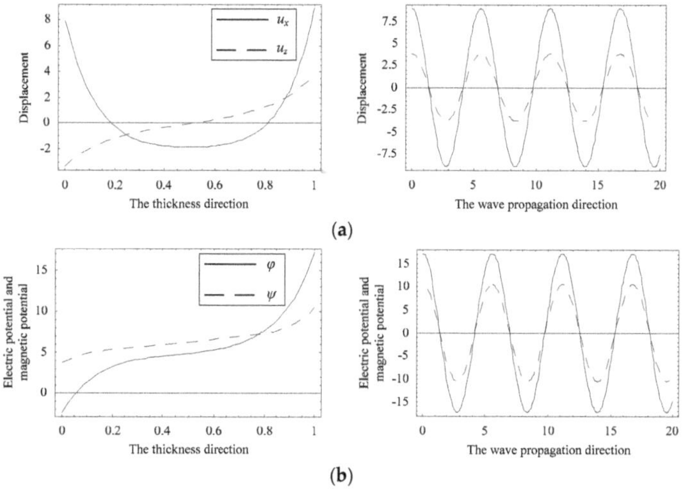

Figure 12. Distributions of the physical quantities when $\Omega = 1.01911$, $\Psi = 0.17938$. (**a**) displacement distribution, (**b**) electric potential and magnetic potential distribution.

3.5. Merits of the Presented Method

Based on the above calculation of wave propagation in a FGPPM plate, we can summarize the following advantages of the presented method, which makes the method attractive.

(1) The complex mathematical issue is reduced to solve an eigenvalue problem, which is capable of accurately determining all the real, imaginary and complex solutions of a transcendental dispersion equation.
(2) The conventional approaches (root-finding routines or finite element simulations) require an iterative search procedure or a far greater coding effort, to find complex roots. The present method can avoid tedious iterative two-variable search and is simple to program. It needs to take a larger polynomial order to obtain solutions of the higher modes, which will cause more computer memory and long time.
(3) The method is easy to implement and can be extended to complex structures such as multilayered or curved structures.

4. Conclusions

This paper presents an analytic method based on polynomial expansions for the determination of the full dispersion spectrum of the guided waves in the FGPPM plate. The correctness of the present method is verified via numerical comparison with available reference results. For the first time, the complete 3D dispersion curves of Lamb wave in a FGPPM plate are illustrated in a wide frequency range. The characteristics of the Lamb waves including the propagating and evanescent modes in various FGPPM plates are investigated. The emphasis on evanescent waves makes this work relevant for applications in the nondestructive evaluation of material or structural properties. Based on the above numerical results, some interesting conclusions can be drawn:

(1) Superior to the conventional methods that necessitate an iterative search procedure to solve the complex roots of a dispersion equation, the presented analytic method can transform the set of differential equations for the acoustic waves into an eigenvalue problem in the form $AX = kX$ to find the complex solutions.
(2) Complex branches of the Lamb wave usually collapse onto the extremum of the real branches. They exhibit both local vibration and local propagation, and some can propagate a quite long distance (more than ten times of the plate thickness). They will turn into the propagating modes with increasing frequency.
(3) Some evanescent modes have a noticeably higher phase velocity than the propagating modes. The phase velocity of the low order evanescent modes is more than four times larger than that of the propagating modes. Also, the wave dispersion of the evanescent mode is quite weak in a certain frequency range.
(4) The magneto-electromechanical coupling factor of the guided wave in a FGPPM plate may be improved by adjusting the graded field. The coupling factor reaches a maximum from 4.4% for the sinusoidal graded field to 9.5% for the cubic graded field. The maximum of the magneto-electromechanical coupling factor for the S0 mode shifts to lower frequencies with increasing the gradient index.

Author Contributions: X.Z. and J.Y. proposed the studied problem and the corresponding solving method; X.Z. and Z.L. conducted the theoretical derivation and the computation; X.Z. and J.Y. analyzed and discussed the results; X.Z. and J.Y. wrote the paper.

Funding: This research was funded by the National Natural Science Foundation of China (No. U1504106), the fundamental research funds for the national outstanding youth project of Henan Polytechnic University (No. NSFRF140301), and the Program for Innovative Research Team of Henan Polytechnic University (T2017-3).

Conflicts of Interest: The authors declare no conflict of interest.

References

1. Wang, J.G.; Chen, L.F.; Fang, S.S. State vector approach to analysis of multilayered magneto-electro-elastic plates. *Int. J. Solids Struct.* **2003**, *40*, 1669–1680. [CrossRef]
2. Tornabene, F. Free vibration analysis of functionally graded conical, cylindrical shell and annular plate structures with a four-parameter power-law distribution. *Comput. Methods Appl. Mech. Eng.* **2009**, *198*, 2911–2935. [CrossRef]
3. Kandasamy, R.; Dimitri, R.; Tornabene, F. Numerical study on the free vibration and thermal buckling behavior of moderately thick functionally graded structures in thermal environments. *Compos. Struct.* **2016**, *157*, 207–221. [CrossRef]
4. Čanađija, M.; Barretta, R.; Sciarra, F.M.D. On functionally graded timoshenko nonisothermal nanobeams. *Compos. Struct.* **2016**, *135*, 286–296. [CrossRef]
5. Barretta, R.; Čanađija, M.; Feo, L.; Lucianod, R.; de Sciarraa, F.M.; Pennac, R. Exact solutions of inflected functionally graded nano-beams in integral elasticity. *Compos. Part B* **2018**, *142*, 273–286. [CrossRef]
6. Rostami, J.; Pwt, T.; Fang, Z. Sparse and dispersion-based matching pursuit for minimizing the dispersion effect occurring when using guided wave for pipe inspection. *Materials* **2017**, *10*, 622. [CrossRef] [PubMed]
7. Chen, J.Y.; Guo, J.H.; Pan, E.N. Wave propagation in magneto-electro-elastic multilayered plates with nonlocal effect. *J. Sound Vib.* **2017**, *400*, 550–563. [CrossRef]
8. Wang, L.G.; Rokhlin, S.I. Recursive geometric integrators for wave propagation in a functionally graded multilayered elastic medium. *J. Mech. Phys. Solids* **2004**, *52*, 2473–2506. [CrossRef]
9. Pan, E.; Han, F. Exact solution for functionally graded and layered magneto-electro-elastic plates. *Int. J. Eng. Sci.* **2005**, *43*, 321–339. [CrossRef]
10. Bhangale, R.K.; Ganesan, N. Free vibration of simply supported functionally graded and layered magneto-electro-elastic plates by finite element method. *J. Sound Vib.* **2006**, *294*, 1016–1038. [CrossRef]
11. Wu, B.; Yu, J.G.; He, C.F. Wave propagation in non-homogeneous magneto-electro-elastic plates. *J. Sound Vib.* **2008**, *317*, 250–264.
12. Cao, X.S.; Shi, J.P.; Jin, F. Lamb wave propagation in the functionally graded piezoelectric-piezomagnetic material plate. *Acta Mech.* **2012**, *223*, 1081–1091. [CrossRef]
13. Singh, B.M.; Rokne, J. Propagation of SH waves in layered functionally gradient piezoelectric-piezomagnetic structures. *Philos. Mag.* **2013**, *93*, 1690–1700. [CrossRef]
14. Xiao, D.L.; Han, Q.; Liu, Y.J.; Li, C. Guided wave propagation in an infinite functionally graded magneto-electro-elastic plate by the Chebyshev spectral element method. *Compos. Struct.* **2016**, *153*, 704–711. [CrossRef]
15. Roshchupkin, D.; Ortega, L.; Snigirev, A.; Snigirev, I. X-ray imaging of the surface acoustic wave propagation in $La_3Ga_5SiO_{14}$ crystal. *Appl. Phys. Lett.* **2013**, *103*, 154101. [CrossRef]
16. Glushkov, E.; Glushkova, N.; Zhang, C.Z. Surface and pseudo-surface acoustic waves piezoelectrically excited in diamond-based structures. *J. Appl. Phys.* **2012**, *112*, 064911–064920. [CrossRef]
17. Auld, B.A. *Acoustic Fields and Waves in Solids*, 2nd ed.; Krieger Publishing Company: Malabar, FL, USA, 1990.
18. Lyon, R.H. Response of an elastic plate to localized driving forces. *J. Acoust. Soc. Am.* **1955**, *27*, 259–265. [CrossRef]
19. Mindlin, R.D.; Medick, M.A. Extensional vibrations of elastic plates. *J. Appl. Mech.* **1959**, *26*, 561–569.
20. Freedman, A. The vibration, with the Poisson ratio, of Lamb modes in a free plate, I: General spectra. *J. Sound Vib.* **1990**, *137*, 209–230. [CrossRef]
21. Quintanilla, F.H.; Lowe, M.J.S.; Craster, R.V. Full 3D dispersion curve solutions for guided waves in generally anisotropic media. *J. Sound Vib.* **2016**, *363*, 545–559. [CrossRef]
22. Yan, X.; Yuan, F.G. Conversion of evanescent Lamb waves into propagating waves via a narrow aperture edge. *J. Acoust. Soc. Am.* **2015**, *137*, 3523–3533. [CrossRef] [PubMed]
23. Yan, X.; Yuan, F.G. A semi-analytical approach for SH guided wave mode conversion from evanescent into propagating. *Ultrasonics* **2018**, *84*, 430–437. [CrossRef] [PubMed]
24. Chen, H.; Wang, J.; Du, J.K.; Yang, J. Propagation of shear-horizontal waves in piezoelectric plates of cubic crystals. *Arch. Appl. Mec.* **2016**, *86*, 517–528. [CrossRef]
25. Datta, S.; Hunsinger, B.J. Analysis of surface waves using orthogonal functions. *J. Appl. Phys.* **1978**, *49*, 475–479. [CrossRef]

26. Li, J.Y.; Dunn, M.L. Micromechanics of Magnetoelectroelastic Composite Materials: Average Fields and Effective Behavior. *J. Intell. Mater. Syst. Struct.* **1998**, *9*, 404–416. [CrossRef]
27. Han, X.; Liu, G.R. Elastic waves in a functionally graded piezoelectric cylinder. *Smart Mater. Struct.* **2003**, *12*, 962–971. [CrossRef]
28. Guo, Y.Q.; Chen, W.Q.; Zhang, Y.L. Guided wave propagation in multilayered piezoelectric structures. *Sci. China Ser. G Phys. Mech. Astron.* **2009**, *52*, 1094–1104. [CrossRef]
29. Ezzin, H.; Amor, M.B.; Ghozlen, M.H.B. Propagation behavior of SH waves in layered piezoelectric/piezomagnetic plates. *Acta Mech.* **2017**, *228*, 1071–1081. [CrossRef]

materials

MDPI

Article

Properties of Love Waves in Functional Graded Saturated Material

Zhen Qu [1,2], Xiaoshan Cao [2,3,*] and Xiaoqin Shen [1,*]

1 School of Science, Xi'an University of Technology, Xi'an 710054, China; quzhen@xaut.edu.cn
2 School of Civil Engineering and Architecture, Xi'an University of Technology, Xi'an 710048, China
3 State Key Laboratory of Transducer Technology, Chinese Academy of Sciences, Shanghai 200050, China
* Correspondence: caoxsh@xaut.edu.cn (X.C.); xqshen@xaut.edu.cn (X.S.)

Received: 8 October 2018; Accepted: 30 October 2018; Published: 2 November 2018

Abstract: In the present study, the propagation of Love waves is investigated in a layered structure with two different homogeneity saturated materials based on Biot's theory. The upper layer is a transversely isotropic functional graded saturated layer, and the substrate is a saturated semi-space. The inhomogeneity of the functional graded layer is taken into account. Furthermore, the gradient coefficient is employed as the representation of the relation with the layer thickness and the material parameters, and the power series method is applied to solve the variable coefficients governing the equations. In this regard, the influence of the gradient coefficients of saturated material on the dispersion relations, and the attenuation of Love waves in this structure are explored, and the results of the present study can provide theoretical guidance for the non-destructive evaluation of functional graded saturated material.

Keywords: functional graded saturated material; inhomogeneity; Love wave; dispersion; attenuation

1. Introduction

The research of the propagation characteristics of Love waves have been found in a wide range of engineering applications, such as seismic engineering, geotechnical engineering, and geophysics. Studies based on the elastic hypothesis have been sufficiently carried out. Since 1956, Biot [1–3] established the constitutive relation and the motion equation of saturated porous media. Based on Biot's work, fruitful results have been yielded thereafter. Deresiewics et al. [4–6] derived the dispersion and attenuation equations of Love waves in the porous media. Wang, Tong, and Santos et al. [7–9] used the iteration method to solve the dispersion equation of porous materials. In addition, Konezak [10] and Ba et al. [11] gave a solution to the propagation of waves in porous layered half-space.

However, the research, which we have mentioned above, mainly focused on the homogeneous hypothesis of media. In the real situation, some saturated materials are always regarded as a layered and inhomogeneous medium, in which the material parameters vary continuously with the medium thickness. On this basis, how to explain the influence of homogeneity on wave propagation characteristics has become a crucial problem. In the recent years, some researchers use analytical methods to solve this problem. For example, Ke et al. [12] and Qian et al. [13] used the iterative method and Wentzel-Kramers-Brillouin (WKB) method, respectively, to deal with the inhomogeneity of materials, but have some limitations. In Ke's work, the inhomogeneity of materials was described just as an exponential function, which we do not think is sufficient. The WKB method is too complicated for calculation. Cao et al. [14–16] used the power series method to solve Love wave and Rayleigh wave propagation problems in the FGM layered composite system.

In this study, the inhomogeneity of the saturated material and solid skeleton is supposed to be transversely isotropic. In addition, the assumption is made concerning the relationship between material thickness and material parameters, of which the latter vary continuously along the depth. Then, the dispersion relations and the attenuation of Love waves are investigated.

2. Statement of the Problem and Governing Equations

The propagation of Love waves in a functional graded saturated media structure is shown in Figure 1. The upper layer is a transversely isotropic inhomogeneous saturated layer with the thickness of H. The surface of this layer is traction free, and the substrate is a homogeneous saturated half-space. Based on the Biot's model of the homogeneous anisotropic saturated porous media, the soil skeleton is considered as a transversely isotropic medium. In terms of the Love waves propagation in the structure shown in Figure 1, the expressions of displacement are given as follows:

$$\begin{cases} u_x = u_y = 0,\ u_z = u_z(x,y,t), \\ w_x = w_y = 0,\ w_z = w_z(x,y,t). \end{cases} \quad (1)$$

Based on the motion equations, presented by Biot [3] in porous media, namely:

$$\begin{cases} \sigma_{ij,j} = \rho \ddot{u}_i + \rho_f \ddot{w}_i, \\ -p_{f,j} = \rho_f \ddot{u}_i + m_{ii}\ddot{w}_i + r_{ii}\dot{w}_i, \end{cases} \quad (2)$$

where p_f is the fluid pressure, and ρ is the density of saturated material, which can be expressed as $\rho = (1-\phi)\rho_s + \phi\rho_f$. ρ_s is the density of solid skeleton, ρ_f is the fluid density, and ϕ is the porosity of the solid. The u_i in the equation is the component of the solid skeleton, and in terms of $w_i = -\phi(u_i - U_i)$, U_i is the displacement of fluid. The comma followed by the subscript i indicates the space differentiation with respect to the corresponding coordinate x, y, and z, the dot "•" represents time differentiation, and the repeated index is the means to summation related to that index. The parameter $m_{ii} = Re[\alpha_i(\omega)]\rho_f/\phi$ and $r_{ii} = \eta/Re[K_i(\omega)]$ are Biot's coefficients put forward by Biot. They are the functions of angular frequency ω and $\omega = ck$. C and k are the velocity and numbers of the waves. Where η is the viscosity of the fluid, and $\alpha_i(\omega)$ and $K_i(\omega)$ are the dynamic tortuosity and permeability. Let u_x, u_y, u_z and w_x, w_y, w_z denote the displacement of the medium. The governing equations for the displacement of medium can be obtained.

Let $\overline{u}_i, \overline{w}_i$ denote the displacement in the substrate layer. The expression of the governing equations for the Love waves propagating in the substrate layer ($x > 0$) are given as follows.

$$\begin{cases} \overline{C}_{44}\frac{\partial^2 \overline{u}_z}{\partial x^2} + \overline{C}_{44}\frac{\partial^2 \overline{u}_z}{\partial y^2} = \overline{\rho}\frac{\partial^2 \overline{u}_z}{\partial t^2} + \overline{\rho}_f\frac{\partial^2 \overline{w}_z}{\partial t^2}, \\ \overline{\rho}_f\frac{\partial^2 \overline{u}_z}{\partial t^2} + \overline{m}_1\frac{\partial^2 \overline{w}_z}{\partial t^2} + \overline{r}_1\frac{\partial \overline{w}_z}{\partial t} = 0, \end{cases} \quad (3)$$

where C_{44} is the coefficient of material parameters. The "$^-$" symbol is used to denote the parameters in the substrate layer.

Similarly, we use $\hat{u}_i$, $\hat{w}_i$ denote the displacement in the upper layer. The governing equations for the Love waves propagating in the upper layer ($-H < x < 0$) can be expressed as follows:

$$\begin{cases} \hat{C}_{44}\frac{\partial^2 \hat{u}_z}{\partial x^2} + \hat{C}'_{44}\frac{\partial \hat{u}_z}{\partial x} + \hat{C}_{44}\frac{\partial^2 \hat{u}_z}{\partial y^2} = \hat{\rho}\frac{\partial^2 \hat{u}_z}{\partial t^2} + \hat{\rho}_f\frac{\partial^2 \hat{w}_z}{\partial t^2}, \\ \hat{\rho}_f\frac{\partial^2 \hat{u}_z}{\partial t^2} + \hat{m}_1\frac{\partial^2 \hat{w}_z}{\partial t^2} + \hat{r}_1\frac{\partial \hat{w}_z}{\partial t} = 0, \end{cases} \quad (4)$$

where the superscript "$'$" indicates the space differentiation with respect to the $x-$ coordinate. The "$\hat{}$" symbol is used to denote the parameters in the upper layer, and these parameters are the functions of the $x - axis$, which needs to be emphasized.

The boundary condition of the present problem should be satisfied as follows: (a) the traction free boundary condition is $\hat{\tau}_{xz}(-H,y) = 0$ at $x = -H$; (b) the stress and displacement are all continuous,

$\overline{\tau}_{xz}(0,y) = \hat{\tau}_{xz}(0,y)$, $\overline{u}_z(0,y) = \hat{u}_z(0,y)$, $\overline{w}_z(0,y) = \hat{w}_z(0,y)$; and (c) the attenuation conditions for Love waves are $\overline{u}_z \to 0$ at $x \to \infty$.

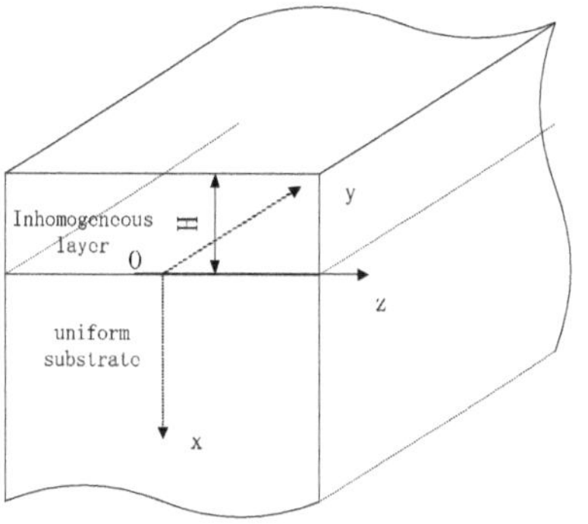

Figure 1. The functional graded saturated material layered structure.

3. Solution of the Problem

In light of the present Love waves propagation problem we have discussed above, the solutions of the governing equations can be supposed as follows:

$$\begin{cases} u_z(x,y,t) = A_z(x)\exp[ik(y-ct)], \\ w_z(x,y,t) = W_z(x)\exp[ik(y-ct)], \end{cases} \tag{5}$$

where $i = \sqrt{-1}$, $k = 2\pi/\lambda$ is the wave number, and c is the phase velocity. $A_z(x)$ and $W_z(x)$ are the amplitudes of the displacement, which will be solved. Furthermore, the "-" symbol and the "^" symbol are used to denote the substrate layer and the upper layer, respectively, so $\overline{A}_z(x)$ is the amplitudes of the displacement of the substrate layer, and $\hat{A}_z(x)$ refers to the amplitudes of the upper layer, respectively.

Firstly, in order to solve the problem in the substrate layer, we combine Equation (5) with Equation (3), and the governing equations can be modified as follows:

$$\begin{cases} \overline{C}_{44}\left(\overline{A}_z'' - k^2\overline{A}_z\right) = -\overline{\rho}c^2k^2\overline{A}_z - \overline{\rho}_f c^2k^2\overline{W}_z, \\ \overline{\rho}_f c^2k^2\overline{A}_z + \overline{m}_1 c^2k^2 W_z + \overline{r}_1 ick\overline{W}_z = 0\,. \end{cases} \tag{6}$$

Then, the Love waves in the substrate layer can be expressed as follows:

$$\begin{cases} \overline{u}_z(x,y,t) = [C_1\exp(i\gamma x) + C_2\exp(-i\gamma x)]\exp[ik(y-ct)], \\ \overline{w}_z(x,y,t) = -\frac{\overline{\rho}_f ck}{\overline{m}_1 ck + \overline{r}_1 i}\overline{u}_z(x,y,t), \end{cases} \tag{7}$$

For the radiation condition of the Love waves, we must have $\mathbf{Im}(\gamma) > 0$ and $\mathbf{Re}(\gamma) > 0$. And, when integrating Equation (7) into the attenuation conditions, we can easily find that $C_2 = 0$.

Secondly, the governing equations in the upper layer can be solved by combing Equation (5) with Equation (4), and the governing equation can be modified as follows:

$$\begin{cases} \hat{C}_{44}\hat{A}_z'' + \hat{C}_{44}'\hat{A}_z' - \hat{C}_{44}k^2\hat{A}_z = -\hat{\rho}c^2k^2\hat{A}_z - \hat{\rho}_f c^2k^2\hat{W}_z, \\ \hat{\rho}_f c^2k^2\hat{A}_z + \hat{m}_1 ck\hat{W}_z + \hat{r}_1 ick\hat{W}_z = 0, \end{cases} \tag{8}$$

In order to solve the variable coefficient Equation (8), we assume the material parameters of the upper layer as the following functional form:

$$\begin{aligned}
&\hat{C}_{44} = \sum_{n=0}^{\infty} a_n^1 \left(\tfrac{x}{H}\right)^n, \hat{\rho} = \sum_{n=0}^{\infty} a_n^2 \left(\tfrac{x}{H}\right)^n, \hat{\rho}_f = \sum_{n=0}^{\infty} a_n^3 \left(\tfrac{x}{H}\right)^n, \\
&\hat{m}_1 = \sum_{n=0}^{\infty} a_n^4 \left(\tfrac{x}{H}\right)^n, \hat{r}_1 = \sum_{n=0}^{\infty} a_n^5 \left(\tfrac{x}{H}\right)^n,
\end{aligned} \tag{9}$$

where the coefficients a_n^i can be determined by the relations between the functions and their Taylor expansions. Then, the solutions of Equation (8) can be assumed to take the similar forms, as follows:

$$\hat{A}_z = \sum_{n=0}^{\infty} s_n \left(\frac{x}{H}\right)^n, \hat{W}_z = \sum_{n=0}^{\infty} t_n \left(\frac{x}{H}\right)^n. \tag{10}$$

According to the integration of Equations (9) and (10) into Equation (8), the two recursive equations for s_n and t_n are presented as follows:

$$\begin{aligned}
&\sum_{i=0}^{n} (i+2)(i+1)a_{n-i}^1 s_{i+2} + \sum_{i=0}^{n} (n-i+1)(i+1)a_{n-i+1}^1 s_{i+1} - (kH)^2 \sum_{i=0}^{n} a_{n-i}^1 s_i \\
&+c^2(kH)^2 \sum_{i=0}^{n} a_{n-i}^2 s_i + c^2(kH)^2 \sum_{i=0}^{n} a_{n-i}^4 t_i = 0,
\end{aligned} \tag{11}$$

$$c^2k^2 \sum_{i=0}^{n} a_{n-i}^3 s_i + c^2k^2 \sum_{i=0}^{n} a_{n-i}^4 t_i + ick \sum_{i=0}^{n} a_{n-i}^5 t_i = 0. \tag{12}$$

We can calculate the coefficients of $(x/H)^n$, s_n, and t_n with n from zero to infinity, using Equations (11) and (12). On this basis, a matrix is described to solve these coefficients.

$$(s_{0j}, s_{1j}) = \mathbf{I}, \tag{13}$$

where j = 3~4 and **I** is a 2 × 2 unit matrix. The solution of Equation (8) can be rewritten as follows:

$$\hat{A}_z = \sum_{j=3}^{4} C_j \sum_{n=0}^{\infty} s_{nj} \left(\frac{x}{H}\right)^n, \hat{W}_z = \sum_{n=0}^{\infty} t_n \left(\frac{x}{H}\right)^n. \tag{14}$$

According to the discussion we have made above, the solution of Equation (4) can be described as follows:

$$\begin{cases}
\hat{u}_z(x,y,t) = \left[\sum_{j=3}^{4} C_j \sum_{n=0}^{\infty} S_{nj} \left(\tfrac{x}{H}\right)^n\right] \exp[ik(y-ct)], \\
\hat{w}_z(x,y,t) = \left[\sum_{n=0}^{\infty} t_n \left(\tfrac{x}{H}\right)^n\right] \exp[ik(y-ct)],
\end{cases} \tag{15}$$

Then, we apply Equations (15) and (7) to the boundary condition of the present problem, and there are a set of homogeneous linear algebraic equations of unknown coefficients C_i, i = 1, 3, 4 obtained. According to the condition for the existence of a non-trivial solution, the determinant of the coefficients matrix Q must be vanished.

$$|\mathbf{Q}| = 0. \tag{16}$$

4. Numerical Results and Discussion

The numerical examples will be given to illustrate the propagation characters of Love waves in the functional graded saturated layer, which are lying on a homogeneous saturated soil half-space. First and foremost, some important hypotheses must be introduced. In light of our problem, we used the following expression [7] to calculate the $\alpha(\omega)$ and $K(\omega)$.

$$\eta[\omega K_i(\omega)]^{-1} = i\phi^{-1}\rho_f\alpha_i(\omega) = i\phi^{-1}\rho_f\alpha_i(\infty)\left[1+\frac{4if_{ci}}{3f}\times\left(1-\frac{3i}{8}\frac{f}{f_{ci}}\right)^{1/2}\right], \quad (17)$$

where $f = \omega/2\pi$ is the wave frequency. $f_{ci} = \omega_{ci}/2\pi = 3\eta\phi\left[8\pi K_i(0)\alpha_i(\infty)\rho_f\right]^{-1}$ is a critical frequency, which was reported by Sharma in 1991 [17]. At the functional graded layer, the material parameters are functions of layer thickness, and these functions can be assumed as follows:

$$g = 1 - \exp(px/H)\,, \quad (18)$$

where the parameter p is the gradient coefficient, which refers to the level of layer inhomogeneity. On this basis, the parameter function of soil thickness can be described as follows:

$$\hat{C}_{44} = \overline{C}_{44}\cdot g, \quad (19)$$

and the other parameters in the upper layer have the similar forms.

In the present paper, the influence of the gradient coefficient on the Love waves dispersion relations and attenuation will be discussed. In detail, from the governing Equation (3), Equation (4), and dispersion relations Equation (16), the wave number k in our problem is a complex $k = k_1 + k_2$. Then, the dispersion relation curves will be drawn as the relation between the phase velocities c and k_1 in convenient, and we designate $\delta = k_2/k_1$ as the attenuation coefficient to evaluate the Love wave attenuation in our problem. In order to solve the complex dispersion equation, we used the method called the minimum modulus value approximation, in order to approximate the suitable solution. The theme of this method is described below. We assume the material parameter of the homogeneous saturated media as follows: $\overline{C}_{44} = 4\,\text{Gpa}, \overline{\phi} = 0.2, \overline{K}_1(0) = 1, \overline{\alpha}_1(\infty) = 1, \eta = 10^{-3}\,\text{pa}\cdot\text{s}$, $\overline{\rho}_s = 30\,\text{kN/m}^3, \overline{\rho}_l = 10\,\text{kN/m}^3, \overline{\rho}_g = 1.2\,\text{kN/m}^3$.

4.1. *Influence of the Gradient Coefficient on Love Wave Dispersion*

In order to describe the influence of the gradient coefficient on the Love wave dispersion, it is necessary to give a solution to the complex Equation (16). First of all, according to the research conducted by Sharma [17] and Wang [7], the Love wave speed has a range in the porous medium that is determined by a critical frequency, f_{ci}. In this paper, we also chose them as the method to calculate the range of the Love wave speed for specific gradient coefficients. Secondly, based on the range of speed, we employed the minimum modulus value approximation method to obtain the suitable solution of Equation (16). The theme of this method should be given as follows: (a) for a given speed range of the specific gradient coefficient p and nth modes of Love wave, we choose four values of (k_1H,k_2H) from $k_1H = 0$, and made them as a square; (b) calculate the determinant of Equation (16); (c) choose the values (k_1H,k_2H), which have the minimum value of determinant and use (k_1H,k_2H) as an angular point to make the new square, which has a half-length of the side of the previous square; (d) repeat the step (c) until the value of determinant reaches zero; and (e) give an increment of k_1H, and repeat the whole procedures. Then, we can draw a dispersion curve of the nth mode of the Love wave. At the same time, the attenuation coefficient $\log[k_2/k_1]$ can also be calculated in the given gradient coefficient p and nth modes of wave.

Figure 2 presents the Love wave dispersion curve of 1st and 2nd modes with the gradient $p = 0.6$. The comparison of the different gradient coefficients ($p = 0.2, 0.6, 0.8$) is shown in Figure 3. The results in Figure 3 suggest that the gradient coefficient p gives a conspicuous impact of Love wave dispersion. And Figure 4 shows the material parameter distributions. With the increase of the gradient coefficient p, the phase velocity of the Love wave decreases obviously, and the influence of the gradient coefficient on the first mode is more intense than that on the second mode. For the first mode of the Love wave, with the increase of dimensionless wave numbers k_1H, the influence of gradient coefficients on phase velocity gradually increases. In terms of the second mode of the Love wave, the influence is smoother than that on the first mode.

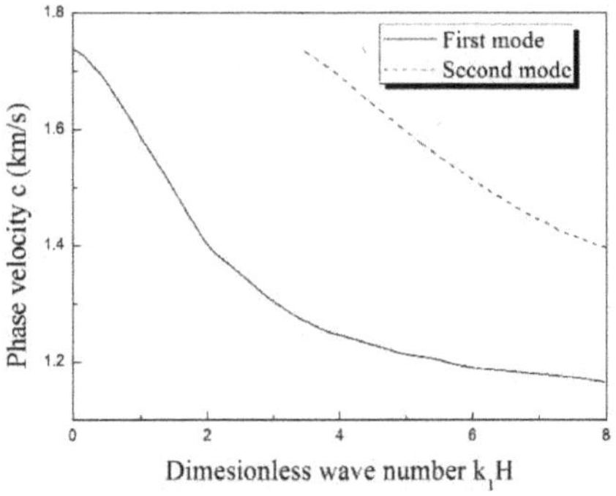

Figure 2. The dispersion curve of the Love wave in the inhomogeneous unsaturated layer lies on homogeneous saturated half-space.

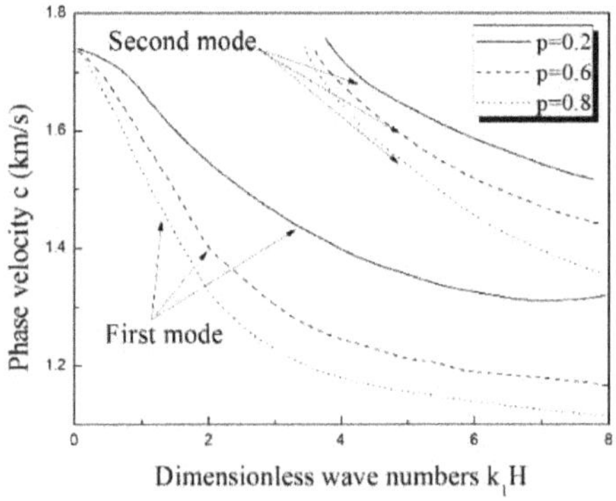

Figure 3. The comparison of the dispersion relation of the Love wave with the different gradient parameter *p*.

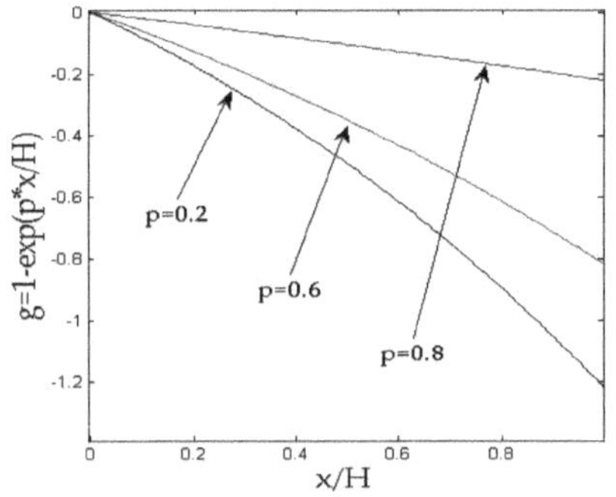

Figure 4. The material parameter distributions.

4.2. *Influence of the Gradient Coefficient on Love Wave Attenuation*

The attenuation of the Love wave is shown in Figure 5. The solid line denotes the first mode attenuation, and the second mode is expressed by the dashed line. As the two modes indicate, the attenuation rapidly increases at first, and then becomes smoother with the increase of dimensionless wave numbers, k_1H. The investigating results of the influence of the gradient coefficient on the attenuation of the Love wave are plotted in Figure 6. In the current study, the discussion mainly focuses on the situation of the Love wave attenuation in the first mode. The solid line refers to the situation of $p = 0.2$, the dashed line describes the $p = 0.6$, and the case of $p = 0.8$ is plotted as the dotted line. It is easily seen that the change of gradient coefficient almost exerts no effect on the Love wave attenuation, and the influence of material inhomogeneity on the attenuation of wave is very little. In this regard, great interest is entailed in the comparison with the rapid influence of inhomogeneity on the dispersion of the Love wave.

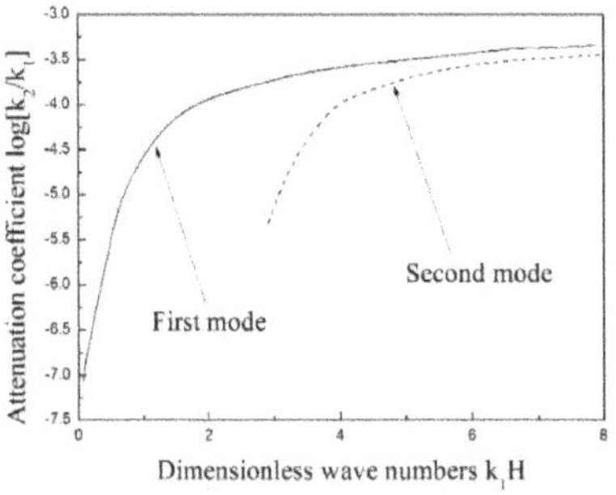

Figure 5. The attenuation curve of the Love wave in the inhomogeneous saturated layer lies on homogeneous saturated half-space.

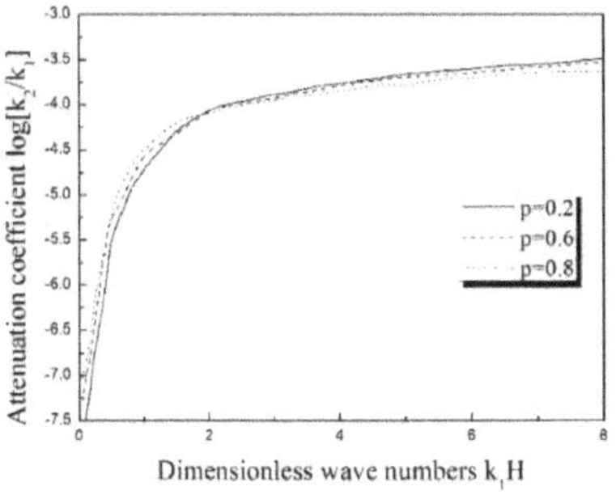

Figure 6. The comparison of the attenuation of the Love wave with the different gradient parameter p.

5. Conclusions

In this paper, based on the Biot's saturated porous medium theory, the influence of inhomogeneity has been theoretically analyzed on the propagation character of the Love wave in a transversely isotropic inhomogeneous saturated layer lying on a saturated half-space. The governing equations of the problem have been solved by the power series method, and the minimum modulus value approximation method is employed to discuss the dispersion equation of the Love wave. The gradient coefficient p has been introduced to describe the inhomogeneity of the saturated media, and we obtained the dispersion and attenuation curve of the Love wave with different gradient coefficients. It is important to note that the gradient coefficient has a great influence on the dispersion of the Love wave, but the effect of the gradient coefficient on the attenuation is less significant.

Author Contributions: Conceptualization, X.C. and X.S.; Data curation, Z.Q.; Formal analysis, Z.Q.; Funding acquisition, X.C. and X.S.; Investigation, Z.Q.; Methodology, Z.Q.

Funding: This research was funded by the National Natural Science Foundation of China (No. 11571275, No. 11572244, and No. U1630144), the Natural Science Foundation of Shaanxi Province (No. 2018JM1014), and the open State Key Laboratories of Transducer Technology (No. SKT1506).

Conflicts of Interest: The authors declare no conflict of interest.

References

1. Biot, M.A. Theory of Elasticity and Consolidation for a porous Anisotropic Solid. *J. Appl. Phys.* **1955**, *26*, 182–185. [CrossRef]
2. Biot, M.A. General solutions of the equation of elasticity and consolidation for a porous material. *J. Appl. Phys.* **1956**, *23*, 91–96.
3. Biot, M.A. Generalized theory of acoustic propagation in porous dissipative media. *J. Acoust. Soc. Am.* **1962**, *34*, 1254–1264. [CrossRef]
4. Deresiewisz, H. The effect of boundaries on wave propagation in a liquid-filled porous solid—II. Love waves in a porous layer. *Bull. Seismol. Soc. Am.* **1961**, *51*, 51–59.
5. Deresiewisz, H. The effect of boundaries on wave propagation in a liquid-filled porous solid—VI. Love waves in double surface layer. *Bull. Seismol. Soc. Am.* **1964**, *54*, 417–423.
6. Deresiewisz, H. The effect of boundaries on wave propagation in a liquid-filled porous solid—XI. Waves in a plate. *Bull. Seismol. Soc. Am.* **1974**, *64*, 1901–1907.
7. Wang, Y.S.; Zhang, Z.M. Propagation of Love waves in transversely isotropic fluid-saturated porous layered half-space. *J. Acoust. Soc. Am.* **1998**, *103*, 695–701. [CrossRef]
8. Tong, L.H.; Liu, Y.S.; Geng, D.X.; Lai, S.K. Nonlinear wave propagation in porous materials based on the Biot theory. *J. Acoust. Soci. Am.* **2007**, *142*, 756–770. [CrossRef] [PubMed]
9. Santos, J.E., Jr.; Corbero, J.; Lovera, O.M. A model for wave propagation in a porous medium saturated by a two-phase fluid. *J. Acoust. Soc. Am.* **2016**, *139*, 639–702. [CrossRef] [PubMed]
10. Konezak, Z. The propagation of Love waves in a fluid-saturated porous anisotropic layer. *Acta Mech.* **1989**, *79*, 155–168. [CrossRef]
11. Ba, Z.N.; Liang, J.W.; Lee, V.W. Wave propagation of buried spherical SH-, P1-, P2- and SV-waves in a layerd poroelastic half-space. *Soil Dyn. Earthq. Eng.* **2016**, *88*, 237–255. [CrossRef]
12. Ke, L.L.; Wang, Y.S.; Zhang, Z.M. Love waves in an inhomogeneous fluid saturated porous layered half-space with linearly varying properties. *Soil Dyn. Earthq. Eng.* **2006**, *26*, 574–581. [CrossRef]
13. Qian, Z.H.; Jin, F.; Kishimoto, K.; Lu, T.J. Propagation behavior of Love waves in a functionally graded half-space with initial stress. *Int. J. Solids Struct.* **2009**, *46*, 1354–1361. [CrossRef]
14. Cao, X.S.; Jin, F.; Jeon, I.; Lu, T.J. Propagation of Love waves in a functionally graded piezoelectric material layered composite system. *Int. J. Solids Struct.* **2009**, *46*, 4123–4132. [CrossRef]
15. Cao, X.S.; Jin, F.; Jeon, I. Calculation of propagation properties of lamb waves in a functionally graded material plate by power series technique. *NDT&E Int.* **2011**, *44*, 84–92.
16. Cao, X.S.; Jin, F.; Kishimoto, K. Transverse Shear surface wave in functionally graded material infinite half space. *Philos. Mag. Lett.* **2012**, *5*, 245–253. [CrossRef]
17. Sharma, M.D.; Gogna, M.L. Wave propagation in anisotropic liquid saturated porous solids. *J. Acoust. Soc. Am.* **1991**, *90*, 1068–1073. [CrossRef]

materials

MDPI

Article

Asymptotic Solution and Numerical Simulation of Lamb Waves in Functionally Graded Viscoelastic Film

Xiaoshan Cao *, Haining Jiang, Yan Ru and Junping Shi

School of Civil Engineering, Xi'an University of Technology, Xi'an 710048, China; 2160720004@stu.xaut.edu.cn (H.J.); ruyan@xaut.edu.cn (Y.R.); shijp@xaut.edu.cn (J.S.)

* Correspondence: caoxsh@xaut.edu.cn

Received: 3 December 2018; Accepted: 10 January 2019; Published: 15 January 2019

Abstract: To investigate Lamb waves in thin films made of functionally graded viscoelastic material, we deduce the governing equation with respect to the displacement component and solve these partial differential equations with complex variable coefficients based on a power series method. To solve the transcendental equations in the form of a series with complex coefficients, we propose and optimize the minimum module approximation (MMA) method. The power series solution agrees well with the exact analytical solution when the material varies along its thickness following the same exponential function. When material parameters vary with thickness with the same function, the effect of the gradient properties on the wave velocity is limited and that on the wave structure is obvious. The influence of the gradient parameter on the dispersion property and the damping coefficient are discussed. The results should provide nondestructive evaluation for viscoelastic material and the MMA method is suggested for obtaining numerical results of the asymptotic solution for attenuated waves, including waves in viscoelastic structures, piezoelectric semiconductor structures, and so on.

Keywords: Lamb wave; functionally graded viscoelastic material; minimum module approximation method; damping coefficient

1. Introduction

Lamb waves, which are a type of plain strain wave in a thin film or a plate with a traction-free boundary, are widely used in nondestructive evaluation. Early research reported on Lamb waves focused on isotropic elastic plates [1]. Since then, scientists have directed more attention to Lamb waves in plates made of various materials, including viscoelastic materials [2], functionally graded materials (FGMs) [3], piezoelectric materials [4], and piezoelectric–piezomagnetic materials [5]. To detect material properties or damage to the structures, much research has been focused on guided waves in composite structures based on numerical and experimental methods [6,7].

FGMs were proposed by scientists as a kind of thermal-protection material in the 1990s [8]. In FGM structures, the material parameters are not constant and vary along one direction continuously. With the development of material technology, the FGM technique has been used not only for common elastic material but also for some smart materials, including piezoelectric [9,10] and piezoelectric–piezomagnetic materials [11]. To evaluate the mechanical properties of FGM structures, researchers have investigated various elastic waves in FGM structures, such as Lamb waves, horizontal shear (SH) waves [12], Love waves [13], and Rayleigh waves [14].

To address wave propagation problems in heterogeneous media, both numerical and analytical methods are employed for solving the wave-governing differential equations with variable coefficients. The main idea of numerical methods is to divide the functionally graded material into multilayer models and to simplify each sublayer as a homogenous layer [15–18]. Scientists have also proposed some analytical solutions for wave propagation problems in different heterogeneous structures. These methods include exact analytical expressions for material parameters following the same exponential function [19], the Wentzel–Kramers–Brillouin (WKB) method for large-wave-number [20] or cutoff problems [21,22], and the special function method for material parameters following some special function [23]. In recent decades, researchers suggested that these equations can be solved by using a power series method [11,24] and a Legendre polynomial method [25,26], which are fit for solving the wave propagation problem in heterogeneous structures in arbitrary cases in which material parameters vary continuously and slowly. The form of the dispersion equations based on these two methods contains series items. Therefore, these dispersion equations should be solved numerically.

It is found that not only elastic materials but also viscoelastic materials in nature have gradient properties. For example, when a material undergoes subsurface aging or subsurface damage, the elastic modulus varies along the thickness of the damaged subsurface region and mechanical gradient characteristics appear [27]. This should also occur for viscoelastic materials. For the wave propagation problem in viscoelastic structures, Lu et al. [28] found that the attenuation of Lamb waves increases with the increase of the thickness of the viscoelastic layer and that the mode is transformed as well. Compared with the propagation characteristics of Love waves in an elastic medium, the energy of Love waves in a Kelvin–Voigt viscoelastic medium is obviously attenuated, as shown by Zhang et al. [29]. SH waves have one displacement component. Yu et al. [30] deduced the dispersion equations for SH waves in orthotropic viscoelastic hollow cylinders. There are few studies on the propagation of Lamb waves with two displacement components in viscoelastic complex structures, and most of them use the Legendre polynomial method [31].

The dispersion equations for wave propagation in a viscoelastic material comprise a set of complex coefficient transcendental equations. To solve the transcendental equations with complex variables, Qian et al. [32] comprehensively analyzed the applicability of the parabolic Newton iteration method, the binary dichotomy method, and the modulus value convergence method. However, when the power series method is employed to solve the wave propagation problem in a functionally graded viscoelastic material (FGVM) structure, the dispersion equation, which is a transcendental equation with complex numbers in series form, is difficult to solve based on the above numerical simulation method. For example, the Newton iteration method requires that the solution be in the form of a display function rather than a series, while the binary dichotomy method and the modulus value convergence method might lead to the existence of spurious solutions.

In this study, we investigate the dispersion and attenuation characteristics of Lamb wave propagation in a thin film made of FGVM, which follows the Kelvin–Voigt model [33]. The governing equations with a displacement function are deduced and the power series asymptotic solution is obtained by using the power series method. Because the series has no explicit expression for the function, we propose the minimum module approximation (MMA) method for solving the complex coefficient dispersion equation. The detailed process of the MMA method, the existence analysis of its solution, and its optimization are given. The reliability of the power series solution is verified by comparison with the exact analytical solution for Lamb wave propagation in a functionally graded viscoelastic film. The dispersion and attenuation characteristics of Lamb wave propagation under different gradient parameters are discussed, and the damping coefficients are analyzed. Conclusions based on these results can provide a theoretical basis for nonhomogeneous viscoelastic structure nondestructive testing.

2. Basic Equation for Lamb Waves in FGVM Film

Consider Lamb waves propagating in an isotropic functionally graded viscoelastic film along the x direction, as shown in Figure 1. The thickness of the film is h. The z direction is along the thickness direction. Let u, v, and w represent the displacement in the x, y, and z directions, respectively. For Lamb waves propagating is this structure, the displacement should satisfy:

$$u = u(x,z,t), v = 0, \ w = w(x,z,t) \tag{1}$$

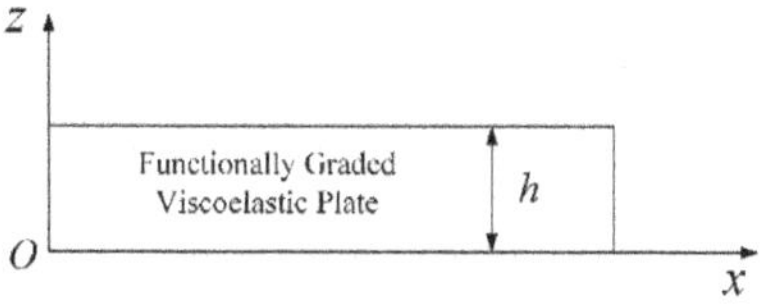

Figure 1. Functionally graded viscoelastic film and coordinate system.

If we let subscripts 1, 2, and 3 represent x, y, and z, respectively, then the stress tensor σ_{ij}, $i, j = 1-3$, can be divided into two parts: the deviation stress $\overline{\sigma}_{ij}$ and the spherical stress tensor σ_{m}, where $\sigma_{\mathrm{m}} = \sigma_{kk}/3$, $k = 1-3$, and the repeated index in the subscript implies summation with respect to that index.

In the Kelvin model, the constitutive equation can be expressed as:

$$\overline{\sigma}_{ij} = 2GS_{ij} + 2\eta\frac{\partial S_{ij}}{\partial t},\ \sigma_{\mathrm{m}} = \sigma_{kk}/3 = KS_{kk} \tag{2}$$

where $i, j, k, l = 1-3$, S_{kl} is the strain tensor, η represents the viscosity coefficient, G is the shear modulus, and K is the bulk modulus.

Equation (2) can be rewritten as:

$$\sigma_{ij} = c_{ijkl}S_{kl} + \overline{c}_{ijkl}\frac{\partial S_{kl}}{\partial t} \tag{3}$$

where c_{ijkl} and $\overline{c}_{ijkl}$ are components of the elastic tensor and viscosity tensor, respectively. In an FGVM, both these elastic parameters as well as the mass density ρ are not constants but are functions of z.

The motion equations have the following form:

$$\frac{\partial \sigma_x}{\partial x} + \frac{\partial \sigma_{xz}}{\partial z} = \rho\frac{\partial^2 u}{\partial t^2},\ \frac{\partial \sigma_{xz}}{\partial x} + \frac{\partial \sigma_z}{\partial z} = \rho\frac{\partial^2 w}{\partial t^2} \tag{4}$$

The relation between the strain and the displacement deduced from Equation (1) is:

$$\varepsilon_x = \frac{\partial u}{\partial x},\ \varepsilon_z = \frac{\partial w}{\partial z},\ \gamma_{xz} = \frac{\partial u}{\partial z} + \frac{\partial w}{\partial x},\ \varepsilon_y = \gamma_{xy} = \gamma_{yz} = 0 \tag{5}$$

By using index reduction, the original fourth-order elastic parameters can be rewritten to second-order elastic parameters. By substituting Equation (5) into the constitutive Equation (3), we obtain the following component forms of the stress:

$$\begin{aligned}
\sigma_x &= c_{11}\tfrac{\partial u}{\partial x} + c_{13}\tfrac{\partial w}{\partial z} + \overline{c}_{11}\tfrac{\partial^2 u}{\partial x\partial t} + \overline{c}_{13}\tfrac{\partial^2 w}{\partial z\partial t} \\
\sigma_{xz} &= c_{44}\left(\tfrac{\partial u}{\partial z} + \tfrac{\partial w}{\partial x}\right) + \overline{c}_{44}\left(\tfrac{\partial^2 u}{\partial z\partial t} + \tfrac{\partial^2 w}{\partial x\partial t}\right) \\
\sigma_z &= c_{13}\tfrac{\partial u}{\partial x} + c_{11}\tfrac{\partial w}{\partial z} + \overline{c}_{13}\tfrac{\partial^2 u}{\partial x\partial t} + \overline{c}_{11}\tfrac{\partial^2 w}{\partial z\partial t}
\end{aligned} \tag{6}$$

where the parameters in isotropic materials satisfy $c_{11} - c_{13} = 2c_{44}$ and $\overline{c}_{11} - \overline{c}_{13} = 2\overline{c}_{44}$.

Substitution of Equation (6) into Equation (4) leads to the following governing equations with respect to the displacement components:

$$\begin{aligned}
& c_{11}\tfrac{\partial^2 u}{\partial x^2} + c_{13}\tfrac{\partial^2 w}{\partial x\partial z} + \overline{c}_{11}\tfrac{\partial^3 u}{\partial x^2\partial t} + \overline{c}_{13}\tfrac{\partial^3 w}{\partial x\partial z\partial t} + c_{44}\left(\tfrac{\partial^2 u}{\partial z^2} + \tfrac{\partial^2 w}{\partial x\partial z}\right) + \overline{c}_{44}\left(\tfrac{\partial^3 u}{\partial z^2\partial t} + \tfrac{\partial^3 w}{\partial x\partial z\partial t}\right) \\
& + \tfrac{\mathrm{d}c_{44}}{\mathrm{d}z}\left(\tfrac{\partial u}{\partial z} + \tfrac{\partial w}{\partial x}\right) + \tfrac{\mathrm{d}\overline{c}_{44}}{\mathrm{d}z}\left(\tfrac{\partial^2 u}{\partial z\partial t} + \tfrac{\partial^2 w}{\partial x\partial t}\right) = \rho\tfrac{\partial^2 u}{\partial t^2} \\
& c_{44}\left(\tfrac{\partial^2 u}{\partial x\partial z} + \tfrac{\partial^2 w}{\partial x^2}\right) + \overline{c}_{44}\left(\tfrac{\partial^3 u}{\partial x\partial z\partial t} + \tfrac{\partial^3 w}{\partial x^2\partial t}\right) + c_{13}\tfrac{\partial^2 u}{\partial x\partial z} + c_{11}\tfrac{\partial^2 w}{\partial z^2} + \overline{c}_{13}\tfrac{\partial^3 u}{\partial x\partial z\partial t} + \overline{c}_{11}\tfrac{\partial^3 w}{\partial z^2\partial t} \\
& + \tfrac{\mathrm{d}c_{13}}{\mathrm{d}z}\tfrac{\partial u}{\partial x} + \tfrac{\mathrm{d}c_{11}}{\mathrm{d}z}\tfrac{\partial w}{\partial z} + \tfrac{\mathrm{d}\overline{c}_{13}}{\mathrm{d}z}\tfrac{\partial^2 u}{\partial x\partial t} + \tfrac{\mathrm{d}\overline{c}_{11}}{\mathrm{d}z}\tfrac{\partial^2 w}{\partial z\partial t} = \rho\tfrac{\partial^2 w}{\partial t^2}
\end{aligned} \tag{7}$$

Lamb waves propagating in a functionally graded viscoelastic film must satisfy not only the governing equations but also the traction-free conditions of the film, which are expressed as:

$$\sigma_z(x,0) = 0,\ \sigma_{xz}(x,0) = 0,\ \sigma_z(x,h) = 0,\ \sigma_{xz}(x,h) = 0 \tag{8}$$

3. Power Series Solution

The solutions of the governing equations can be expressed as:

$$u = U(z)\exp[\mathrm{i}k(x-ct)], w = -\mathrm{i}W(z)\exp[\mathrm{i}k(x-ct)] \tag{9}$$

where i is the imaginary unit, k and c are the wave number and wave velocity, respectively, and $U(z)$ and $W(z)$ are the unknown amplitudes of the displacement.

Substitution of Equation (9) into Equation (7) leads to:

$$\begin{aligned} &\hat{c}_{44}U'' + \hat{c}'_{44}U' + \left(\rho\Omega^2 - \hat{c}_{11}k^2\right)U + k(\hat{c}_{11} - \hat{c}_{44})W' + \hat{c}'_{44}kW = 0 \\ &\hat{c}_{11}W'' + \hat{c}'_{11}W' + \left(\rho\Omega^2 - \hat{c}_{44}k^2\right)W - k(\hat{c}_{11} - \hat{c}_{44})U' - \left(\hat{c}'_{11} - 2\hat{c}'_{44}\right)kU = 0 \end{aligned} \tag{10}$$

where $\hat{c}_{44} = c_{44} - \mathrm{i}\Omega\overline{c}_{44}$, $\hat{c}_{11} = c_{11} - \mathrm{i}\Omega\overline{c}_{11}$, $\Omega = ck$ is the frequency, and the prime symbol (′) represents differentiation with respect to thickness z. Both $\hat{c}_{44}$ and $\hat{c}_{11}$ are functions of z and Ω. Suppose that the material parameters vary along the thickness direction slowly, so that, for a certain Ω, the material parameters of the isotropic FGVM film can be expressed as follows:

$$\hat{c}_{11} = f_1\left(\frac{z}{h}\right),\ \hat{c}_{44} = f_2\left(\frac{z}{h}\right),\ \rho = f_3\left(\frac{z}{h}\right) \tag{11}$$

Suppose that the material parameters can be expressed in the power series form:

$$f_i\left(\frac{z}{h}\right) = \sum_{n=0}^{\infty} a_n^{\langle i\rangle}\left(\frac{z}{h}\right)^n\ (i = 1,2,3) \tag{12}$$

Therefore, the solutions of Equation (10) can also be expressed in a power series form as:

$$U(z) = \sum_{n=0}^{\infty} s_n\left(\frac{z}{h}\right)^n,\ W(z) = \sum_{n=0}^{\infty} t_n\left(\frac{z}{h}\right)^n \tag{13}$$

By substituting Equations (11)–(13) into Equation (10) and equating the coefficient of $(z/h)^n$ to zero, the following recursive equations can be obtained:

$$\begin{aligned} &\sum_{l=0}^{n}(l+2)(l+1)a_{n-l}^{\langle 2\rangle}s_{l+2} + \sum_{l=0}^{n}(n-l+1)(l+1)a_{n-l+1}^{\langle 2\rangle}s_{l+1} + (kh)^2\sum_{l=0}^{n}\left(a_{n-l}^{\langle 3\rangle}c^2 - a_{n-l}^{\langle 1\rangle}\right)s_l \\ &+(kh)\sum_{l=0}^{n}(l+1)\left(a_{n-l}^{\langle 1\rangle} - a_{n-l}^{\langle 2\rangle}\right)t_{l+1} + (kh)\sum_{l=0}^{n}(n-l+1)a_{n-l+1}^{\langle 2\rangle}t_l = 0 \\ &\sum_{l=0}^{n}(l+2)(l+1)a_{n-l}^{\langle 1\rangle}t_{l+2} + \sum_{l=0}^{n}(n-l+1)(l+1)a_{n-l+1}^{\langle 1\rangle}t_{l+1} + (kh)^2\sum_{l=0}^{n}\left(a_{n-l}^{\langle 3\rangle}c^2 - a_{n-l}^{\langle 2\rangle}\right)t_l \\ &-(kh)\sum_{l=0}^{n}(l+1)\left(a_{n-l}^{\langle 1\rangle} - a_{n-l}^{\langle 2\rangle}\right)s_{l+1} - (kh)\sum_{l=0}^{n}(n-l+1)\left(a_{n-l+1}^{\langle 1\rangle} - 2a_{n-l+1}^{\langle 2\rangle}\right)s_l = 0 \end{aligned} \tag{14}$$

where s_0, s_1, t_0, and t_1 are undetermined coefficients. For $l \geq 2$, all of the s_l and t_l are linear functions of s_0, s_1, t_0, and t_1.

To simplify calculating the relation between s_l, t_l and s_0, s_1, t_0, t_1, let:

$$\left(s_{0j}, s_{1j}, t_{0j}, t_{1j}\right) = \mathbf{I} \tag{15}$$

where j = 1–4 and **I** is a 4 × 4 unit matrix. Therefore, the solution of Equation (10) can be rewritten as:

$$U(z) = \sum_{j=1}^{4} C_j \sum_{n=0}^{\infty} s_{nj} \left(\frac{z}{h}\right)^n, \; W(z) = \sum_{j=1}^{4} C_j \sum_{n=0}^{\infty} t_{nj} \left(\frac{z}{h}\right)^n \tag{16}$$

where the constants C_j (j = 1–4) are to be determined. For $n = 0$ and 1, s_{nj} and t_{nj} are defined by expression (15); for other values of n, s_{nj}, and t_{nj} can be determined by solving Equation (14).

By substituting (16) into the boundary conditions, we then obtain the following linear algebraic equations for determining constants C_j (j = 1–4):

$$\begin{gathered}
-\left(\hat{c}_{11}^{0} - 2\hat{c}_{44}^{0}\right)khxC_1 + \hat{c}_{11}^{0}C_4 = 0 \\
\hat{c}_{44}^{0}C_2 + kh\hat{c}_{44}^{0}C_3 = 0 \\
\sum_{j=1}^{4}\left\{\sum_{n=0}^{\infty}\left[-kh\left(\hat{c}_{11}^{h} - 2\hat{c}_{44}^{h}\right)s_{nj} + \hat{c}_{11}^{h}(n+1)t_{(n+1)j}\right]\right\}C_j = 0 \\
\sum_{j=1}^{4}\left\{\sum_{n=0}^{\infty}\left[\hat{c}_{44}^{h}(n+1)s_{(n+1)j} + \hat{c}_{44}^{h}kht_{nj}\right]\right\}C_j = 0
\end{gathered} \tag{17}$$

The sufficient and necessary condition for the existence of a nontrivial solution is that the determinant of the coefficient matrix must vanish. Therefore, for the dispersion relation for Lamb waves, there exists:

$$\left|T_{ij}\right| = 0 \tag{18}$$

where

$$\begin{gathered}
T_{11} = -kh\left(\hat{c}_{11}^{0} - 2\hat{c}_{44}^{0}\right),\; T_{14} = \hat{c}_{11}^{0},\; T_{22} = 1,\; T_{23} = kh, \\
T_{3j} = \sum_{n=0}^{\infty}\left[-kh\left(\hat{c}_{11}^{h} - 2\hat{c}_{44}^{h}\right)s_{nj} + \hat{c}_{11}^{h}(n+1)t_{(n+1)j}\right],\; T_{4j} = \sum_{n=0}^{\infty}\left[(n+1)s_{(n+1)j} + kht_{nj}\right]
\end{gathered}$$

where j = 1–4, and other items of T_{ij} equal zero. The superscripts 0 and h represent the material parameters at the bottom and upper surfaces, respectively.

Owing to the existence of the complex relation, Equation (18) is a complex coefficient transcendental equation. In this paper, we suppose that k and the wavelength λ are both real numbers. The wave velocity c contains both real and imaginary parts. The real part of the wave velocity represents the phase velocity, and the imaginary part is related to the attenuation characteristic of the wave.

4. MMA Method and Optimization

4.1. MMA Method

For solving an equation with a complex variable, we should obtain a solution for which both the real part and the imaginary part of the equation should be zero. Consider the complex equation:

$$f(z) = 0 \Rightarrow f(x,y) = 0, \tag{19}$$

where x and y are the real and imaginary parts, respectively, of the complex variable z, and f, which is a function of z, is also complex. We suppose $z = a + ib$ is the solution of Equation (19). It should satisfy the conditions:

$$\mathrm{Re}[f(a,b)] = 0 \text{ and } \mathrm{Im}[f(a,b)] = 0, \tag{20}$$

where $\mathrm{Re}[f(x,y)]$ and $\mathrm{Im}[f(x,y)]$ represent the real and imaginary parts of the function $f(x,y)$.

Let

$$G(x,y) = \{\mathrm{Mod}[f(x,y)]\}^2 = \{\mathrm{Re}[f(x,y)]\}^2 + \{\mathrm{Im}[f(x,y)]\}^2, \tag{21}$$

where $\text{Mod}[f(x,y)]$ is the module of the function $f(x,y)$ and the square of the module is expressed by the function $G(x,y)$.

Suppose that the solution of Equation (19) exists and is unique in the region $(x_0, \overline{x}_0)$, $(y_0, \overline{y}_0)$. The model in Figure 2 is used to illustrate the solution steps. The first loop step is shown in Figure 2a: We divide the solution region into $n \times n$ grids, obtaining $(n+1)^2$ nodes in total, and calculate the module of each node by using Equation (21) and find the node (a_1, b_1) satisfying:

$$G(a_1, b_1) = \min\{G(x_0, y_0), G(x_0 + i\Delta x_0, y_0 + j\Delta y_0)\} \quad (i, j = 1, \ldots, n). \tag{22}$$

where

$$\Delta x_0 = \frac{\overline{x}_0 - x_0}{n}, \; \Delta y_0 = \frac{\overline{y}_0 - y_0}{n}.$$

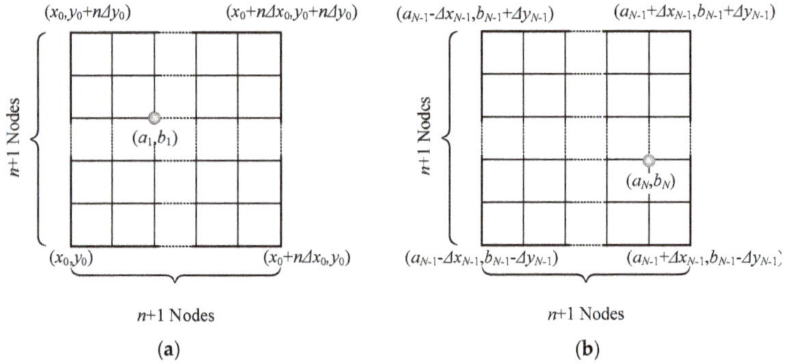

Figure 2. Steps of the MMA method. (**a**) First loop step, (**b**) *N*th loop step.

For the second loop step, based on n, we obtain the four points $(a_1-\Delta x_0, b_1-y_0)$, $(a_1+\Delta x_0, b_1-y_0)$, $(a_1+\Delta x_0, b_1+y_0)$, and $(a_1-\Delta x_0, b_1+y_0)$, remesh the square determined by the four points as the vertices into $n \times n$ grids once again, and also calculate the module of each node to find the minimum.

For the Nth loop step (Figure 2b), we repeat the above step, obtaining:

$$G(a_N, b_N) = \min\{G(x_{N-1}, y_{N-1}), G(x_{N-1} + i\Delta x_{N-1}, y_{N-1} + j\Delta y_{N-1})\} \; (i, j = 1, \ldots, n) \tag{23}$$

where (a_N, b_N) is the node with the minimum module after N time steps. The approximate solution of Equation (19) $z = a_N + ib_N$, can then be obtained.

4.2. Optimization of the MMA Method

The MMA method can be applied to solve equations with complex variables. To optimize the method, the following function q is introduced:

$$q = \left(\frac{4}{n^2}\right)^{\frac{1}{(n+1)^2}} \tag{24}$$

where q represents the average percentage of solution area reduction with each calculation and n^2 and $(n+1)^2$ are the number of grids and nodes of the solution region, respectively.

To optimize the MMA method, we should find the n value needed to satisfy that q reaches its minimum. In other words, $\ln q$ should also be the minimum. Therefore, n should satisfy:

$$\frac{\mathrm{d} \ln q}{\mathrm{d}n} = 0 \tag{25}$$

The solution of Equation (25) is $n = 3.77$. Because the number of grids needs to satisfy the condition of positive integers, we take the approximate solution at $n = 4$.

In practical numerical analysis, we always predict with a certain level of uncertainly the region over which the equation with complex variables has a solution. If, in the first step, (a_1, b_1) lies on the boundary of the region, $(x_0, \overline{x}_0)$, $\left(y_0, \overline{y}_0\right)$, the solution might not lie in the region. In this case, n should be selected to be a larger number to certify the existence of the solution. However, a spurious solution might exist if the calculation region is too large. Normally, the solutions of these problems are always irrational. This means that we can find the solution as the module infinites approaches zero. We should check for the convergence of solution by testing the ratio of the module reduction in several continuous steps. For example, we can calculate the ratio of the module for every three loops and judge the convergence to avoid a spurious solution.

It is worth noting that, if the minimum modulus is located at the boundary in the first calculation, there might be no solution in the computational domain. If the minimum modulus is not at the boundary after mesh refinement, the solution exists in the computational domain. Otherwise, the computational domain needs to be enlarged and recalculated. To verify the existence of the solution and avoid a spurious solution, we suggest that n should be selected as 6–8 in the first loop step in practical calculations.

5. Numerical Results and Discussion

5.1. Comparison with the Exact Analytical Results

To verify the validity of the power series method, the exact analytic solution and the asymptotic solution of the power series are compared when all material parameters vary with the same exponential function. The exact analytical solution can be obtained for waves propagating in the special FGVM thin film. The governing equations can be simplified to ordinary differential equations with constant coefficients, and the analytical solution can be obtained directly.

We suppose that the material parameters follow:

$$\lambda = \lambda_0 e^{p(z/h)},\ \mu = \mu_0 e^{p(z/h)},\ \eta = \xi\mu = \xi\mu_0 e^{p(z/h)},\ \rho = \rho_0 e^{p(z/h)} \tag{26}$$

where λ_0, μ_0, and ρ_0 are material parameters of the film lower surface at z = 0, η is the viscosity coefficient, ξ is a constant and is selected as 10^{-5}, and p represents the gradient parameter. The analytical solution in this condition can be selected as a reference for the solution of the power series reported in this paper.

The material parameters used in this paper can be deduced as:

$$\begin{aligned} \hat{c}_{11} &= \left(\lambda_0 + 2\mu_0 - \tfrac{4}{3}\mathrm{i}\Omega\xi\mu_0\right)e^{p(z/h)} = \beta_1 e^{p(z/h)} \\ \hat{c}_{44} &= \left(\mu_0 - \mathrm{i}\Omega\xi\mu_0\right)e^{p(z/h)} = \beta_2 e^{p(z/h)} \\ \rho &= \rho_0 e^{p(z/h)} = \beta_3 e^{p(z/h)} \end{aligned} \tag{27}$$

The displacement amplitude can also be expressed in an exponential function form as:

$$U_e(z) = Ae^{\alpha(z/h)},\ W_e(z) = Be^{\alpha(z/h)} \tag{28}$$

By substitution of Equation (28) into Equation (10) the homogeneous linear equations for the undetermined coefficients A and B can be deduced as:

$$\begin{aligned} &\left[\beta_2\alpha^2 + p\alpha\beta_1 + \left(\beta_3 c^2 - \beta_1\right)(kh)^2\right]A + \left[kh(\beta_1 - \beta_2)\alpha + khp\beta_2\right]B = 0 \\ &\left[kh(\beta_2 - \beta_1)\alpha + p(2\beta_2 - \beta_1)\right]A + \left[\beta_1\alpha^2 + p\alpha\beta_1 + \left(\beta_3 c^2 - \beta_2\right)(kh)^2\right]B = 0 \end{aligned} \tag{29}$$

Equations (29) comprise a set of linear homogeneous equations with respect to A and B. From the necessary and sufficient conditions for the existence of a nontrivial solution, we obtain that the determinant of the coefficient matrix is equal to zero:

$$\begin{vmatrix} \beta_2\alpha^2 + p\alpha\beta_1 + \left(\beta_3 c^2 - \beta_1\right)(kh)^2 & kh(\beta_1 - \beta_2)\alpha + khp\beta_2 \\ kh(\beta_2 - \beta_1)\alpha + p(2\beta_2 - \beta_1) & \beta_1\alpha^2 + p\alpha\beta_1 + \left(\beta_3 c^2 - \beta_2\right)(kh)^2 \end{vmatrix} = 0 \tag{30}$$

Considering that Equation (30) is a fourth-order equation, we suppose that the solution is α_j (j =1–4). The relation between A and B is derived by calculating Equation (29) as follows:

$$B_i = f_i A_i (i = 1\text{–}4) \tag{31}$$

The displacement amplitude solution of Equation (10) can be rewritten as:

$$U = \sum_{j=1}^{4} A_j e^{\alpha_j (z/h)} \quad W = \sum_{j=1}^{4} f_j A_j e^{\alpha_j (z/h)} \tag{32}$$

Similarly, by considering the boundary conditions, we then obtain the dispersion equation:

$$\left|T_{ij}\right| = 0 \tag{33}$$

where

$$T_{1j} = (\beta_1 - 2\beta_2)k + \beta_1 f_j \alpha_j,\ T_{2j} = \alpha_j - kf_j$$
$$T_{3j} = \left[(\beta_1 - 2\beta_2)k + \beta_1 f_j \alpha_j\right] e^{\alpha_j h},\ T_{4j} = \left(\alpha_j - kf_j\right) e^{\alpha_j h} (j = 1 - 4)$$

In numerical analysis, the normalized wave velocity $\hat{c}$ and the dimensionless wave number kh are applied for describing the wave propagation property. The normalized dimensionless wave velocity $\hat{c}$ satisfies:

$$\hat{c} = c/c_{sh} \tag{34}$$

where $c_{sh} = \sqrt{G/\rho}$, which is the bulk shear wave velocity. The Poisson ratio is a constant and satisfies $\nu = 0.25$.

To evaluate the accuracy and precision of the power series and MMA methods, the relation between the normalized wave velocity and the dimensionless wave number for Lamb waves propagating in the special FGVM thin film is plotted in Figure 3. When $p = 0$, the FGVM thin film becomes a homogenous viscoelastic thin film. Figure 3a,b present the real and imaginary parts of the normalized wave velocity, respectively. It is found that the solution obtained by using the power series method agrees well with the exact analytical solution.

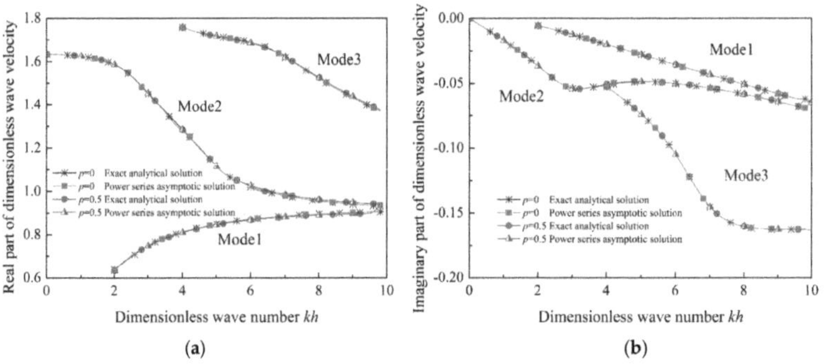

Figure 3. Wave velocity of Lamb waves in homogenous viscoelastic film and in special FGVM film. (**a**) Real part, (**b**) imaginary part.

By checking the results of the change of phase velocity, we find that there exists little difference between the wave velocity curves for Lamb waves in the homogenous viscoelastic thin film and in the special FGVM thin film. Normally, the dispersion curves of Lamb waves in homogenous film can be determined by bulk shear wave velocity and the Poisson ratio. Both the bulk shear wave velocity and the Poisson ratio in homogenous thin film are same as those in the special FGVM thin film. It can be used to explain that the dispersion curves of the two cases are almost identical. This implies that we cannot measure the gradient parameters by variation of the wave velocity

We further study the wave structure of Lamb wave propagation in different viscoelastic thin films. The normalized displacement amplitude is defined as:

$$\hbar_u = |U|/|U(0)| \cdot \text{sign}\{\text{Re}(U)/\text{Re}[U(0)]\},\ \hbar_w = |W|/|U(0)| \cdot \text{sign}\{\text{Re}(W)/\text{Re}[U(0)]\} \quad (35)$$

where $U(0)$, which represents displacement component at z = 0, is selected to be 1 in the numerical analysis, Re is the real part of the complex number, and the sign function satisfies

$$\text{sign}(x) = \begin{cases} 1, & x \geq 0 \\ -1, & x < 0 \end{cases} \quad (36)$$

The normalized displacement amplitude of the first two modes at $kh = \pi$ and $kh = 2\pi$ are plotted in Figure 4. The curves obtained from the exact solution and these obtained by using the power series method coincide completely. In a homogenous viscoelastic thin film, the wave structure is symmetric or antisymmetric. However, because of the asymmetric properties of the FGVM thin film, the displacement amplitudes are not symmetric.

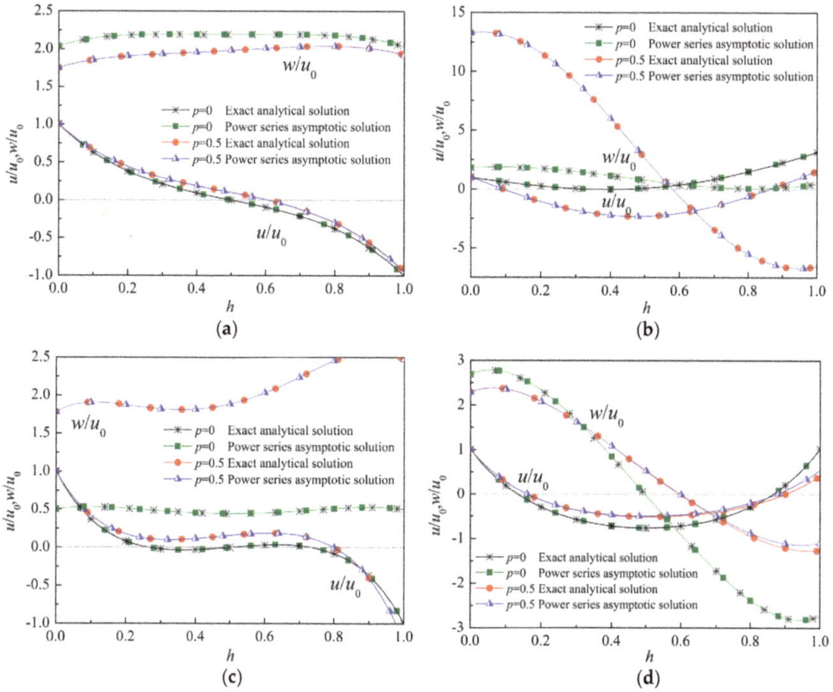

Figure 4. Wave structure. (**a**) $kh = \pi$, Mode 1, (**b**) $kh = \pi$, Mode 2, (**c**) $kh = 2\pi$, Mode 1, (**d**) $kh = 2\pi$, Mode 2.

To further investigate the influence of the gradient property on the displacement, we denote the ellipticity of particles on lower and upper surfaces as $\chi_0 = |w_0/u_0|$ and $\chi_h = |w_h/u_h|$, respectively. The relation between the gradient parameter and the ellipticity of particles at $kh = \pi$ and $kh = 2\pi$ is plotted in Figure 5. It is found that the influence of the gradient property on the ellipticity of a particle on the surface is more obvious than that on the wave velocity.

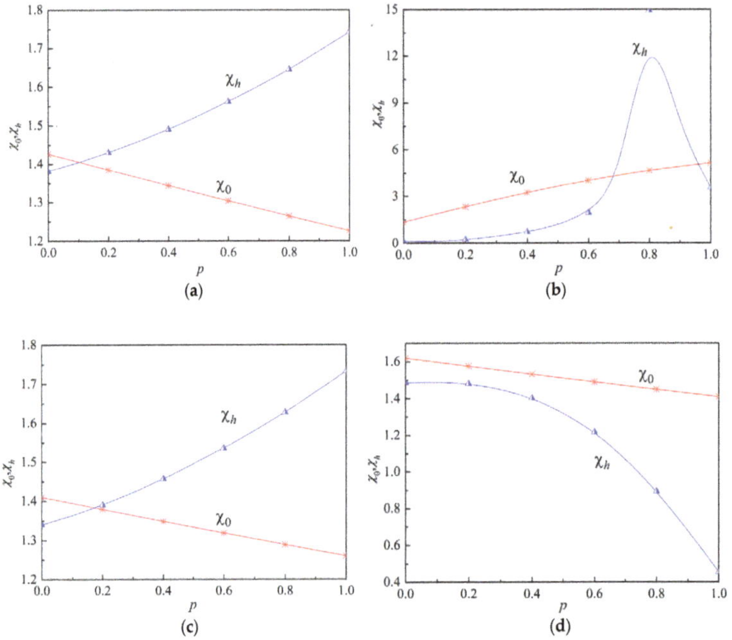

Figure 5. Ellipticity of particles on lower and upper surfaces versus the gradient parameter. (**a**) $kh = \pi$, Mode 1, (**b**) $kh = \pi$, Mode 2, (**c**) $kh = 2\pi$, Mode 1, (**d**) $kh = 2\pi$, Mode 2.

5.2. Material Parameters Varying Linearly

For numerical analysis with the theoretical model described above, we assumed that the Lamé parameters λ and μ, mass density ρ, and viscosity coefficient η in the functionally graded viscoelastic film varied as follows:

$$\lambda = \lambda_0 + p_1\lambda_0\frac{z}{h},\ \mu = \mu_0 + p_2\mu_0\frac{z}{h},\ \rho = \rho_0 + p_3\rho_0\frac{z}{h},\ \eta = \xi\mu \tag{37}$$

where λ_0, μ_0, and ρ_0 are material parameters of the film lower surface at z = 0; η is the viscosity coefficient; ξ is a constant and is selected as 10^{-5} (except in Section 5.2.3); and p_1, p_2 and p_3 are the gradient parameters of λ, μ, and ρ, respectively $(0 \leq p_1, p_2, p_3 < 1)$.

The material parameters used in this paper can be deduced as:

$$\hat{c}_{11} = \lambda + 2\mu - \frac{4}{3}\mathrm{i}\Omega\eta,\ \hat{c}_{44} = \mu - \mathrm{i}\Omega\eta \tag{38}$$

5.2.1. All Material Parameters Varying Identically

Suppose that all material parameters vary along the thickness direction linearly and identically, i.e., $p_i = p(i = 1, 2, 3)$. The wave velocity plotted as a function of wave number when p = 0, 0.3, 0.5, and 0.7 is shown in Figure 6. By comparing the results for the case in Section 5.1, a similar conclusion can be reached. If all the material parameters vary along the thickness direction identically, the wave velocity curves almost coincide.

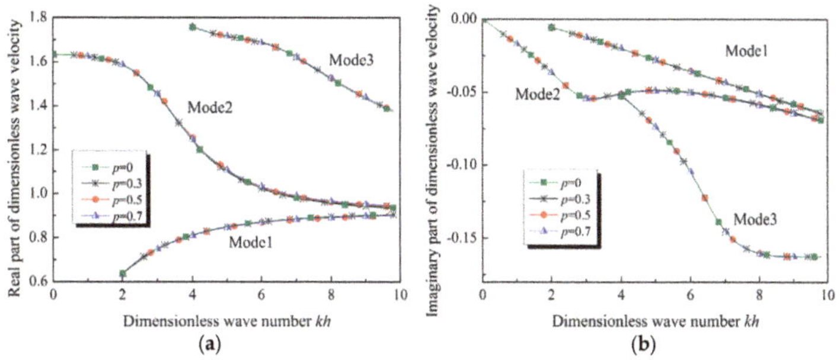

Figure 6. Wave velocity of Lamb waves in homogenous viscoelastic film and in FGVM film (with all material parameters varying linearly and identically). (**a**) Real part, (**b**) imaginary part.

In these cases, the bulk wave velocity, including the shear wave velocity c_{sh} and the longitudinal wave velocity c_L, where $c_L = \sqrt{(\lambda + 2\mu)/\rho}$, are constants. This implies that, if the bulk wave velocities are constants, the wave velocity of Lamb waves in the FGVM thin film are almost similar to that in a homogenous film.

Similarly, the wave structures are also plotted in Figure 7. It is found that the gradient parameter has an obvious influence on the wave structure. It is also shown in Figure 7 that the ellipticity of particles on the upper and lower surfaces is different owing to the gradient property of the FGVM film. Considering testability, we suggest that the ellipticity of a particle on the surface can be applied for measuring the gradient parameter when all material parameters vary identically.

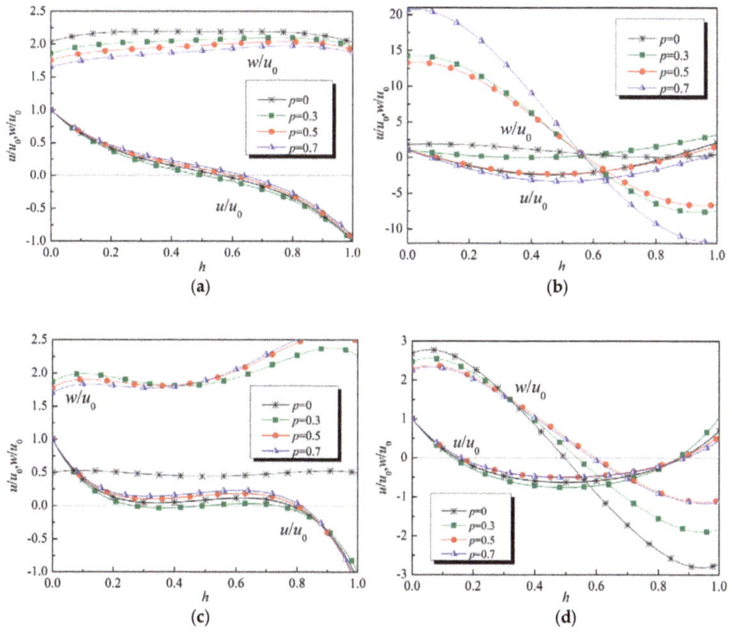

Figure 7. Wave structure of Lamb waves (with all material parameters varying linearly and identically). (**a**) $kh = \pi$, Mode 1, (**b**) $kh = \pi$, Mode 2, (**c**) $kh = 2\pi$, Mode 1, (**d**) $kh = 2\pi$, Mode 2.

5.2.2. Material Parameters Varying Independently

To investigate the influence of the elastic modulus and density gradient on dispersion and attenuation characteristics of Lamb waves in gradient viscoelastic film, we chose three types of films for which the gradient parameters are:

$$\text{A}: p_1 = p_2 = p_3 = 0;\ \text{B}: p_1, p_2 = 0,\ p_3 = 0.2;\ \text{C}: p_1 = 0.2,\ p_2 = 0.2,\ p_3 = 0$$

Film A is a homogeneous film, which can be used for referencing the propagation characteristics in the gradient film. The elastic modulus in film B is a constant, and the density increases along the thickness of the film and the density of the lower surface is the same as that of the homogeneous film. Conversely, the density in film C is a constant, and the elastic modulus varies along the thickness direction linearly.

The real and imaginary parts of the wave velocity in the three types of films are shown in Figure 8. When the mass density increases, the real part and the absolute value of the imaginary part of the wave velocity of each mode are less than these in the homogenous film; when the elastic modulus increases, the real part and the absolute value of the imaginary part of the dimensionless wave velocity of each mode are larger than these in the homogenous film.

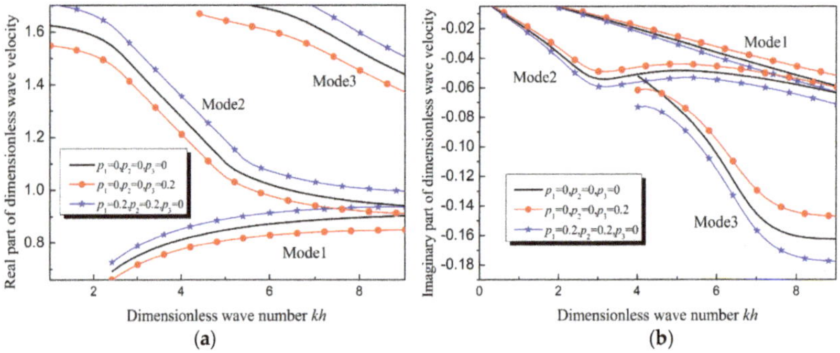

Figure 8. Dimensionless Lamb wave velocity in functionally graded viscoelastic films of different gradient parameters: (**a**) Dispersion curves of Lamb waves in the three films, (**b**) attenuation curves of Lamb waves in the three films.

5.2.3. Relative Viscosity Coefficient Varying Independently

In engineering application, the relative viscosity coefficient might vary along one direction because of the environment. However, in these cases, the mass density and the elastic parameters might not change. To reveal the influence of the gradient relative viscosity coefficient on the wave property, we suppose that the relative viscosity coefficient ξ varies along the thickness direction and that other parameters including λ, μ, and ρ are constants. The material parameters are:

$$\lambda = \lambda_0,\ \mu = \mu_0,\ \rho = \rho_0,\ \eta = \xi\mu,\ \xi = 10^{-5} p_4(z/h) \tag{39}$$

The wave velocity is plotted as function of wave number in Figure 9. In Figure 9a, the curves for the real part of the phase velocity almost coincide. This suggests that the influence of the gradient relative viscosity coefficient on the dispersion curves is too slight to measure. However, obvious differences can be observed in Figure 9b, which describes the relation between the imaginary parts of the wave velocity. The absolute value of the imaginary part of the dimensionless wave velocity increases with the increase of the gradient relative viscosity coefficient. The physical meaning of the imaginary parts of the wave velocity will be discussed in the next section.

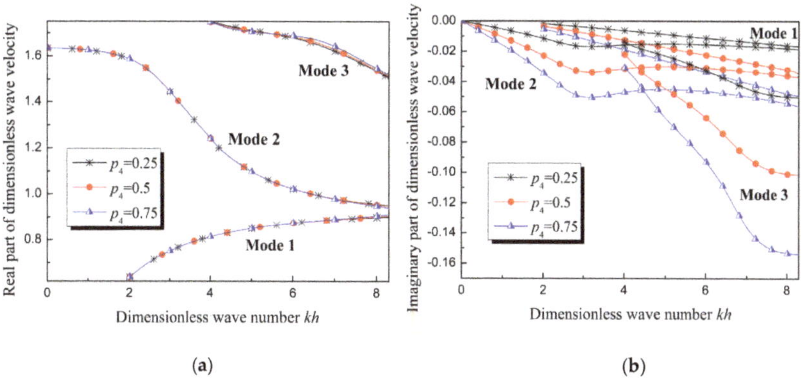

Figure 9. Wave velocity of Lamb waves in homogenous viscoelastic film and in FGVM film (with only the relative viscosity coefficient varying linearly and identically). (**a**) Real part, (**b**) imaginary part.

5.3. Influence of Gradient Parameter on Wave Attenuation

Equation (18) is a complex equation. As the wave number is a real number, the wavelength is also a real number, the obtained wave velocity is complex, and the product Ω of the wave velocity c and the wave number k is complex, which can be expressed as follows:

$$c_p = \mathrm{Re}(c), \Omega = \omega + \mathrm{i}\widetilde{\omega} = ck \tag{40}$$

where ω and $\widetilde{\omega}$ are the real and imaginary parts of Ω, respectively, ω is frequency, and $\widetilde{\omega}$ is related to the attenuation of the wave amplitude. From Equation (40), we have:

$$\omega = c_p k \tag{41}$$

In viscoelastic materials, the wave propagation process is essentially a quasi-periodic motion, and the period of the particle displacement is determined by the phase velocity. The period is expressed as follows:

$$T = \frac{2\pi}{k\mathrm{Re}(c)} = \frac{2\pi}{\omega} \tag{42}$$

To analyze the attenuation trend, we define the amplitude ratio of the adjacent period as the damping coefficient γ, given by:

$$\gamma = \exp\left[-2\pi\frac{\mathrm{Im}(c)}{\mathrm{Re}(c)}\right] = \exp\left(-\frac{2\pi\widetilde{\omega}}{\omega}\right) \tag{43}$$

In this study, the normalized product of frequency and thickness $\hat{\omega}h$ is selected to be the abscissa. If

$$\hat{\omega} = \frac{\omega}{c_{sh}}$$

is satisfied, then

$$\hat{\omega}h = \frac{\omega}{c_{sh}}h = \frac{c_p}{c_{sh}}kh \tag{44}$$

The influence of the gradient properties on the damping coefficient is plotted in Figure 10. The damping coefficient increases with the increase of the frequency. When material parameters are constants, or material parameters vary along the thickness direction with the same exponential function, or both Lamé parameters and mass density vary linearly, the damping coefficient of Mode 1 and Mode 2 is similar, as shown in Figure 10a,b. However, when the Lamé parameters and mass density do not vary identically, a difference in the damping coefficient can be observed, as shown in Figure 10b. When the mass density increases, the damping coefficient increases at high frequency ($\hat{\omega}h > s0$). This implies that, at high frequency, if the mass density increases along the thickness direction, then the Lamb waves in the FGVM thin film should attenuate more quickly than those in a homogenous material. Conversely, if the Lamé parameters increase along the thickness direction, then the attenuation of Lamb waves in the FGVM thin film should be weakened. If only the relative viscosity coefficient increases along the thickness direction, then the attenuation of Lamb waves will become more serious, as shown in Figure 11. As the gradient coefficient increases, the damping coefficient increases and the attenuation tendency becomes obvious.

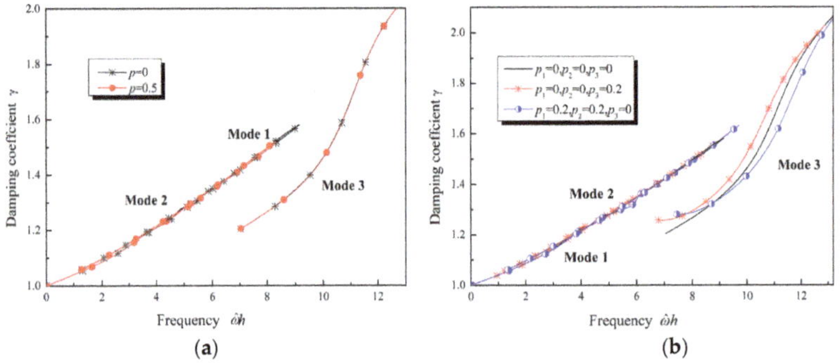

Figure 10. Damping coefficient of Lamb waves with Lamé parameters and density varying along the thickness direction: (**a**) Material parameters varying following the same exponential function; (**b**) material parameters varying following a linear function.

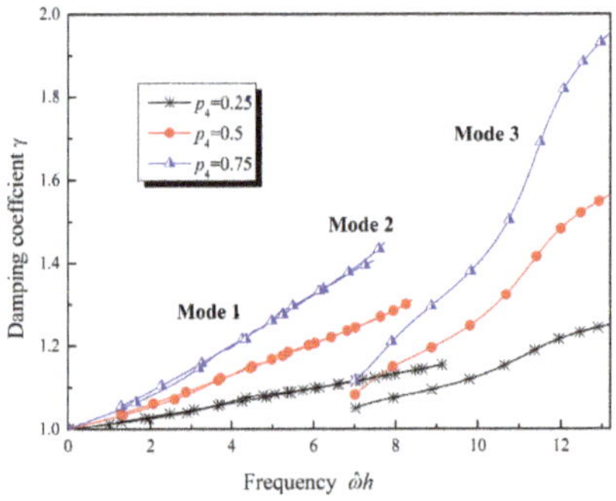

Figure 11. Damping coefficient of Lamb waves with the relative viscosity coefficient varying.

6. Conclusions

The power series method can be employed for solving the governing differential equations for Lamb wave propagation in FGVM thin films. The MMA method is proposed to solve the transcendental equations in the form of a series with complex coefficients. It is suggested that the meshing number should be selected as 6–8 in the first loop step and 4 in other loops. The numerical results obtained by these methods agree well with the exact analytical solution.

When Lamé parameters and mass density vary along the thickness direction identically, the influence of the gradient properties on the wave velocity is slight but that on the wave structure and the ellipticity of particles on the surface is obvious. This suggests that the ellipticity of particles on the surface should be selected to measure the gradient property if the bulk wave velocities are constants in the FGVM thin film. When Lamé parameters and mass density vary along the thickness direction independently, the variation of the phase velocity can be used for testing the gradient parameters.

However, when the relative viscosity coefficient is a variable and both Lamé parameters and mass density are constants, the gradient property will not affect the phase velocity. The attenuation tendency becomes obvious with the increase of the gradient relative viscosity coefficient.

The method proposed herein and the results obtained should provide theoretical guidance for ultrasonic nondestructive testing of heterogeneous viscoelastic materials and enable the safe evaluation of surface acoustic wave devices.

Author Contributions: Conceptualization, X.C.; Data curation, H.J. and Y.R.; Funding acquisition, X.C. and J.S.; Methodology, X.C. and Y.R.; Software, X.C. and H.J.; Supervision, X.C.; Writing—original draft, X.C., H.J. and Y.R.; Writing—review & editing, X.C. and J.S.

Funding: This research was funded by the National Natural Science Foundation of China (Nos. 11572244, 11872300). It also was supported by NSAF (No. U1630144) and the Open Subject of State Key Laboratories of Transducer Technology (No. SKT1506).

Conflicts of Interest: The authors declare no conflict of interest.

References

1. Achenbach, J.D. *Wave Propagation in Elastic Solids*; Tongji University Press: Shanghai, China, 1992; pp. 232–248.
2. Lu, Y.; Zhu, Y.; Zhu, Z. The propagation of Lamb waves in a plate with viscoelastic layer coating. *Acta Acust.* **2006**, *31*, 355–362.
3. Cao, X.S.; Jin, F.; Jeon, I. Calculation of propagation properties of Lamb waves in a functionally graded material (FGM) plate by power series technique. *NDT E Int.* **2011**, *44*, 84–92. [CrossRef]
4. Amor, M.B.; Ghozlen, M.H.B. Lamb waves propagation in functionally graded piezoelectric materials by Peano-series method. *Ultrasonics* **2015**, *55*, 10–14. [CrossRef] [PubMed]
5. Wu, X.H.; Shen, Y.P.; Sun, Q. Lamb wave propagation in magneto-electro-elastic plates. *Appl. Acoust.* **2007**, *68*, 1224–1240. [CrossRef]
6. De Luca, A.; Caputo, F.; Khodaei, Z.S.; Aliabadi, M.H. Damage characterization of composite plates under low velocity impact using ultrasonic guided waves. *Compos. Part B Eng.* **2018**, *138*, 168–180. [CrossRef]
7. De Luca, A.; Perfetto, D.; De Fenza, A.; Petrone, G.; Caputo, F. Guided waves in a composite winglet structure: Numerical and experimental investigations. *Compos. Struct.* **2019**, *210*, 96–108. [CrossRef]
8. Koizumi, M. The concept of FGM. *Ceram. Trans. FGM.* **1993**, *34*, 3–10.
9. Qian, Z.H.; Jin, F.; Lu, T.J.; Kishimoto, K. Transverse surface waves in functionally graded piezoelectric materials with exponential variation. *Smart Mater. Struct.* **2008**, *17*, 065005. [CrossRef]
10. Cao, X.; Jin, F.; Jeon, I.; Lu, T.J. Propagation of Love waves in a functionally graded piezoelectric material (FGPM) layered composite system. *Int. J. Solids Struct.* **2009**, *46*, 4123–4132. [CrossRef]
11. Kuo, H.Y.; Bhattacharya, K. Fibrous composites of piezoelectric and piezomagnetic phases. *Mech. Mater.* **2013**, *60*, 159–170. [CrossRef]
12. Han, X.; Liu, G.R. Effects of SH waves in a functionally graded plate. *Mech. Res. Commun.* **2002**, *29*, 327–338. [CrossRef]
13. Kiełczyński, P.; Szalewski, M.; Balcerzak, A.; Wieja, K. Propagation of ultrasonic Love waves in nonhomogeneous elastic functionally graded materials. *Ultrasonics* **2016**, *65*, 220–227. [CrossRef]
14. Cao, X.S.; Jin, F.; Wang, Z.K. On dispersion relations of Rayleigh waves in a functionally graded piezoelectric material (FGPM) half-space. *Acta Mech.* **2008**, *200*, 247–261. [CrossRef]
15. Yuan, L.; Shen, Z.H.; Ni, X.W.; Lu, J. Numerical calculation of laser induced surface wave in material with changes of near-surface properties. *Infrared Laser Eng.* **2007**, *36*, 328–331.
16. Sun, H.X. Numerical simulation of laser-generated Rayleigh wave by finite element method on viscoelastic materials. *Acta Phys. Sin.* **2009**, *58*, 6344–6350.
17. Cai, C.; Liu, G.R.; Lam, K.Y. A transfer matrix approach for acoustic analysis of a multilayered active acoustic coating. *J. Sound Vib.* **2001**, *248*, 71–89. [CrossRef]
18. Du, J.K.; Ye, D. SH waves in laminated structure of functionally gradient piezoelectric material. *J. Solid Rocket Technol.* **2005**, *28*, 133–136.
19. Collet, B.; Destrade, M.; Maugin, G.A. Bleustein–Gulyaev waves in some functionally graded materials. *Eur. J. Mech.* **2006**, *25*, 695–706. [CrossRef]

20. Qian, Z.; Jin, F.; Wang, Z.; Kishimoto, K. Transverse surface waves on a piezoelectric material carrying a functionally graded layer of finite thickness. *Int. J. Eng. Sci.* **2007**, *45*, 455–466. [CrossRef]
21. Shen, X.; Ren, D.; Cao, X.; Wang, J. Cut-off frequencies of circumferential horizontal shear waves in various functionally graded cylinder shells. *Ultrasonics* **2018**, *84*, 180–186. [CrossRef]
22. Li, X.Y.; Wang, Z.K.; Huang, S.H. Love waves in functionally graded piezoelectric materials. *Int. J. Solids Struct.* **2004**, *41*, 7309–7328. [CrossRef]
23. Vlasie, V.; Rousseau, M. Guided modes in a plane elastic layer with gradually continuous acoustic properties. *NDT E Int.* **2004**, *37*, 633–644. [CrossRef]
24. Cao, X.S.; Jin, F.; Wang, Z.K. Bleustein-Gulyaev(B-G) waves in functionally graded piezoelectric layered structures. *Sci. China* **2009**, *52*, 613–625. [CrossRef]
25. Dahmen, S.; Amor, M.B.; Ghozlen, M.H.B. Investigation of the coupled Lamb waves propagation in viscoelastic and anisotropic multilayer composites by Legendre polynomial method. *Compos. Struct.* **2016**, *153*, 557–568. [CrossRef]
26. Yu, J.G.; Ratolojanahary, F.E.; Lefebvre, J.E. Guided waves in functionally graded viscoelastic plates. *Compos. Struct.* **2011**, *93*, 2671–2677. [CrossRef]
27. Paehler, D.; Schneider, D.; Herben, M. Nondestructive characterization of sub-surface damage in rotational ground silicon wafers by laser acoustics. *Microelectron. Eng.* **2007**, *84*, 340–354. [CrossRef]
28. Yu, J. Viscoelastic shear horizontal wave in graded and layered plates. *Int. J. Solids Struct.* **2011**, *48*, 2361–2372.
29. Zhang, Z.; Sun, C.; Wu, D. Love wave forward modeling in Kelvin-Voigt viscoelastic medium. In Proceedings of the Annual Meeting of Chinese Geoscience Union, Beijing, China, 20–23 October 2014; p. 1453.
30. Zhang, X.M.; Wang, Y.Q.; Yu, J.G. Guided circumferential SH wave in orthotropic viscoelastic hollow cylinders. *Eng. Mech.* **2013**, *30*, 78–81. [CrossRef]
31. Lefebvre, J.E.; Zhang, V.; Gazalet, J.; Gryba, T.; Sadaune, V. Acoustic wave propagation in continuous functionally graded plates: An extension of the Legendre polynomial approach. *IEEE Trans. Ultrason. Ferroelectr. Freq. Control* **2001**, *48*, 1332–1340. [CrossRef]
32. Li, N.; Qian, Z.; Wang, B. Study on computational methods of dispersion curves in complex wavenumber range. *Chin. J. Appl. Mech.* **2016**, *33*, 365–370.
33. Yang, T.Q. *Theory of Viscoelasticity*; Huazhong University of Science and Technology Press: Wuhan, China, 1990.

MDPI
St. Alban-Anlage 66
4052 Basel
Switzerland
Tel. +41 61 683 77 34
Fax +41 61 302 89 18
www.mdpi.com

Materials Editorial Office
E-mail: materials@mdpi.com
www.mdpi.com/journal/materials

www.ingramcontent.com/pod-product-compliance
Lightning Source LLC
Chambersburg PA
CBHW051725210326
41597CB00032B/5605
9783039216581